Brain-Computer Interface Technologies

Claude Clément

Brain-Computer Interface Technologies

Accelerating Neuro-Technology for Human Benefit

Springer

Claude Clément
Campus Biotech
Wyss Center for Bio and Neuroengineering
Genève, Geneve, Switzerland

ISBN 978-3-030-27851-9 ISBN 978-3-030-27852-6 (eBook)
https://doi.org/10.1007/978-3-030-27852-6

This Springer imprint is published by the registered company Springer Nature Switzerland AG
The registered company address is: Gewerbestrasse 11, 6330 Cham, Switzerland

To our patients, who give us so much motivation to continue.
To all my colleagues at the Wyss Center for Bio and Neuroengineering and to all the people I worked with in the past 30 years in the field of active implantable devices.
Geneva, June 3, 2019

Preface

If we were able to sneak along our spinal cord and nerves, or to slip through by the interface of our ears or eyes, we would enter in the limitless cosmos of the billions of neurons living in our body which are making us as we are. The ancients were thinking that our heart was the center of our emotions. It is not true. The heart is the machine our body needs to preserve life, but what characterizes human beings, personal features, sensations, emotions, and feelings resides in our nervous system where, in a majestic dynamic ballet, interconnections change, neurons die and appear, and areas damaged by a traumatism receive help from other sections of the brain. The way our synapses interleave is in a continuous evolution, and we ignore the rules governing these changes.

For a long time, we could not do much more than "listen" to the brain by collecting tiny electrical signals at the surface of the scalp using the well-known electroencephalograms (EEG). These receivers, kept distant from the brain by the skin, scalp, skull, and dura-mater, hear only a remote murmur: the choir of billion neurons.

Since the end of the 1980s, the advent of technologies in the field of active implants has allowed us to place electrodes on or in the brain. We are now able to hear, in detail, what the brain is saying. In our thirst of understanding everything, neurosciences were first trying to grasp the overall complexity of the brain, even to model or simulate it. We realize now that we should not compare the brain to a computer. Connections between neurons are not governed by a binary system but rather by multidimensional nonlinear relations of chaotic nature. This is the miracle: from chaos appear motor actions, perceptions, emotions, feelings, memories, and ideas. Today, and probably for many more years, we are not able to "program" soft human particularities like love, attraction for another individual, survival instinct, or duty to reproduce. Science does not explain falling in love, genius, or creation of a unique piece of art.

Nevertheless, we have discovered that electrical signals, injected at appropriate locations, may inhibit, modify, or influence the relations between the brain and its environment. Clinical tests have shown that implanting electrodes in the nervous system may treat a large variety of conditions, including cognitive, affective, and psychiatric disorders. It raises fundamental questions in terms of ethics and society.

Is "emotionally augmented human" a sustainable concept? Do we want to annihilate our differences and our personal characteristics?

We know that we will never reach the confines of our universe. We should also realize that we should remain modest in our conquest of the brain. Finding rationality in the never-ending dance of neurons is maybe a vain challenge. Shall we keep untouched those mysteries which make us unique and unpredictable?

Genève, Switzerland Claude Clément

Contents

About the Author

Claude Clément born in 1955, is from the French-speaking part of Switzerland. He has first worked in R&D for the watch industry (Swatch Group) as head of the transducers and actuators development group. He entered the world of medical technologies by heading the diversification activities of Swatch in the field of wearable programmable drug delivery pumps. Afterward, he spent 27 years in the field of active implantable medical devices, as director of Manufacturing Engineering at Intermedics (now Boston Scientific), as plant manager of the Swiss Operations at Medtronic, and later as a consultant for major companies, mainly in the field of pacemakers, and for various highly innovative start-ups. Starting 1996, he put in place and ramped up the highly automated factory of Medtronic in Lake Geneva area. This plant is the world's largest site for the assembly of active implantable medical devices, producing large volumes of pacemakers, defibrillators, and neurostimulators. Until 2014, he was CEO of MyoPowers, a start-up company developing an electromechanical implant to treat severe incontinence. Beginning of 2015, he joined the Wyss Center for Bio and Neuroengineering as CTO. He is or was founder, chairman, or board member of several start-ups and small businesses. He is chairman of the BioAlps Association, a diversified life science cluster in Western Switzerland. He holds a master's degree in Electrical Engineering from the Swiss Federal Institute of Technology (EPFL) in Lausanne and an MBA from HEC at the University of Lausanne (Switzerland).

Abbreviations and Acronyms

510k	Premarket submission to the FDA, substantially equivalent to a legally marketed device
5G	Fifth generation of cellular network technology
AB	Advanced Bionics
AC	Alternating Current
AD	Alzheimer's Disease
ADHD	Attention Deficit Hyperactivity Disorder
AI	Artificial Intelligence
AIMD	Active Implantable Medical Device
ALD	Atomic Layer Deposition
ALS	Amyotrophic Lateral Sclerosis
AMF	Alfred Mann Foundation
BAHA	Bone-Anchored Hearing Aid
BCI	Brain Computer Interface
BD	Big Data
BGA	Ball Grid Array
BMI	Brain Machine Interface
BSc	Boston Scientific
CBP	Chronic Back Pain
CDRH	Center for Devices and Radiological Health (FDA)
CE	Conformité Européenne
CHUV	Centre Hospitalier Universitaire Vaudois
CI	Cochlear Implant
CLIS	Completely Locked-In Syndrome
CMOS	Complementary Metal Oxide Semiconductor
CNS	Central Nervous System
CoC	Chip-on-Chip
CoGS	Cost of Goods Sold
CoNQ	Cost of Non-quality
COTS	Component of the Self
CRM	Cardiac Rhythm Management

CSEM	Centre Suisse d'Electronique et de Microtechnique
CSF	Cerebrospinal Fluid
CT	Computed Tomography
DARPA	Defense Advanced Research Projects Agency
DBS	Deep Brain Stimulation
DC	Direct Current
EAP	Expedited Access Pathway
EC	Ethics Committee
ECAPS	Evoked Compound Action Potential Signal
ECG	Electrocardiogram
ECoG	Electrocortical Grid
EEG	Electroencephalogram
EFS	Early Feasibility Study
EMD	Electromagnetic Disturbance
EMG	Electromyogram
EOL	End of Life
EPFL	Ecole Polytechnique Fédérale de Lausanne
EtO	Ethylene Oxide
FBSS	Failed Back Surgery Syndrome
FCB	Flip Chip Bonding
FCC	Federal Communications Commission
FDA	Food and Drug Administration
FES	Functional Electrical Stimulation
FI	Fecal Incontinence
FIH	First in Human
fMRI	Functional Magnetic Resonance Imaging
FPGA	Field Programmable Gate Array
FT	Feedthrough
FtO	Freedom to Operate
GES	Gastric Electrical Stimulation
GNS	Gastric Nerve Stimulation
HDE	Humanitarian Device Exception
HF	High Frequency
IC	Integrated Circuit
ICD	Implantable Cardiac Defibrillator
IDE	Investigational Device Exemption
IMMG	Intramuscular Myogram
IMS	Intramuscular Stimulation
IoE	Internet-of-Everything
IoMT	Internet-of-Medical-Things
IoT	Internet-of-Things
IP	Intellectual Property
IPA	Isopropyl Alcohol
IPG	Implantable Pulse Generator
IR	Infrared Light

ITU	International Telecommunication Union
LCP	Liquid Crystal Polymer
LED	Light-Emitting Diode
LIFUS	Low-Intensity Focused Ultrasounds
M2M	Machine-to-Machine Communication
MDD	Major Depressive Disorder
MDR	Medical Device Regulation
MDT	Medtronic
MEA	Microelectrode Array
MEG	Magnetoencephalogram
MIL-STD	Military Standard
MMI	Mind-Machine Interface
MPE	Maximum Permissible Exposure
MRI	Magnetic Resonance Investigation
NB	Notified Body
NESD	Neural Engineering System Design
NI	Neural Interface
NIR	Near Infrared Light
NNP	Networked Neuroprosthetics System
OAB	Overactive Bladder
OCD	Obsessive-Compulsive Disorder
OEM	Original Equipment Manufacturer
OR	Operation Room
OSA	Obstructive Sleep Apnea
PBS	Phosphate Buffered Saline
PCA	Patient-Controlled Analgesia
PCB	Printed Circuit Board
PD	Parkinson's Disease
PET	Positron Emission Tomography
PI	Principal Investigator
PM	Personalized Medicine/Program/Project Manager
PMA	Pre-market Approval
PMA-S	Pre-market Approval Supplement
PMS	Pain Management System/Post-market Surveillance
PNS	Peripheral Nerve Stimulation
PVD	Physical Vapor Deposition
QA	Quality Assurance
QMS	Quality Management System
RA	Regulatory Affairs
RAA	Reactive Accelerated Aging
RF	Radio Frequency
RGA	Residual Gas Analysis
RI	Retinal Implant
RNS	Responsive Neurostimulator System
ROS	Reactive Oxygen Species

RR	Rate Responsive
SAR	Specific Absorption Rate
SCI	Spinal Cord Injury
SCS	Spinal Cord Stimulation
SEM	Scanning Electronic Microscope
SNR	Signal-to-Noise Ratio
SNS	Sacral Nerve Stimulation
SoC	System-on-Chip
TACS	Transdermal Alternating Current Stimulation
TDCS	Transdermal Direct Current Stimulation
TENS	Transdermal Electrical Nerve Stimulation
TESS	Targeted Epidural Spinal Stimulation
TNS	Tibial Nerve Stimulation
UE	European Union
UEA	Utah Electrode Array
UI	Urinary Incontinence/Urge Incontinence
US	Ultrasounds
UV	Ultraviolet Light
V&V	Verification and Validation
VCSEL	Vertical Cavity Surface Emitting Laser
VNS	Vagal Nerve Stimulation
WB	Wire Bonding
WLAN	Wireless Local Area Network
WP	Work Package
YAG	Yttrium Aluminum Garnet

Chapter 1
Introduction

The objective of this book is to provide a general overview, in easy language, not scientific, of neuro-technologies, in the context of translational medicine, from concept to human clinical applications. Deep explanations on the physiological and clinical aspects of neurological disorders are not the purpose of this book. An abundant literature is available for a more scientific and medical understanding.

The subtitle of the book is *How to build the brain-computer interface of the future*. The keyword is *build*, and the emphasis will be put on the translational development, from concept to patient, with a special focus on how to practically execute projects in the field of active implantable medical devices applied to neurological indications. *Build* also concretely means that our intent is to design, manufacture, and commercialize devices which will provide improvements in the quality of life of patients suffering from neurological disorders of various origins, from birth defects, accidents, diseases, degeneration, or age-related degradation.

1.1 Brain-Computer Interface (BCI)

There are several definitions of brain-computer interfaces, which sometime have other names like brain-machine interface (BMI), mind-machine interface (MMI), or neural interface (NI). The most global definition of BCI is a direct interaction between the neural system and electronic systems. Some authors limit the use of the term "BCI" to bidirectional communications with the brain only. The term BCI made its first appearance at the University of California in the 1970s.

Other notions, like neuromodulation and neuroprosthetics, may somehow overlap with the terminology BCI. As this book is focused on technologies, we will not have any restrictive definition of what a BCI is. We will cover the technical challenges of any system intended to enter in contact with the entire nervous system and senses (see Fig. 1.1a).

C. Clément, *Brain-Computer Interface Technologies*,
https://doi.org/10.1007/978-3-030-27852-6_1

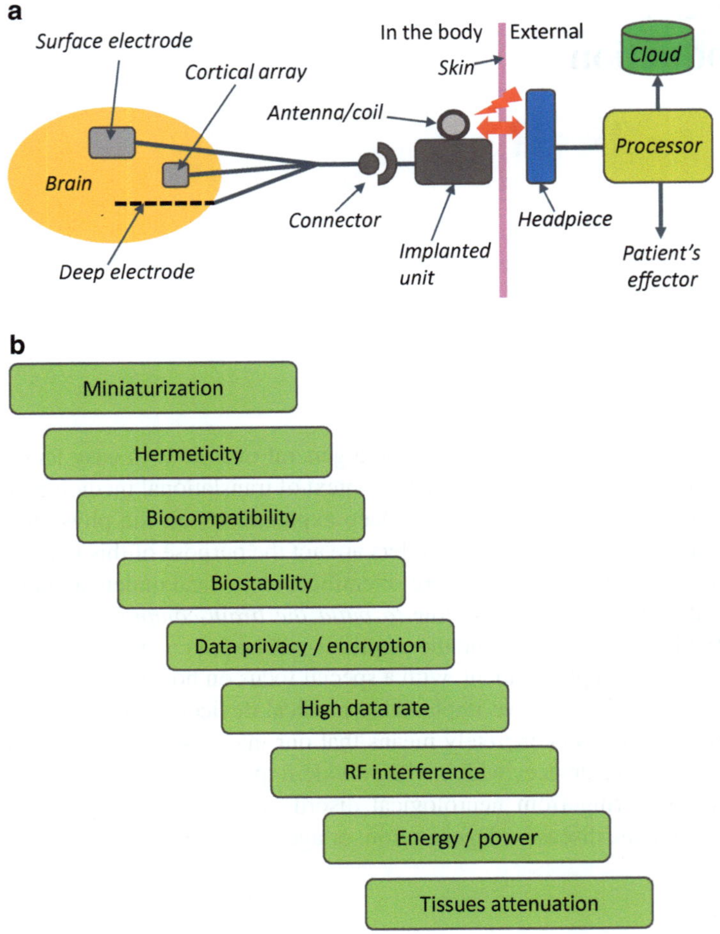

Fig. 1.1 (**a**) Global description of a bidirectional BCI. (**b**) Main challenges in building BCI systems

We will see along the various chapters of this book that interfacing with the brain is a very complex task, mainly due to the nature of the human body. A thorough easy-to-access article on BCIs has been published by the economist [12]. It is a good introduction to understand the global context. At page 7 of this document, the quote says: *"The brain is not the right place to do technology."* We'll explain this statement in this book and cover the main challenges (see Fig. 1.1b) involved in the building of devices interfacing with the brain and the nervous system.

In a first stage, BCI systems are unidirectional, limited to "reading" the brain (see Fig. 1.2). There are plenty of possible configurations of BCI for collecting signals from the brain.

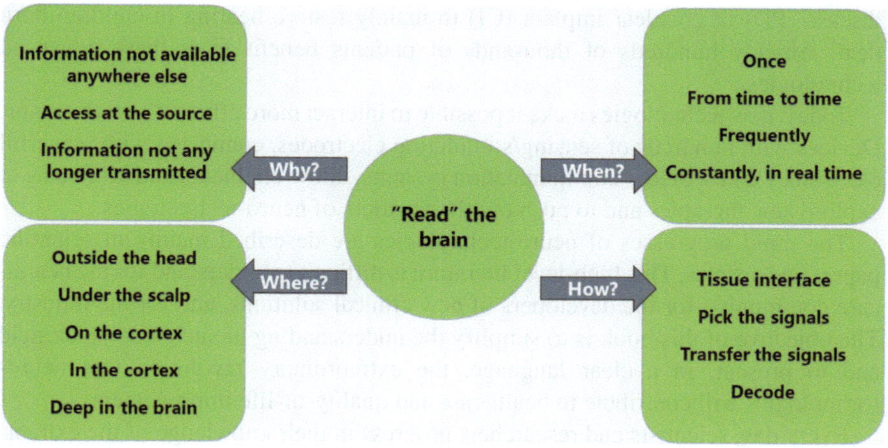

Fig. 1.2 Reading the brain

1.2 Technology Versus Science

Technologies are now available and make it possible to interact with the human brain and nervous system. This book covers the past achievements, the current work, and the future perspectives of BCI and other interactions between medical devices and the human nervous system. Repairing and rehabilitating patients suffering from neurologic impairments, from paralysis to movement disorders and epilepsy, are described in detail, from a pragmatic point of view. Whenever possible, we try to interact with the nervous system without breaking the skin barrier. Nevertheless, such severe disorders often require an invasive solution, based on an implanted device. This book explains the unique and special environment of active implants electrically interfacing with the brain, spinal cord, peripheral nerves, and organs.

BCI should be understood as a wide concept:

- B: Brain, but also central nervous system, spinal cord, vagal nerve, peripheral nervous system, senses, and various organs.
- C: Computer, but also "machines" (BMI: brain-machine interface), implanted electronics, and external electronics.
- I: Sensing and/or stimulating from electrodes or tissue interfaces.

Active implantable medical devices (AIMDs) have been available from the 1960s, mainly to treat cardiac disorders. Pacemakers and implantable defibrillators are now very mature, reliable, and efficient devices, several of them being implanted in patients every minute all over the planet. Using similar technologies, based on hermetically sealed electrical stimulators and sensing devices, the industry of active implants started to address other unmet medical needs at other locations in the body, like deep brain stimulation (DBS) cancelling the symptoms of the Parkinson's disease (PD) or cochlear implant (CI) to mainly restore hearing in children born deaf. Already hundreds of thousands of patients benefit from these advanced technologies.

Today, new technologies make it possible to interact more efficiently with organs. Devices with hundreds of sensing/stimulating electrodes, connected with powerful electronics and wireless communication systems, allow engineers and clinicians to explore new therapies and to push out the frontiers of neuro-technologies.

The rapid progresses of neuro-technologies are described mainly in scientific papers and articles. This high-level literature is difficult to understand for the health-care community, for the developers of new clinical solutions, and for the industry. The objective of this book is to simplify the understanding of such a complex field and to present, in a clear language, the extraordinary revolution that neuro-technologies will contribute to healthcare and quality-of-life improvement.

Every day, scientists and researchers progress in their knowledge of the extraordinary complexity of our nervous system, rising hopes and expectations for better therapies, more accurate diagnostics, and coverage of unmet medical needs.

This constantly improving grasp of the interactions between cells, neurons, brain circuits, and organs paves the way to new technological solutions. The main goal of this book is to describe how to translate the considerable progress of neurosciences, in devices, tools, interfaces, software, and other technological steps, which will give patients a better life.

Experts in translational neuro-medicine must be bilingual. They need to understand the language of neuroscientists and to be able to translate it properly in technological needs, specifications, and human factors. Working together, scientists and engineers have the power to assess the technical limitations, the physics of implants in the human body, and the realistic long-term perspectives.

This book will provide down-to-earth global analysis of neuro-technology for human benefit, including science, technology, regulatory, clinical, reimbursement, patient's acceptance, surgical aspects, and long-term perspectives. We will review the evolution of the AIMD industry, moving from cardiac to neuro-applications. A critical analysis on the pioneer implantable neuro-indications will also show that many people already benefit from neuro-technologies. Reviewing "who-is-doing-what" in this field will confirm the statement that "the next decades are going to be the age of neuro-technologies."

1.3 This Is Not Science Fiction

Neuro-technology is not science fiction. Since the 1980s, millions of people have benefitted from implants not related to cardiac disorders. Every day, in the streets or public transportation systems of large cities, you meet somebody who has a neuro-device implanted, but you do not even notice it. This is a proof that the neuro-industry has already succeeded in repairing people to a level that the other bypassers do not know anything of the problem. Let's quickly mention some successful therapies and corresponding devices related to the nervous system. A deeper review of some of them can be found in Sect. 3.2.

1.3.1 Cochlear Implants (CI)

Interfacing directly with neuroreceptors of the inner ear was the first commercial achievement of neuro-technologies. CIs are mainly implanted in children born with a nonfunctioning conduction of the sound waves from the eardrum to the cochlear, often due to malformation of the middle ear. CIs are also indicated for the treatment of severe deafness of adults and elderly people. A tiny electrode is introduced in the cochlear and stimulates the natural neuroreceptors of the inner ear. The electrode is connected to an implanted electronic in a hermetic housing which receives signals from an external sound processor positioned on the scalp, at the rear of the ear. Natural sounds are picked by a microphone and processed by the external unit. CI will be described in more details in Sect. 3.3.1.

1.3.2 Deep Brain Stimulation (DBS)

Available since the end of the 1980s, DBS systems consist of electrodes placed in specific areas deep in the brain, mainly to treat movement disorders like Parkinson's disease, dystonia, or essential tremors. The leads are tunneled under the scalp and then along the neck to be connected to an implantable pulse stimulator (IPG) located in the pectoral area. The electrical signals applied in the brain block the symptoms characteristic to PD like uncontrolled movements and tremor of the upper limbs. Details on DBS will be covered in Sect. 3.3.2.

1.3.3 Spinal Cord Stimulation (SCS)

SCS represents about 50% of the overall market of neurological implants. Electrical signals are sent to selected area of the spinal cord, mainly for the treatment of chronic back pain. Paddle electrodes are connected to an IPG located in the back. Electrical stimulation blocks the pain signals at the root of the nerves and prevents them to reach the brain. Technical aspects of SCS can be found in Sect. 3.3.3.

1.3.4 Sacral Nerve Stimulation (SNS)

Stimulation of the sacral nerve permits to treat mild to moderate forms of urinary and fecal incontinence. Sacral nerves control functions of the pelvic area. Stimulating them with electrodes placed nearby, connected to an IPG, provide remote control of the bladder and sphincters. More details on urinary incontinence in Sect. 3.3.5.

1.3.5 Vagal Nerve Stimulation (VNS)

The vagal nerve is the second "communication neurohighway," after the spinal cord. It includes afferent and efferent fibers. Stimulating it permits some control on epilepsy, treatment-resistant major depressive disorders (TR-MDD), and other treatments of organs related disorders. Stimulation of the vagal nerve is done either by placing a cuff electrode around the nerve, connected to an IPG, or by transcutaneous stimulation.

1.3.6 Various Devices

In addition, several devices have been developed and approved to treat diseases related to the nervous system. Some examples:

- Programmable implantable drug delivery pumps for intrathecal injection (in the cerebrospinal fluid (CSF)) to treat chronic pain, end-of-life pain, and tremors.
- Gastric nerve stimulation (GNS) is aiming to treat obesity by electrical stimulation of the upper part of the stomach.
- Retinal implants have proven efficient to give some visual perception to totally blind patients (more details in Sect. 3.3.4).
- Tibial nerve stimulation (TNS) has shown potential to treat mild urinary incontinence by stimulating the tibial nerve, by external or implanted stimulation.
- Functional electrical stimulation (FES) is already used by some groups to directly apply electrical stimulation to nerves or muscles for the restoration of simple movements for paralyzed patients.

1.4 Pioneers, Doers, and Dreamers

1.4.1 Pioneers

We will see later in this book that most of the technical developments related to electrical interactions with the human body find their origins in cardiac applications. It is known since centuries [1] that muscles and nerves react to electrical stimulation. Implantable systems could only be realized when transistors, integrated electronics, and small batteries became available in the late 1950s. First came the pacemakers and about 30 years later implantable defibrillators which needed much more sophisticated electronics. Then, in the late 1980s, the first neuromodulation devices appeared: deep brain stimulation and spinal cord stimulation. At the same period, CIs made their way to market. Early neuro-devices are not strictly speaking BCI but rather stimulators interacting with the nervous system. In this sense, spending some time to understand how they developed is also part of the objectives of this book: how to build the BCI of the future.

1.4.1.1 Pacemakers

Early pacemakers [2], in the later 1950s, were simple pulse generators, with fixed pulse rate, basic non-programmable electronics, and mercury batteries potted in epoxy or silicone rubber. Long-term reliability was poor, as epoxy encapsulation did not provide long-term hermeticity. Nevertheless, those simple devices opened the door to an entire industry by providing acceptable life-supporting solutions to thousands of people with serious cardiac disorders.

In the 1970s, the first laser welded hermetic titanium-encapsulated pacemakers paved the way for high reliability implants with sophisticated, programmable, and integrated electronics. Hermetic sealing achieved two major steps in the field of implantable devices:

- Protection of the patient in case of battery leakage.
- Protection of the implanted electronics from moisture and body fluids.

The pacemaker industry has set the fundamental grounds of active implants. Early devices were not hermetically encapsulated, meaning that sooner or later, electronic components will be exposed to moisture. At that time, the electronics of the implants were based on discreet components like simple transistors, resistors, and capacitors, assembled with a comfortable distance between them. In this configuration, diffusion of moisture through the plastic encapsulation was not critical.

When electronics became more integrated, with thousands of transistors on integrated circuits (ICs) and short distance between components, simple epoxy or silicone encapsulations were not enough to provide long-term reliability. Total hermeticity was required to avoid exposure of sensitive electronic components to moisture and oxygen. Laser seam welding of titanium housing provided the solution for long-term reliable high-tech implants. Feedthroughs are key components to build hermetic packaging. They consist in one or several conductive wires sealed in an insulator, itself brazed in the packaging. These wire connections allow communication between the electronics in the package and the tissue interfaces. These technologies could then be applied to other indications.

The pacemaker industry is now a mature technological field with very high reliability. About 1.5 million pacemakers are implanted every year.

1.4.1.2 Implantable Cardiac Defibrillators (ICDs)

The first ICDs appeared in the early 1990s. Compared to pacemakers which generate low-voltage pulses to stimulate, resynchronize, or assist the heart, ICDs are designed to provide high-voltage high-energy electrical shocks in case of sudden cardiac arrest, severe tachycardia, or ventricular fibrillation. ICDs include advanced electronics and high-voltage circuits which require hermetic encapsulation. As ICDs are life-supporting devices, they cannot rely on rechargeable batteries, which might be depleted when needed. The primary battery being at low voltage (3.5 V), a complex voltage multiplier rises it to about 700 V, necessary to generate high-energy shock, in the range of up to 40 J. This large amount of energy cannot be continuously stored in the implant. Therefore, when electrodes detect a situation of fibrillation or a heart stop, the multiplier starts loading a capacitor with the appropriate energy for the shock. It takes 10–20 seconds before the ICD is ready to fire.

Modern ICDs have been miniaturized and are now used in large numbers of cardiac indications, combining regular stimulation and defibrillation. Hundreds of thousands ICDs are implanted every year.

1.4.1.3 Cochlear Implants

As mentioned earlier, CIs have been a major contributor to the evolution of active implants. They are the first neurological active implanted devices to have reached a large population. Unlike pacemakers and ICDs, CIs are battery-less devices. The implanted electronics get its energy through transdermal inductive magnetic coupling of an implanted coil and an external coil. The acoustic signal is transmitted through the same inductive coupling.

In the last 30 years, about 700–800 thousand CIs have been implanted in children with congenital deafness or in older patients with severe hearing disorders.

1.4.1.4 Deep Brain Stimulation

In the late 1980s, DBS became the first therapy interacting directly with the brain. Patient's benefits were amazing even if the understanding of the effects of electrical stimulation on the thalamus were then not totally understood. Today, more than 200,000 patients are well treated, having no visible symptoms of Parkinson's any longer. Compared to more drastic surgery, like tissue ablation, DBS has the advantage of being controllable and reversible.

1.4.1.5 Spinal Cord Stimulation

A few years after DBS, it was understood that stimulation electrodes could be placed on the spinal cord where afferent nerve merges to it. Applying mild current at this location allows a substantial reduction of the perception of pain, for example, in cases of chronic back pain (CBP) or lower limbs pain. Compared to other methods for treating pain, like drugs, SCS has the advantage of have no side effect and to be reversible.

1.4.2 Doers

The pioneer indications are going on growing and serving more and more patients. More recently several products made their way to the market to treat other patients' needs. There is currently a formidable energy focused on applying technology to treat neurological disorders. Some projects are leveraging the technologies of the pioneers to treat new indications. Other groups are pushing the former technologies further with the objective to meet medical needs which were not so far reachable. Here below, you'll find a brief description of recent (last two decades) and ongoing initiatives with promising outcomes.

1.4.2.1 Spinal Cord Stimulation

SCS has been described as a pioneer technology, but, because of its success, it also belongs to this chapter. SCS is the largest indication in the field of neuro-technologies. Its impact in terms of quality of life and societal benefit is clear. The therapy is expected to improve. New projects, using high-frequency stimulation, show promising results, even if the scientific rational is not yet fully understood. Controlling pain through electrical stimulation is a high-potential target. A lot of progress is expected in this domain.

1.4.2.2 Sacral Nerve Stimulation

Like SCS, SNS is a therapy which is mainly unknown of by the population, but hundreds of thousand patients have already benefited of SNS for a better control of urinary incontinence. Originally, the indication was limited to mild forms of urge incontinence (UI) and overactive bladder (OAB). Medtronic was a pioneer in this indication [3]. Today, we see new companies like Axonics [4] and Nuvectra [5] entering in this field and an extension of indications in the direction of fecal incontinence. So far, SNS is not able to treat severe forms of incontinence, like post-prostatectomy incontinence and older women severe incontinence, which remain real unmet medical needs.

1.4.2.3 Vagal Nerve Stimulation

Cyberonics (now LivaNova) [6] was first to attempt stimulating the vagal nerve in order to control epilepsy. It demonstrated that a lot can be achieved by interfacing with the vagal nerve. VNS is still one of the only therapy available for the treatment of some forms of epilepsy. There are several other initiatives aiming to stimulating the vagal nerve for other indications. In neurology, it has been shown that VNS might be efficient to treat forms of depression, like major depressive disorders (MMD), without understanding all the brain mechanisms associated with these results.

Other applications of VNS, not strictly neurological, have been developed, for example, for the treatment of morbid obesity through gastric electrical stimulation (GES). Original work in this direction has been done by EnteroMedics [7] which has now merged with ReShape Lifesciences [8] providing a gastric band for the same purpose. VNS as proposed by EnteroMedics failed in proving to be superior to other solutions.

1.4.2.4 Retinal Implants

Three to four companies are achieving tremendous successes in their endeavor to provide some sense of vision to blind people. Retinal implants are still limited to hundreds of pixels. It is small compared to the performance of a healthy retina. But, getting some basic visual perception is an enormous improvement for blind people. From the ongoing work, we can anticipate substantial achievements. Several teams are currently progressing fast on other neuro-interfaces to restore vision, where electrodes are not located in the eyes but rather on the optic nerve or on the visual cortex.

1.4.2.5 Peripheral Nerve Stimulation (PNS)

Stimulation of nerves outside the brain and spinal cord domain has shown a high potential. Several companies are working in the field of PNS with exciting successes. Among them SNS can be considered as PNS. Other therapies, like gastric nerve stimulation (GNS), for fighting obesity also belong to the PNS group. FES and TNS, already described above, are addressing specific medical needs with several approved devices.

1.4.2.6 Intelligent Prosthesis for Amputees

Various ongoing projects are aiming to connect intelligent prosthesis with the remaining nerves at the root of the lost limbs, either to be able to activate the prosthesis directly from the patient's nerves or to provide a sensory feedback (haptic) from sensors placed in the prosthesis and connected to the nerves. The number of patients who benefit from these devices is still limited, but large progresses will come soon, especially when lower limbs amputees would become eligible.

1.4.2.7 Diagnostic and Monitoring of Epileptic Patients

The common approach to assess occurrence, intensity, frequency, and localization of epileptic seizure is to use electroencephalography (EEG). Unfortunately, EEG caps cannot be worn for extended periods of time. Home-based accurate long-term monitoring is not available yet, with the exception to NeuroPace RNS system (see Sect. 3.4.7). It consists in an implantable recorder, inserted in a craniotomy, and connected to 8–16 electrodes (paddle cortical electrodes or penetrating electrodes). Several groups are currently developing less invasive implantable system for medium- to long-term diagnostic and monitoring of epileptic patients, with objectives of being able to forecast or event predict seizures. An example is UNEEG [9], a Danish company part of the Widex Group [10], a hearing aid supplier.

1.4.2.8 BCI for Sensing Motor Areas of the Cortex

Since more than a decade, the BrainGate Initiative [11] gathers five US institutions in a consortium which leads the way of research and development in the domain of reading movement intentions of paralyzed patients. Sensing the cortical activity is mainly done through the so-called blackrock array or Utah array (see Fig. 1.3), a microelectrode array (MAE) [12]. This tiny tissue interface of up to 100 fine electrodes penetrates about 1.5 mm in the motor cortex.

So far, the electrodes are connected to a bundle of thin gold wires and a transdermal connector called pedestal (see Fig. 1.4). The pedestal is attached to the skull.

Fig. 1.3 Utah or blackrock array. *(Courtesy: Blackrock Microsystems LLC)*

Fig. 1.4 Utah array connected to a transdermal pedestal. (*Courtesy of Blackrock Microsystems LLC)*

Until now, around 15–20 paralyzed patients have had 1 or 2 Utah arrays inserted on their motor cortex. The greatest challenge resides in the real-time decoding of the movement intentions. Early work enabled a paralyzed patient to successfully move, by his/her thoughts only, a cursor (2D) on a screen, click on icons, use a speller, and conduct other tasks similarly to the activation of a computer mouse. Later, it became possible to decode and extract information corresponding to more complex movements, up to a dozen degrees of freedom. "Move, Reach, and Grasp" movement intentions of the arm have been decoded successfully, allowing the activation of a robot arm for simple tasks like drinking from a bottle or taking food in a bowl with a fork. Recently, the robot arm was replaced by direct FES stimulation of the paralyzed patient's arm.

Current work (see Sect. 7.3.6) is using the same type of BCI to regain contact with people with completely lock-in patient syndrome (CLIS).

1.4.2.9 Others

Several developments related to innovative devices for interfacing with the nervous system are going on around the planet. To cite just a few:

- Simulation of the spinal cord for reactivate walk in paralyzed patients or for rehabilitation after stroke.
- Stent-like electrodes placed in brain blood vessels for sensing brain signals.
- Stimulation of the inner ear to repair vestibular disorders.
- Stimulation of the optic nerve or on the visual cortex to treat blindness.
- Steerable DBS for a more accurate treatment of PD.
- Use DBS for other syndromes like obsessive-compulsive disorders (OCD), chronic depression, migraine, Tourette's syndrome, obesity, addictions, epilepsy, etc.
- PNS to treat amputees' phantom pain.
- Mirror restoration of unilateral facial paralysis.
- Neurofeedback for tinnitus.
- Hypoglossal nerve stimulation to treat sleep apnea.
- Gastric nerve stimulation for gastroparesis, nausea, and vomiting.
- Brain re-synchronization for dyslexia or certain speech disorders.
- Wired and wireless networks of implants for FES.
- ....

The ongoing development efforts in neuro-technologies will have a considerable impact on health and quality of life. Some improvements are done step-by-step. Some will be disruptive and revolutionary. Learning from the work of the pioneers is essential to execute good research and development today. Anticipating the trends and changes in our environment induces us to also listen to the "dreamers."

1.4.3 Dreamers

Pioneers and "doers" in neuro-technologies were or are mainly physicians, health-care specialists, surgeons, engineers, regulators, scientists, and researchers. Their focus is on improving therapies and diagnostics, with patients in the center.

A new category of players appeared recently: dreamers. Their goals are to use BCI for nonmedical applications. They usually do not have a full understanding of the specificities of the human body. They also underestimate the technical challenges related to interfacing with the brain.

They are successful, wealthy, and young entrepreneurs who founded and grew immense companies, mainly in communication, Internet, software, online commerce, or electrical cars. Their capacity to reinvent entire industries is amazing. As such, doers should listen to dreamers and grasp opportunities whenever possible.

Dreamers want to push BCI beyond its current stage, aimed to repairing people with neurological disorders. They want to distribute BCI over the entire volume of the brain, with thousands of tiny "grains" of electronics communicating in a wireless network. Among other dreams, they would like to extend the use of BCI for the augmentation of the natural capacities of our brain. Nonmedical applications, like a new sort of B2B (in this case brain-to-brain communication), or connecting your phone to your brain, or driving your car directly from your brain, are, for most of us utopia, but not for the dreamers.

At the end of this book, we will come back to the societal, economical, and ethical aspects of nonmedical BCI.

1.5 The Age of Neuro-Technologies

As described before, in the field of active implantable devices, the end of the twentieth century has been the age of cardiac rhythm management (CRM). The beginning of the twenty-first century is definitively the age of neuro-technologies.

1.5.1 Convergence of Technologies

We will see later that interfacing with the brain is technically difficult, in terms of materials, energy, handling tiny signals, encapsulation, miniaturization, and communication. CRM devices have less channels, are principally stimulating and not sensing, have less sophisticated electronics, and are in body locations where size is not as critical as in the head. In this less demanding environment, cardiac devices were able to provide efficient therapies using simple technologies, starting in the 1960s and growing significantly in the 1970s–1980s.

Progress in integrated microelectronics, constant reduction of electrical consumption, and achievements in miniaturization are now opening the way to multichannel, large bandwidth, and real-time BCI. Other technical barriers refrain the progress of implanted electronics: energy, multiple hermetic feedthroughs, and implantable connectors. Innovative firms are proposing alternatives to improve the performances of these strategic building blocks. In Chap. 4, we are going to identify the limitations induced by the body itself. More details on technical barriers are to be found in Sect. 7.2.

1.5.2 Limitations of the Pharma- and Bio-Industries

Regarding the treatment of neurological diseases or disorders, only a very few potent new drugs have been developed during the last decades. Many pharmaceutical groups and biotechs are using substantial resources in the search of solutions for neurodegenerative diseases, like Alzheimer's disease, dementia, or some psychiatric disorders. After several decades of considerable efforts, results are disappointing. It may be time to search for solutions outside of the traditional pharmaceutical approach, for example, among neuro-technology concepts.

For other indications, like epilepsy, a large percentage of patients are not responsive to available drugs. The likelihood of finding better drugs is limited. There too, neuro-technologies might give some hope, especially for patients not responsive to drugs. We also believe that new neuro-devices might be of great help to provide better diagnostic and follow-up to drug-responsive epileptic patients, allowing them a better titration of their drug, more efficacies of the treatment, and less side effects.

Some of these limitations in the development of new or better drugs could be superseded by technological means, like electrical stimulation, close-loop therapies, localized programmable drug delivery, or combinations of them. In the near future, we will see pharmaceutical companies and technological innovators work hand in hand in search of solution to unmet neurological needs, in a promising win-win situation. Pharmaceutical companies understand patients' needs, the medical and clinical aspects. Technological firms will provide a new set of tools to tackle the disease.

1.5.3 Unmet Medical Needs

As mentioned before, several neurological disorders or diseases have no or poor solutions of treatment. With the aging of the population, the load of neurological deficiencies on our societies becomes unbearable.

It has been shown in the recent past that technologies could improve, sometimes substantially, the quality of life of many patients. The example of young kids recovering earing capacities with the help of a cochlear implant shows us that solutions can be found. Thirty years ago, few people would have thought that DBS could be such a major improvement in the life of parkinsonians. Maybe, in a foreseeable future, epilepsy seizures could be predicted or even treated by a BCI system. Today, victims of strokes get limited improvement through long rehabilitation programs. Maybe, in a near future, BCI will enable quicker and more substantial recovery. Progress of technologies will for sure reduce the number of unmet medical needs and drastically improve the outcomes of poorly met medical needs.

References

1. https://en.wikipedia.org/wiki/Electrical_muscle_stimulation
2. Jeffrey K (2001) Machines in our hearts. The Johns Hopkins University Press, Baltimore and London. ISBN 0-8018-6579-4
3. https://en.wikipedia.org/wiki/Sacral_nerve_stimulation
4. http://www.axonicsmodulation.com/product/
5. https://nuvectramedical.com/corporate/about-us/

6. https://en.wikipedia.org/wiki/LivaNova
7. https://www.prnewswire.com/news-releases/enteromedics-announces-affordable-vbloc-therapy-program-for-weight-loss-patients-300480925.html
8. https://www.reshapelifesciences.com/
9. https://www.uneeg.com/
10. https://www.widex.com/
11. https://www.braingate.org/
12. Technology quarterly, Brain-Computer Interfaces, The Economist, January 6th, 2018

Chapter 2
From Concept to Patient

2.1 Translational Medicine

There are several definitions of translational medicine. The description best fitted to the purpose of this book is a methodology which assures that research findings, innovations, ideas, and concepts in the improvement of human health reach patients in need of better therapies. Translational medicine is a global process of transforming ideas in medical products, diagnostic tools, and better healthcare.

2.1.1 From Ideas to Products

Product development is mainly a sequential process, with various phases which must be concluded properly before entering the next phase. In some occurrences, some parallel development can be done with the aim to accelerate the process, but this induces additional risks.

A strict and structured development methodology is the best way to successfully reach the ultimate goals of a translational process: a good product which helps many patients to get a better life. All medical devices which are on the market today went through a rigorous development process, which leaves no room to doubts and uncertainties.

The key of an efficient development is to start the project on solid grounds (see Fig. 2.1): understanding of patients' and healthcare providers' needs, clear specifications, thorough risk analysis, respect of human body limitations, and good command of technologies.

Several methods are used to structure a project. Most common are the sequential "waterfall model" (see Fig. 2.2) and the "V-model" (see Fig. 2.3).

Development models are not good or bad. The main point is to follow a strict methodology and be certain that everything is clearly documented from the very

C. Clément, *Brain-Computer Interface Technologies*,
https://doi.org/10.1007/978-3-030-27852-6_2

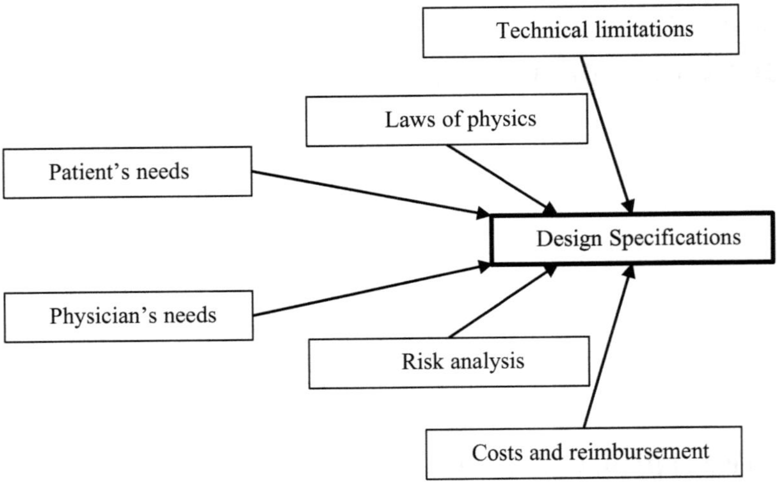

Fig. 2.1 Start the project on solid grounds

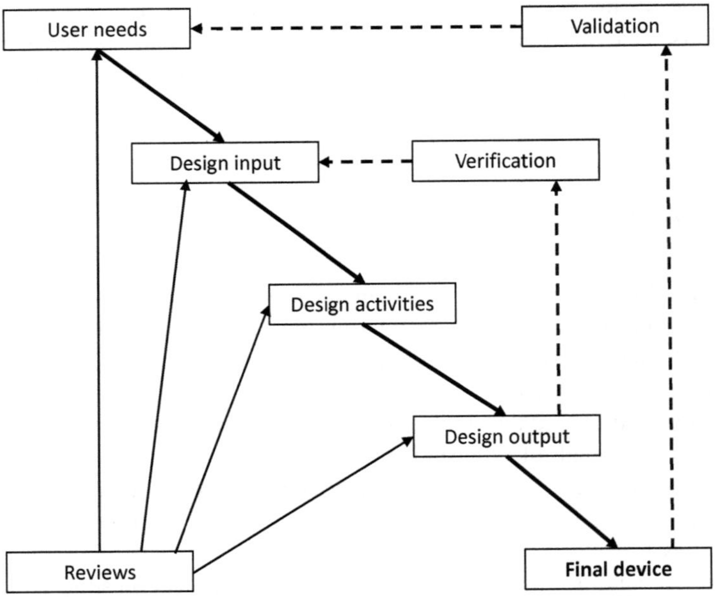

Fig. 2.2 Waterfall model

beginning of the project. Even early day efforts, like brainstorming, conceptual studies, review of the literature, search for patents, and competition analysis, should be traceable in written reports, dated, signed, and properly archived. Failure to document properly each development steps is one of the principal root causes of projects falling in the valley of death.

Developers and manufacturers of active medical devices have a large experience on how to carry a project, from idea to product (see Figs. 2.4 and 2.5), in a way

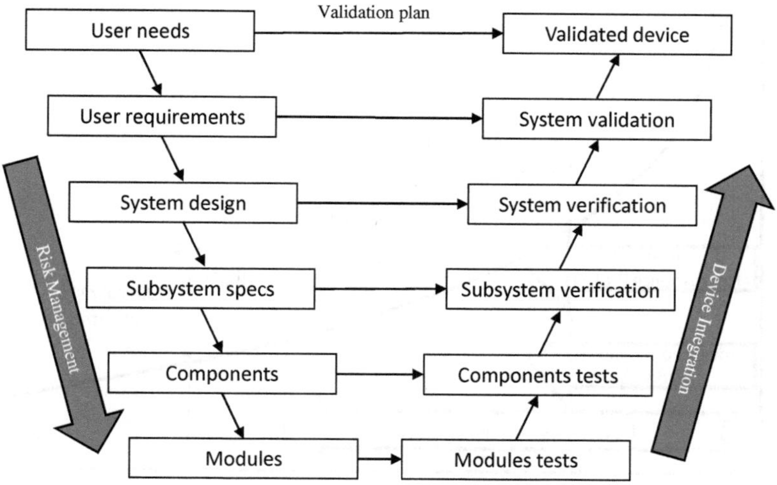

Fig. 2.3 V-Model

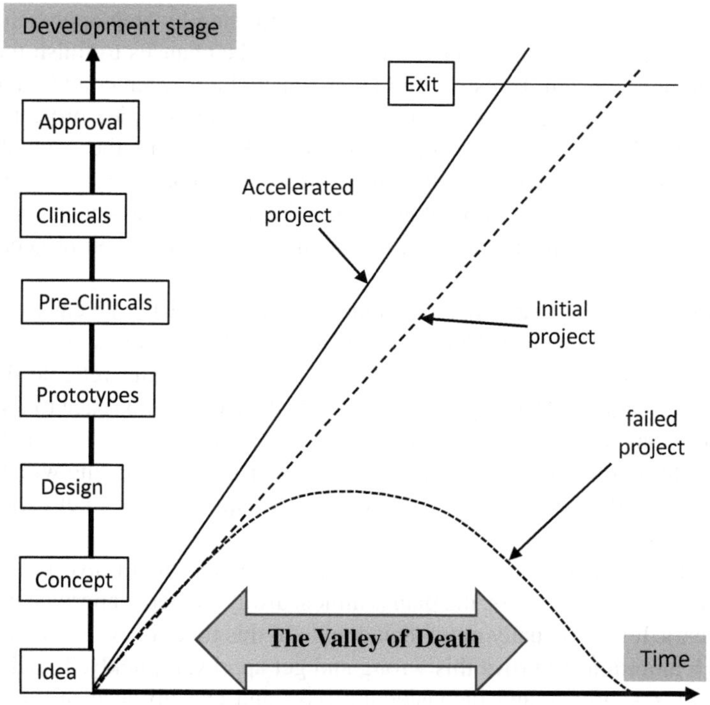

Fig. 2.4 Project evolution

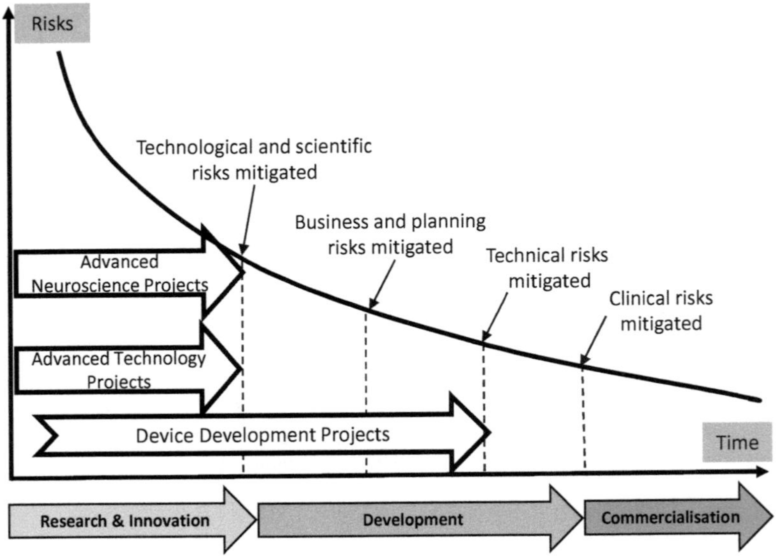

Fig. 2.5 Risk mitigation and life cycle

which optimizes the chances of success. A careful step-by-step approach, solving one problem at a time, takes longer but increases the chances to finish the project without major redesign. Investors and other stakeholders often complain about the long development cycles of active medical devices. For complex projects, it takes usually more than 10 years and costs more than 100 M$ from the original concept to the commercial product. Development teams trying to take shortcuts, like attempting to solve several issues in parallel, or skipping proper assessments and tests, or including too many innovations at a time, are not able to carry their project until the end.

It is often believed that long development cycles of AMIDs are caused by heavy regulations and bureaucracy. This is doubtfully true. The reason is to be searched in the ultimate goals of transitional medicine: provide safe and reliable devices to thousands of patients. Developing human grade devices lets no space to risk taking, approximations, and partial tests. When we go to humans, the very first implant must be safe. This is especially true for AMIDs which are interfacing with the brain. Performance and efficacy can be improved in subsequent phases, but patient safety is not negotiable.

At the beginning of a complex project, under the pressure of investors, management, and other stakeholders, the plan is almost always to get to approval within less than 5 years. It takes usually more than double of this time. I have seen many teams with high ambitions to prove this wrong and get approval quickly. To achieve this, they all took shortcuts and inconsiderate risks and end up deep in the Valley of Death. At the bottom of the Valley of Death, we find the cadavers of very nice ideas, which were pushed to their limits by greedy and unexperienced entrepreneurs.

Looking for AIMD projects which went through the full pre-market approval (PMA) route in the last two decades, none of them took less than 10 years from concept to approval; none of them was below 100 M$. Many companies try to get approval earlier and with reduced investment, but none of them succeeded. Why? I wanted to say: *"Nobody knows,"* but, in fact, they failed precisely because they wanted to be fast and cheap. The main global reasons for failure to be fast and cheap are described on Fig. 2.6.

2.1.2 Valley of Death

Very often, innovative projects, even promising ones, fall in what we call the "Valley of Death." It is the portion of the development life cycle between the middle or end of the design phase and First-In-Human (FIH). These projects die *"at the door of the clinic."*

Why is it the case? Here are some of the reasons, which, individually or combined, may lead projects to be trapped in the Valley of Death:

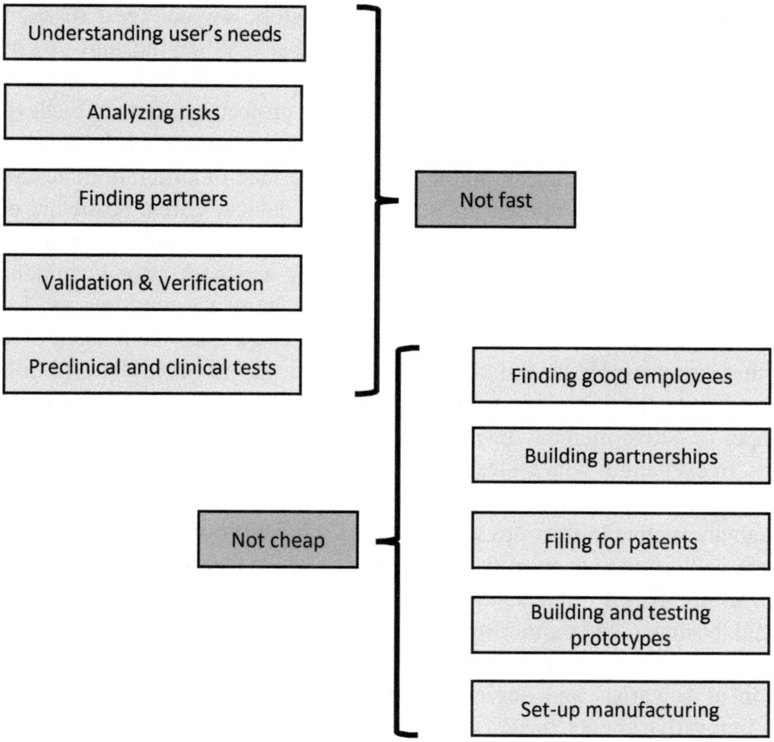

Fig. 2.6 Not fast, not cheap

- Start-up created too early.
- Not enough resources (financial and human) to complete the de-risking phase.
- Failure to plan carefully the development process.
- No enforcement of deadlines.
- Overspending and deviations from the original budget.
- Conceptual and feasibility phases not done thoroughly.
- Too many technological risks taken simultaneously.
- Deep redesign late in the development process.
- Poor understanding of the constrains of the AIMD industry.
- No plan B.
- Overoptimism (in terms of timing and financial resources).
- ....

Neurological devices are complex projects, requiring more time and more financial resources than conventional projects. Start-ups rarely succeed in carrying such complex projects to commercialization. Some early exit strategy must be prepared for having survival alternatives and avoiding falling in the Valley of Death. Start-up founders often refuse to give up their projects, do not pull out in time, and do not let larger, more experienced companies continue the work. The principal root cause of the death of a start-up is the blindness of its founder(s).

Start-ups are certainly not the best structure to conduct development of complex neuro-technological projects. Conventional investors are not used to such long cycles and large investments, with so distant payback. When the start-up enters in difficulties, they often give up and let the company fall in the Valley of Death.

It is extremely rare that a complex implantable project succeeds to be developed in an academic environment. As mentioned above, cycles are too long and grants not enough to support translational initiatives. The lack of structure in academia is not adapted to the development of human grade device. Often, academic entities initiate the project and carry it up to preclinical stage. Unfortunately, in most cases, this early work is conducted without traceability and with poor documentation, finally turning to be worthless for translation to human grade devices. Lacking hands-on experience, academics who attempt to design a device make all the mistakes that many people have done before. A good way to avoid these pitfalls is to spend valuable time searching for *"what-went-wrong-and-why."* Being a skilled developer of active medical devices takes decades of hard work, of learning from failures, of exchanges with colleagues, and of continuous learning. A young postdoc is not likely to immediately be a good developer.

As an alternative to start-ups and academia, philanthropic and nonprofit foundations are well fitted to support the translation of neuro-technological projects. Such organizations dispose of considerable financial means and have access to large financial, business, and partnering networks. An example is the Wyss Center for Bio and Neuroengineering in Geneva, Switzerland [1]. By having hired a unique combination of scientists and engineers, of experienced program managers, and of young innovative talents, the Wyss Center is well positioned to protect projects from falling in the Valley of Death. Furthermore, the experience and bandwidth of such a foundation provide unequaled opportunities to accelerate the development of complex neuro-projects.

2.1.3 Multicultural Approach

I usually say that, for doing good translational medicine, one must be *bilingual*, speaking both the language of science and the language of technology. It goes even further: one must understand both cultures.

To be able to lead sophisticated projects, like BCI for movement restauration of paralyzed patients, the development team must have a very large spectrum of competencies. These diverse talents, skills, and knowledge must be able to cross-fertilize in a synergetic multicultural approach. Each of the groups described in Fig. 2.7 are necessary for success, but only the combination of them will be enough.

Start-ups do not have the luxury to hire competent representatives of each of these groups. Purely academic organizations lack engineering and regulatory skills. Large multinational medical device companies do not invest in early stage projects and prefer that the de-risking work is done by somebody else. These are the reasons why we believe that the development model described in this book is particularly well adapted to BCI and other complex neuro-interfaces.

Multiculturalism is not limited to the aggregation of many diverse professions. It is also the mixture of nationalities, cultural backgrounds, ages, genders, and experience. An efficient development team should be able to build on the differences and is more critical toward bad habits and preconceived approaches. When exploring disruptive solutions in the fields of technology, things must be done differently. Neuro-technologies are now erupting from nowhere. It is an opportunity to generate breakthrough ideas.

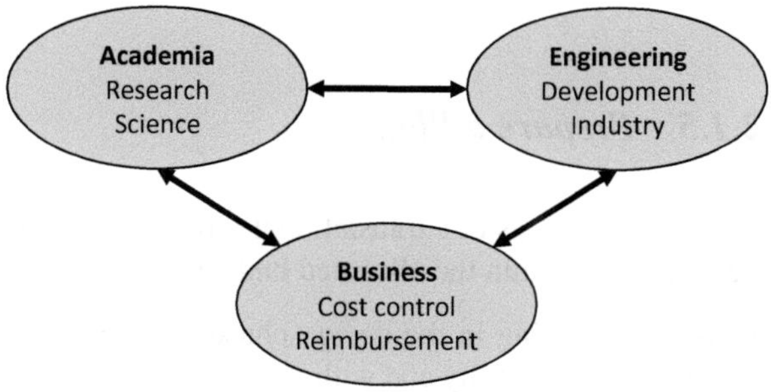

Fig. 2.7 Combination of skills

2.1.4 Setting Priorities

Developing neuro-devices for human applications requires keeping focus on the end goal: *treat patients in a safe and efficient way*. It implies that the project must be based on reasonable specifications, which are reachable with the resources, skills, and experience available, within a predictable time frame.

Translational projects usually fail when deadlines are constantly pushed out. It is better to set modest objectives rather than shooting to the moon. In case of unexpected difficulties, it is acceptable to downgrade some specifications for keeping a deadline. Reasonable objectives lead to success, even if not a revolution. A step-by-step development strategy fosters a succession of successes, which, in the long run, become a revolution. Simple pacemakers took several decades to become flexible programmable intelligent cardiac rhythm management tools. The first steps were addressing the most urgent but addressable needs. They were then life-supporting devices with not much programmability. Today, most pacemakers are providing quality-of-life improvements, fully tailored to the specific need of the patient.

For BCI, we should follow the same philosophy: first solve what is urgent to be solved, even if the results may be far away from our ultimate dreams. Revolutions will have to wait. For example, restoring simple movements of one arm of a tetraplegic patient is a huge step. If the gain in mobility and independence is very modest, the result is a great addition to self-esteem and dignity. Even if it is not permanently available, the capacity of being able, from time to time, to feed yourself or brush your teeth is well compensating the humiliation of being constantly dependent on assistance.

On the same wavelength, paralyzed people get along rather well with their wheelchairs. Architectural barriers are being removed, and wheelchairs have facilitated access almost everywhere. Often, tetraplegics are not only unable to move but suffer from other disorders such as urinary incontinence. Many of them would like to first have a device to control their bladder. Being able to walk again may be a second priority. I heard tetraplegic patients say: *"leave me my wheelchair but do something for my incontinence."*

Engineers and scientists are not always listening to patients and doctors. Good translational medicine is achieved when patients' major priorities are finally met. Helping a few people today with simple solutions is better than to have unreachable ambitions.

2.1.5 Prepare a Plan

The key of success of a translational project is planning (see Fig. 2.8). For a plan to be reachable within the allocated budget, several conditions must be met:

- Before entering in the design phase, conceptual and feasibility phases must be conducted carefully and without concessions. It is capital to allocate enough time

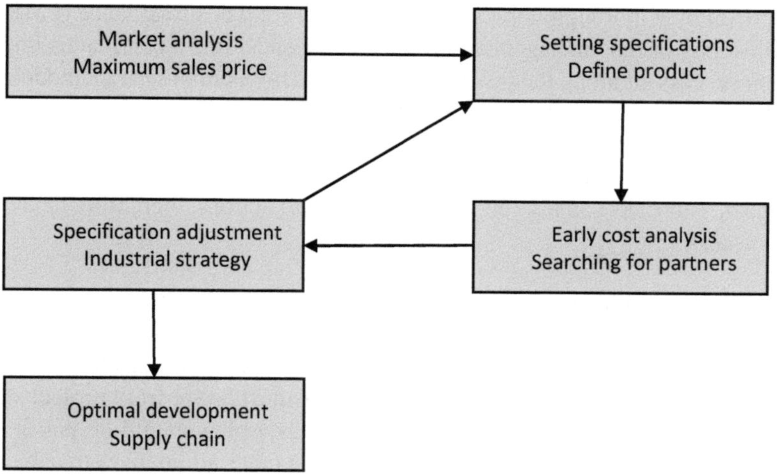

Fig. 2.8 Development plan

to this preliminary phase. It is not lost time. It is not an expense, but an investment. Making the right choices now will save a lot of time later. Many projects fail because feasibility was not demonstrated. Entering the design phase without being sure that there will be a design out is a huge risk to take.

- In a first step, a gross description of the concept, for example, in the form of some variations of 3D printed mock-ups, should allow early discussions with patients and their doctors. This preliminary feedback from the end users is a key to success.
- A market analysis must be done before entering the design phase. It will quantify the market size, the potential penetration of the future product, identify current and future competitors, and allow a market segmentation related to the possible sales price. Having a gross understanding of reimbursement is important.
- In this early phase, the project manager must be able to assess if the estimated ex-factory cost is in line with the sales price tolerated by the market (as evaluated during the market analysis phase), with enough gross margin. Starting a project with predictable thin margin is a main cause of failure. Never forget that "the market is always right." Therefore, if the target sales price does not leave enough room for safe margins, then the overall specifications of the product must be downgraded.
- A thorough risk analysis must be put in place at the beginning of the project. Annex 1 covers the basic principles of risk management in the field of active medical devices. Regular reviews and updates of the risk analysis should induce mitigations of unacceptable risks. These mitigations may have an impact on design input and specifications. Risk analysis is a development tool and a dynamic one. Risk analysis must be started at the very beginning of the feasibility study phase, constantly upgraded through the various development phases, then continuously updated until approval, and followed-up carefully after

approval by a thorough implementation of post-market surveillance (PMS). The philosophy of risk management is that it never ends. It is a continuous improvement process all along the product life cycle. In my opinion, the methodology of risk management should be extended to nontechnical factors, like competition, IP, costs, partnerships, supply chain, regulatory, and so on. Alike technical risks, "soft" risks may be quantified in terms of probability of occurrence, severity, and impact. Soft risks can also be mitigated by smart actions, increasing the chances of success.

- Whenever specifications are changed, for example, to meet deadlines or to fix problems, a systematic analysis of the impacts of the changes should be done. It is current that "fixing a problem" induces many new problems, etc.
- Detailed milestones with reasonable deadlines must be set from project start. In case of trouble to meet a deadline, a contingency plan should be put in place. Deadlines should not be moved out until all the other alternatives to achieve the milestone on time have been tried.
- The project should be split in work packages (WPs), each having objectives, a deadline, allocated resources, and a budget. Some WPs may overlap or be run in parallel.
- WPs should be allocated to sub-teams, with one WP leader being responsible for reporting and outcomes.
- The program manager (PM) manages the project, coordinates the activities, synchronizes the WPs, enforces deadlines, tracks the budget, but does not do development or engineering work.
- The plan should include contingencies and alternative routes for critical items.
- A visual representation (e.g., a Gantt chart) of the overall planning should be maintained and be kept available to the project team and management.

Development of devices goes through several sequential phases, with reviews and close loops to modify, correct, and improve specifications. The journey from the idea to the clinical application is long and tortuous (see Fig. 2.9):

A good plan also includes priorities. As a metaphor, let's say that the project is to have a fast train through the Alps. The priority is to build bridges and tunnels, minimize curves, and assure that the tracks are solid and reliable. Choosing the type of locomotive can be done later. When you dig a tunnel and your budget is running short when you are half the way through, you have only two options:

- You find additional money and get the tunnel completed.
- You do not get money, and you have a cave instead of a tunnel.

There are multiple ways to follow the project development, assess progresses, quantify necessary resources, and track interactions between work packages. Sophisticated software packages allow detailed tracking but consume a lot of time for keeping all data up to date. At the beginning of a project, simple tools like Gantt charts on spreadsheets are recommended to have a good view on the long term. Figure 2.10 shows an example of a gross overall planning for the development of a BCI:

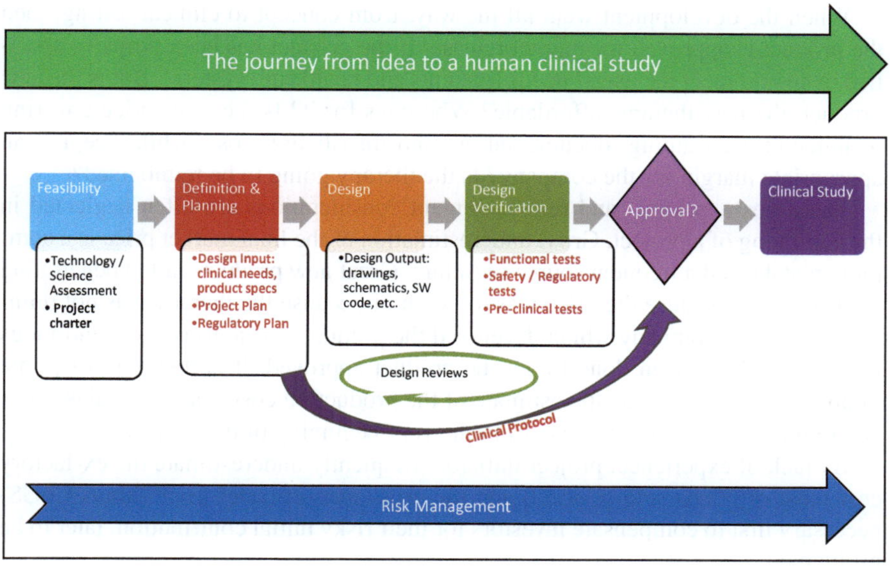

Fig. 2.9 From idea to clinical study

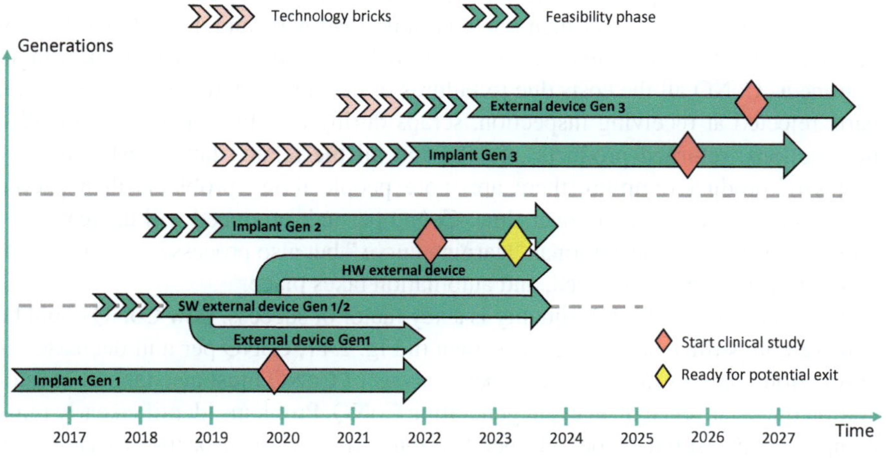

Fig. 2.10 Overall development plan for a family of products, succession on three generations of products with some steps done in parallel

2.1.6 Think About Costs

Besides the importance of having a clear plan and a solid development budget, a thorough understanding of the final product cost is capital for the success of the project. Translational initiatives "must go to the end," meaning reaching the market and serving patients.

When the development went all the way, from concept to clinical testing, then the product is approved for market release. If the product has been properly specified, it should meet the users' needs (patients and healthcare system). But is the new product, the new therapy, affordable? Who pays for it? Is the sales price covering manufacturing, training, documentation, and distribution costs while keeping an appropriate margin for the company? Is the therapy going to be reimbursed?

These costs, pricing, and reimbursement considerations are often neglected in the beginning of a project. Gross underestimation of the final market price is a common mistake and a frequent source of failure. Great new products failed penetrating their markets, because the price was too high or because the therapy was not reimbursed. For the company which developed the product, it is a disaster, as enormous investments have been done to get the product approved. It is therefore of prime importance to have a realistic estimate of the production costs, targeted sales price, and reimbursement expectations from the very beginning of the project.

By lack of experience, project initiators frequently underestimate the ex-factory cost (cost of goods sold (CoGS)) and the gross margin (sales price minus CoGS) necessary first to compensate investors for their risky initial contribution, later to be profitable.

Estimating CoGS, one very often forgets about a lot of factors going far beyond the cost of parts and labor. Processes, clean room assembly, engineering support, quality controls, scraps, reworks, testing, sterilization, factory costs, and energy are usually more expensive, per manufactured unit, than parts and direct labor. During the first years of production, the cost of non-quality (CoNQ) may be extremely high. I define as CoNQ all the costs due to failures or deviations in the production cycle: parts rejected at receiving inspection, scraps during assembly and tests, reworks, field actions, returned products, recalls, and all the related labor costs. In early stages of product ramp-up, there are more people fixing problems than people assembling devices. Slowly over time, CoNQ per unit produced will decrease, not only because of the conventional "learning curve," but also processes are improved, better machines are introduced, and automation takes place.

In the industry of AIMDs, quality is a key factor of success. Low CoNQ could be a major competitive advantage. As shown in Fig. 2.11, CoNQ per unit decreases as production volumes increase. A proper analysis of CoNQ must also take in consideration the occurrence of an event generating CoNQ. Problems identified early (e.g., component measured as out of specification at receiving inspection) have much lower costs than issues occurring late in the production cycle (like a failure detected in a finished sterilized product). As costs (per case) of late failures increase exponentially (see Fig. 2.12), the entire manufacturing process should be designed to avoid dramatic failures, like field actions, recalls, or ex-plantations.

Unlike other industries, the main contribution of process automation in the assembly and test of AIMDs is a reduction of CoNQ. If a fully automated final test system prevents shipping potentially bad units, savings in field actions will be considerable. Two decades ago, scrap rates in manual assembly plans were measured in percent. Today, automated assembly factories manage scrape rate in permille or even better.

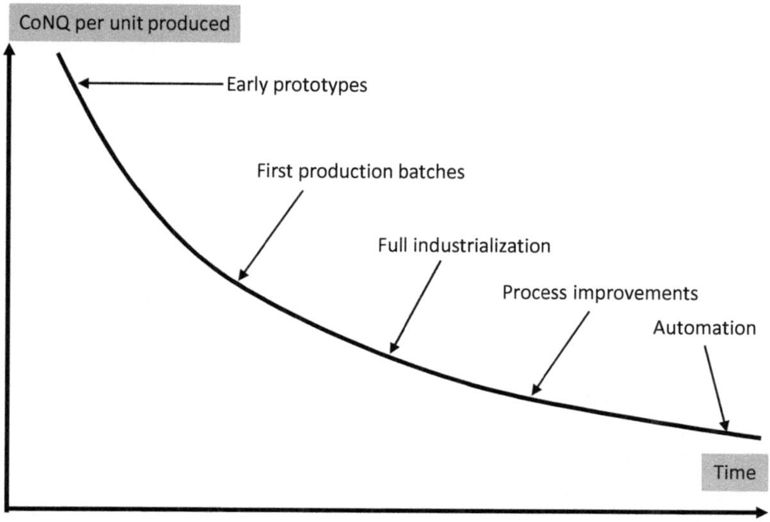

Fig. 2.11 Evolution of the costs of non-quality

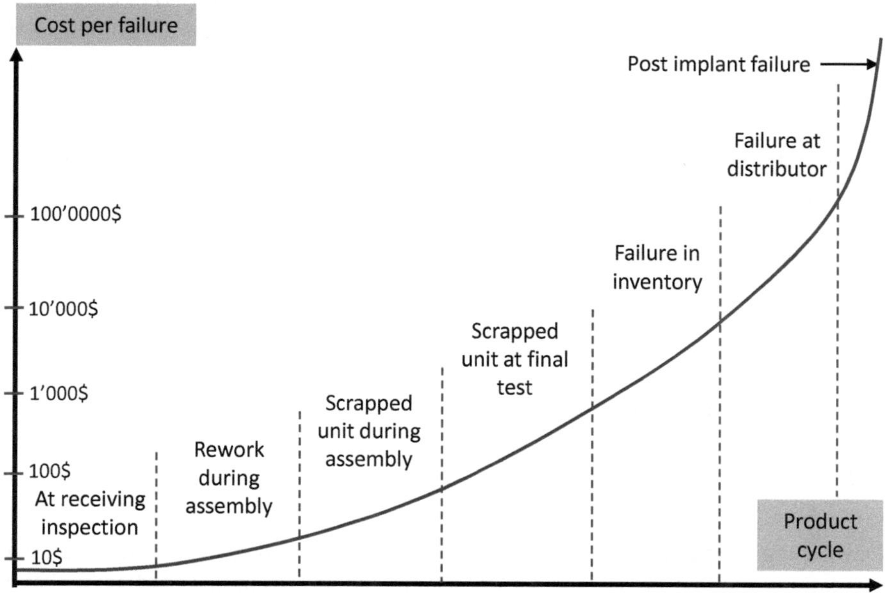

Fig. 2.12 Evolution of the cost (per case) of failure

The production philosophy of AIMDs includes multiple tests of all the units during the manufacturing cycle. Such a thorough screening induces high production costs but reduces radically the probability of having a failure in the field. More testing in production is an investment for the future.

Major medical device companies with large production volumes usually count with gross margins >80% (e.g., CoGS = 20 then sales price >100). It seems excessive, but such a fat margin is necessary to absorb R&D investments, regulatory costs, IP costs, training, inventory, heavy margins of distributors, and risk taking. Producing and selling AIMDs is a risky business. Investors and stakeholders protect their interests by adding a risk premium to sales prices. Start-up and smaller businesses may manage to survive with smaller margins during the few first years, but the smaller the margin, the longer it will take to payback the initial investment.

In the medical device industry, due to long development cycles, high regulatory, and clinical costs, a lot of cash is absorbed every year until the product is approved. But approval is not the end of the adventure! Approval is just the authorization to put a product on the market. It does not secure reimbursement nor commercial success. Several additional years will be needed to reach profitability. Depending on how profitable the business is, again a few years are expected to reimburse investors, amortize R&D and IP costs, and develop distribution channels. A project is not yet successful when the product is approved for sales, far from this. When the business becomes profitable, investors are still concerned about their investment. After a few years of dividends or substantial increase of the company value, then the story is a success.

These very long-term perspectives must be clear for everybody (not only investors but also the designers of the product) from day 1. BCI projects are complex and difficult. Negative cash-flow over a long period of time will create a deep exposure and a late payback.

Some BCI initiatives will never be profitable as an independent business. Such projects still have a large research content, are aimed to the advancement of neurotechnologies, have humanitarian objectives, or similar ideal goals. In these cases, the focus on costs is eluded by more noble incentives. Supporters of such initiatives are philanthropy, foundations, or grants. Nevertheless, even if profitability is not the main driver, high final product costs may seriously limit the success of a therapy.

2.2 Understanding Our Environment

2.2.1 Regulated Environment

Developing, manufacturing, and putting AIMDs on the market are driven by national regulations and international standards. Companies active in this field must obey a certain number of rules, follow guidance, get certifications, get audited for compliance, provide evidence of product safety, demonstrate clinical relevance, and get approval to put devices on the market.

Many developers of AIMDs have insufficient understanding of the constrains induced by regulations and standards. Most of academic groups starting a development of an AIMDs are underestimating the importance of defining a proper "regulatory path" early on. Even start-ups do not allocate appropriate resources to regulatory matters.

Regulations and applicable standards are "laws," which will be enforced by competent authorities, notified bodies (NB) in Europe, and federal agencies or administrations like the Food and Drug Administration (FDA) in the USA. These authorities have received the mission from their respective governments to protect the population from inadequate devices, to assure patient safety, to prevent import of non-compliant products, and to provide post-market surveillance. As every law, medical regulations are often perceived as restrictive. But, as in other fields of activity, laws provide the fundamental protections for proper business conduct. Sometimes we do not like laws, but they are laws.

Medical devices are regulated by a complex system, with substantial differences from country to country. From the beginning of a project, we must have a clear view about where the product will be first put on the market and where it will be extended later. For example, there are slight differences between regulations applicable in Europe and in the USA. Classification could also differ between the two continents. In our world of globalization, it becomes also important to extend regulatory analysis to the other parts of the world like Asia, Australia, Canada, and South America. Until recently, for AIMDs, it was usually quicker to get CE marking in Europe than a full PMA approval in the USA. Today, the situation tends to reverse, as the FDA has introduced new accelerated and facilitated routes for special projects. These new tools often apply to BCI indications. Exemption routes, like the Humanitarian Device Exemption (HDE), limited to 4000 patients per year, are well adapted to complex neuro-projects. On the opposite, the European Union (EU) is currently introducing a new Medical Device Regulation (MDR) [4] with additional constrains. In a near future, we may even expect that European companies may file for approval first in the USA, later in Europe.

The field of implanted BCI is one of the most difficult in terms of regulatory compliance. Even if BCI systems are rarely life supporting, they interact with the brain, our most valuable organ. Implanted BCI are class III devices, the most regulated category. This domain is also relatively new and therefore lacks some basic standards, norms, and guidelines. For historical reasons, the bulk of standards ruling class III devices are based on cardiac applications. The specificities of the nervous system and the way we interface with it are generating new needs for standards and development rules. As an example, implants placed on or inserted partly in the skull are more exposed to impact than devices in the abdomen or in the chest. For the first above-the-neck devices, cochlear implants, a specific impact test standard has been defined by the industry. But so far, it has not been adapted to other skull implants. Therefore, BCIs are frequently in a regulatory "no-man's land." Fortunately, The FDA has recently taken the initiative to provide some guidance in the field of BCIs. The FDA commissioner has issued a statement [5] regarding the needs of providing clear regulations in this field. A non-binding draft guidance [6] has been issued early 2019 which is the cornerstone of a future frame for the industry of neurological AIMDs. This guidance is currently being reviewed by experts in this field. A summary of this draft is discussed in Annex 3.

Another under-regulated domain is wireless communication with implants. Compared to cardiac devices, neuro-technological applications exchange huge volumes of information, requiring very large bandwidth and high frequencies. Radio-frequency (RF) bands have been allocated to medical applications long before the emergence of wireless BCI. It appears today that other bands must be allocated for a proper coverage of the BCI needs. A problem arises from the fact that medical devices and RF bands are regulated by two different authorities. Radio communications are ruled worldwide by the International Telecommunication Union (ITU) [2], but enforcement is under national regulations, like in the USA the Federal Communications Commission (FCC) [3]. Structurally, there are some mismatches between nations. As an example, in Europe, if a product gets the CE mark, it can be commercialized in all the countries recognizing the CE-mark system (EU countries plus a few others [7]). But what happens if the frequency band used by the device is not authorized in one or another member country? The CE-mark system is based on harmonized standards, with one exception: frequency bands. BCI designers must have a good understanding of the fact that the regulatory environment is not fully settled for neuro-applications.

Having the optimal regulatory strategy is a critical factor of success. Or, in other words, a poor command of regulatory matters is a cause of failure. Asking regulatory experts for early opinion and guidance is an investment, not an expenditure.

2.2.2 Users' Needs

Quality of life and health improvement of patients in need are the ultimate goals of translational projects. The main user is the patient. Secondary users are physicians, nurses, other healthcare professionals, and patients' families and friends. All these users have specific needs, sometimes conflicting but often complementary. Understanding the users' needs and translating them in clear specifications are a fundamental step of project development. How many products were fully developed, approved and produced, and found, at the end of an expensive process, that they were not meeting users' expectations? How many great products have been desperately looking for a market?

Overlooking users' needs is a common mistake of medical products designers. Engineers have a tendency of thinking that their brilliant ideas correspond to the dreams of the users. Technical features and functions should be the answers to specific needs expressed by users. But they should never be included in a product simply because they are technically attractive to engineers, superior to competition, or good looking. Features and functions not needed by users add unnecessary costs and risks to the project.

There are several ways to identify and prioritize user's needs:

- Surveys and interviews of patients, including members of the family.
- Interviews of neurologists, neurosurgeons, anesthetists, and nurses.
- Analysis of the satisfaction of patients implanted with products from competition.

As described under Sect. 2.1.4, engineers often have a preconception of what they think patients would need. They may be attracted by technical features which do not correspond to preferences and priorities of patients. For this reason, patients' needs must be clearly identified before specifications are established.

It must be noted that user's needs are the compilation of a diverse population of patients and doctors. Priorities may differ from one patient to another. A well-specified product will not only take care of the majority or median patient, but it should also provide flexibility (e.g., through programmability) to also meet the needs of minority or atypical patients.

In certain cases, some groups of patients may be excluded from the use of a device and a given therapy. For example, the device or therapy may not be adapted to children, pregnant women, patient with cardiac disorders, and so on. Excluding too many groups or large groups of patients is a drawback for the acceptance of the future device.

Exclusion of categories of patients is often linked to technical barriers. For example, a BCI implant communicating through an optical channel may perform differently depending on the type of skin or may tolerate only short communication distance, excluding fat patients. Excluding patients because the technology under-performs are ethically unacceptable. Covering a large spectrum of patients is also part of the user's needs.

Another example of the tight relations between user's needs and technology is compatibility of the device with magnetic resonance imaging (MRI). Patients suffering from neuro-diseases have a high likelihood to be frequently diagnosed using MRI equipment. It may be a critical tool for checking the evolution of their disease. In consequence, MRI compatibility may be considered as a user's need for some categories of patients. If this is the case, designers must strive to make the device MRI compatible. See Sect. 4.11.3 for more details on MRI.

Electromagnetic compatibility between several active devices implanted in the same patient may, in some specific cases, become part of user's needs. Coexistence might be a must for certain patients. Is it sustainable to exclude a paralyzed patient from the benefits of a BCI if he/she has already an implanted defibrillator? The answer is clearly NO. Therefore, user's needs might include requirement regarding coexistence. See Sect. 4.11.4 for more details on coexistence.

2.2.3 Human Factors

For decades, development was principally driven by technical specificities and medical considerations. Today, a lot of importance is given to soft criteria, called "ergonomics," "human factors," or "usability engineering." These notions describe interactions between people and medical devices, consisting in three main actors:

- Users: patients, doctors, nurses, and family.
- Interface or user interface (see Fig. 2.13): hardware and software facilitating the interaction.

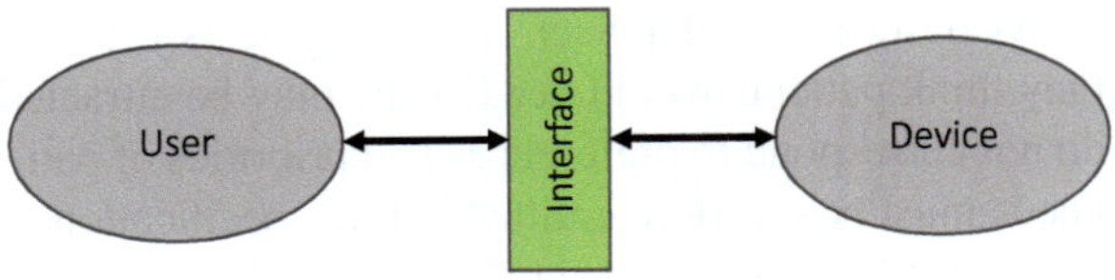

Fig. 2.13 Interfacing between user and device

- Devices: medical systems providing a therapy or a diagnostic.

Developing an optimum interface is one of the main ways to succeed in the development of a medical device:

- For the users:
  - Facilitates daily use.
  - Provides continuous information on the device's operation.
  - Improves acceptance of the therapy.
  - Accelerates training.
  - Improves compliance.
  - Empowers patients.
  - Minimizes errors and misunderstandings.
  - Communicates efficiently alarms and errors.
  - Reduces needs of user manuals.
- For support:
  - Fast and unambiguous diagnostic of device malfunctioning.
  - Displays device's status (battery charge, connections, etc.).
  - Facilitates repairs and maintenance.
  - Reduces nurse interventions.
  - Accelerate programming and adjustments.
- For manufacturers:
  - Increases reliability.
  - Accelerates field actions in case of trouble.
  - Reduces occurrence and severity of adverse events.
  - Gathers long-term records on performance and reliability.
  - Provides statistics on operation and usage.

During the design of a medical system, the user interface is in the center of the concept. Designers must fully understand nontechnical matters which will later be translated in specifications:

- Patient needs, feelings, and perception regarding the therapy.
- Doctors needs with regard on how to install and set up the system.
- Nurse and family expectations regarding supporting the patient.
- Patient interactions with the user interface (display, input keys, language, acoustic feed-back, alarms, language, etc.)
- Access to help and support.

Fig. 2.14 Patient interface

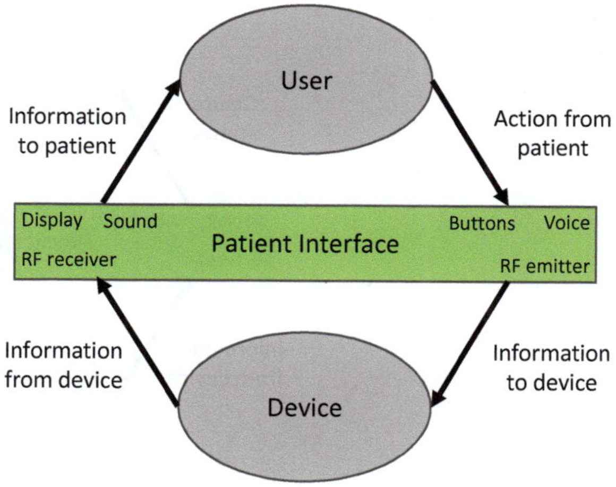

In the field of neuro-diseases, patients often have limited movement capabilities, sub-optimal vision and hearing, or reduced cognitive functions. The user interface must be designed to be fully adapted to these limitations. Misunderstanding this is one of the most common reasons of failure of treatment. Patients will not use or will miss full benefit of the system if the user interface is too difficult to handle.

Complex systems (programmable, implantable, remote controlled, etc.) often require two user interfaces:

- Patient interface (see Fig. 2.14):
  - Provides feedback to the patient regarding the treatment.
  - Displays status of the device (battery, functioning, etc.).
  - Rings alarms and warning signals.
  - Allows simple patient's controls (start-stop, +/–, etc.).
- Physician interface (see Fig. 2.15):
  - Allows downloading device memory.
  - Permits adjustment of parameters and reprogramming.
  - Let access to deep device diagnostic and maintenance.

Usually, the patient interface is small, light, battery operated, and nicely designed. The physician interface is more complex, sometimes similar of a laptop computer. They may or may not use the same communication medium with the device.

In the example of an implanted device with bidirectional wireless radio communication, the user interfaces configurations differ from patients to physicians.

The main standard for usability is IEC 62366 (international standard), EN 62366 in its European format.

In Europe, the new Medical Device Regulations (MDR/2017/745) requires application of human factors and usability engineering, as stated in the Annex 1 of

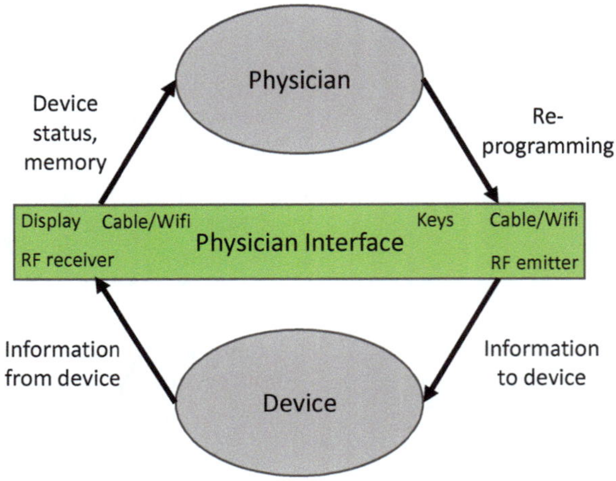

Fig. 2.15 Physician interface

the document, *General Requirements for Safety and Performance Requirements*, Section 5:

> In eliminating or reducing risks related to use error, the manufacturer shall:
>
> (a) reduce as far as possible the risks related to the ergonomic features of the device and the environment, in which the device is intended to be used (design for patient safety), and,
>
> (b) give consideration to the technical knowledge, experience, education, training and use environment, where applicable, and the medical and physical conditions of intended users (design for lay, professional, disable or other users).

2.2.4 Implantology

I call "implantology" the art of implanting a device in the human body. Too often, the implant is designed without much consideration to the surgical technics, to the needs of special tools, to the duration of the procedure, or to the comfort of surgeons.

I noticed several times that there was an imbalance between the level of sophistication of a device and unelaborated surgical tools, ancillaries, and procedures. A good example is the insertion of a MEA in the motor cortex. This very fragile and expensive tissue interface must be punched about 1.5 mm in the cortex, through the arachnoid. Today, the MEA is placed over the desirable location and hits by a simple pneumatically activated hammer, with high risks of damaging the array or having it sliding aside. Using such a gross tool is increasing the risks of inappropriate lesions of the cortex or having to scrap the array. If the array is attached to expensive electronics, it may mean scrapping tens of thousand dollars. Surgeons should have their task facilitated by tools in line with the technology of the implant. Well-planned surgical procedures with appropriate tools increase the chances of successful implantation. This is especially important when the skull must be opened, for example, for the insertion of electrodes in or on the brain. Procedures should optimize the probability to "*be right at the first attempt.*" Brain tissues are so fragile that doing several insertion trials will ultimately results in irreversible damages.

Let's come back to the case of inserting a MEA in the motor cortex. The surface on the brain moves slightly due to heart pulses and respiration. These movements at about 1 Hz and 0.1 Hz have an amplitude limited to a few tenths of millimeters, but, at the scale of the tiny tips of the MEA, it is like shooting on a moving target. In my opinion, insertion tools of the future will have movement sensors for automatic compensation of these small displacements.

Other tissue interfaces are introduced in the body by tunneling them between layers of body tissues. Placement of DBS electrodes in the brain and tunneling the extension cable under the scalp and along the neck is a well-established method. DBS electrodes are stiff and rather robust. SCS electrodes are less rigid, formed like flat paddles. They may be damaged if not inserted carefully. Often, the tunnel is first opened by appropriate tools, and then the electrodes are pulled or pushed in the tunnel.

More recent developments have generated thin, fragile, and sometimes flexible or even stretchable electrodes, which require specially adapted tools. Using conventional insertion tools may damage the body interface or the cable. Tunneling tools with force feedback should be developed for delicate leads introduction.

A large field of opportunities is opening in the adaptation of the surgical equipment for optimum placement of neurological devices. Developers of these new tools should take great care to ergonomics, human factors, and usability, in the perspective of surgeons. Designers of great implants often forget about surgeons. As patients, neurologists, and nurses, surgeons are users of the device. Their needs must be properly assessed and met. Inappropriately inserted implantable devices are likely to later cause an adverse event. Often, damage is caused, but not identified, during surgery.

Designer should not forget that neurosurgeons are usually conservative in the way they operate. They like established procedures and mature technologies. The current explosion of new technologies like tiny fragile intracortical arrays, intrafascicular electrodes inserted in tiny nerves or access to sensory organs, is rapidly changing the surgical environment. New skills are needed. Surgeons must be listened to, trained, and equipped in consequence. The best implant will fail if it cannot be implanted properly.

In a near future, neurological procedures will become so complex that there will be an important need for surgical simulation platforms, where surgeons can get a virtual training before moving to humans. Surgical robot may also contribute to achieve difficult electrodes introduction.

The time spent in an operation room (OR) is very expensive, in the range of 50 to 100$ per minute. Improving surgical procedures and facilitating insertion may have a substantial impact on the overall cost of the therapy.

2.2.5 Think About Patients and Healthcare Players

We saw the importance of understanding well the users' needs and taking in account human factors. In addition, engineers should try to see therapies and devices with the eyes of patients. Depending on the patient's disease or disability, he/she may have unexpected requirements or wishes. For example, the head is the only part of the body which a paralyzed person still controls. He/she often is much concerned if neurologists and surgeons propose invasive and risky procedures around his/her head. I often hear patients saying: "do not mess with my head, this is the only thing still under my control." Aesthetic criteria may also be of high value for these patients.

Another example is DBS for patients suffering from Parkinson's disease. The first generations of IPGs were having primary non-rechargeable batteries. The rather high-energy consumption imposes insertion of a new IPG with fresh batteries, approximately every second year. With the noble goal of trying to extend the duration between two surgical interventions, engineers designed new types of IPGs for DBS, with rechargeable batteries. The objective was to extend the lifetime of the IPG and has replacement surgery only after 8–10 years. The drawback is that the battery needs to be recharged frequently (e.g., every 1–2 weeks). But the engineers forgot to think about patients. What engineers perceived as an improvement was a problem for patients: anytime the patient recharges the device, he/she is reminded that he/she is a parkinsonian! In many cases, patients prefer having a surgical intervention every second year and try to forget about their disease in-between.

Designers of AIMDs also often forget about the caregivers and family. This is of high importance regarding the functionalities and features of the external units, like headpiece, behind-the hear interface, remote control, patient unit, charger, or displays. All these users' interfaces (users being not only patients but anyone who takes care of the patient, at the hospital or at home) must be easy to operate by non-technical people. Far too often, these external devices have complex menus, with hermetic language, on poorly readable screens. Professional caregivers are familiar with patient interfaces and usually get an adequate training focused on the device. But nonprofessionals, like family members, who play a key role in the well-being of their loved ones are often lost in the complexity of operating these interfaces. It induces risks of errors or sub-optimal performance of the interface.

For optimal use of the implant, the external interfaces must be well adapted to users. Not the average user, but the most difficult categories of patients and family members of old age, with bad eyes, with restricted cognitive capabilities, and limited comprehension of electronic equipment. Not everybody has a smartphone and is able to scroll through complex menus. User interfaces must be accessible to older people. Simple and understandable icons and graphics, clear alarm signals, large push bottom, and messages in the language of the user are important for patient acceptance.

BCI users have serious disorders. They are not only suffering from the consequences of their disease but also from how they are perceived by other people. A patient with Parkinson told me that people were speaking loud and slowly to him, as if he had trouble to understand or was mentally retarded! Movement disorders have nothing to do with cognitive functions, but some people do not understand it. The device we provide to the patient shall help him/her to get a better quality of life but also to be perceived in a better way. Giving patients a certain level of control is a capital factor of empowerment. The remote controls and other external devices should be seen by the patient as a tool, an improvement, or even an opportunity. It should definitively not be a barrier, a constrain, or a burden. How many times have we seen patients afraid to use their control device? How many do not trust the device? How many hates it?

The aspect, form factor, shape, and look of the remote control should also be adapted to the type of disease. A person suffering from urinary incontinence and benefitting from SNS wants to have a small and discreet remote control. He/she does not want the device to be seen, identifying the holder as incontinent. But the control should always be at hand. A remote control in a wristwatch or in a necklace will be far better accepted than a bulky box. The big thing for a parkinsonian having a DBS is that he/she does not look any longer as a parkinsonian. In consequence, the remote control should be as transparent as the symptoms. Remote controls are the only tangible connection to an invisible inaccessible implant.

2.2.6 Do Not Listen to Engineers

I dare being critical with engineers: I'm one of them. Engineers have the tendency to be attracted by new technologies, for the sake of technologies. They often forget to see the big picture. Is this attractive function really needed? Or is it just something which pleases me (as an engineer)?

Overengineering is a classical mistake in a medical device development. Performances beyond specifications have no value but a high cost. In addition, they induce unnecessary risks. Having multiple options, sensors which will be rarely used, features of doubtful utility, overload the project. Designers should frequently reassess each feature and ask the fundamental question "Is it a **must** or simply a **nice-to-have**?"

Engineers underestimate the impact of costs. We are, even in developed countries, more and more limited by the costs of healthcare. Meeting unmet medical needs is a grand goal, but is our society able and willing to pay for it?

Usually, engineers start designing before having the overall picture or before specifications are settled. It is good for creativity, but, if not controlled, it leads to chaos. Engineers must be forced by the program manager to think first and design later. Before entering the design phase, they should spend time on nontechnical matters like understanding the field, the users' needs, the environment, the competition, and the regulations.

Engineers have insufficient knowledge of human anatomy, of patient's psychology, of regulatory and legal affairs, and of costs control and design for manufacturability. In complex therapies, for example, BCI, these factors are much more important than technical skills. A good designer of medical device is a person with a large spectrum of knowledge. Good teamwork may compensate the lack of breadth of individual specialists.

Development teams include several engineers. Each of them wants to have his/her own innovation included in the project, something which will leave a legacy, their "baby." This creates unsound competition. A good project has one main innovation only. Other features are in support of the big innovation. Program managers should keep this in mind. Integrating several major innovations in a single project is a sure way to fail. At least one of the innovative features will be late or turn out being not achievable. This will wreck the entire project.

2.2.7 Check What Others Are Doing

Before jumping in a project, convinced that the first idea is the best possible one, it is better to step back and take a deep breath. The first idea is rarely the best one. "Spending" time to look around is frequently a good way to save time later. Looking around, listening carefully, and trying to understand the reasons why others took in unexpected way debating or even arguing must be considered as an investment. Many other organizations are developing devices similar as yours. Some are on very different topics, but aspects of their projects could inspire us. Never underestimate competitors. Instead, respect competition. Frequently, competition is ahead of you, maybe working on a device not as great as yours, but they have already learned from mistakes you have not even made yourself.

A good understanding of "who-is-doing-what-why-how" is of extreme importance. I usually add "how, where, why, and when" to really get the overall picture of the environment. The chances to win a battle are poor if you do not know who the enemy is, where it is, and what kind of weapons he detains. This looks like trivial, but how many companies failed from ignoring these basic rules?

If you know "who is doing what," you will also know "who has already done some elements of my project." Reinventing the wheel is out of reach for small project teams. Reinventing takes more time, costs more, and adds risks, compared to trying to get the already existing element, subassembly, feature, software, or design. Smart development teams do not immediately search for solutions (see Fig. 2.16). First, they search for where the solutions are and find ways to get access to them. If no solution to a given challenge can be found in the environment, then search for your own solution.

It is even worth sometimes to search several decades back in the past. Companies might have had good ideas a long time ago but were not able to turn them to reality, because the technologies were not ready then. Old patents are a great source of inspiration. Often, some disclosures were highly innovative, but could not be executed, for various reasons. Discovering these jewels might give you a competitive advantage. Of course, these ancient ideas are not patentable again, as already disclosed. But, if the patent has already fallen in the public domain, your freedom to

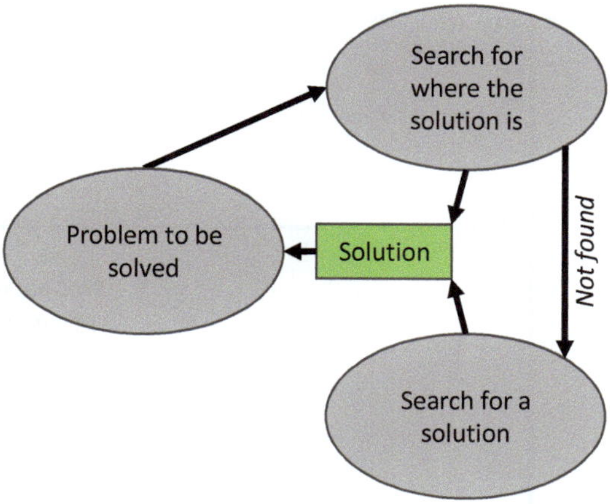

Fig. 2.16 Searching for a solution

operate (FtO) is secured. Adapting old concepts to newer technologies is also a way to finally meet unmet needs.

2.2.8 Search Reasons of Successes and Failures

Analyzing the reasons why competitive or similar products have succeeded is a fundamental exercise to be done before starting your own project. In addition to technical, engineering, manufacturing, and IP characteristics, we also should understand the strategies and tactics chosen by competitors to get their products on the market. Understanding their regulatory path, clinical approach, reimbursement strategy, marketing segmentation, and pricing are as important as technical features. Who were the development partners of my competitors? Why did they select these suppliers? Why did they use distributors instead of their own sales force? Why did they first market their product in country X?

The funding and financial structure of a successful competitor is also a valuable source of information. It may also be an "eye-opener." Find out how much money and how long time it took them to get their product approved. From my experience, for a class III complex AIMD, I know that we should count >100 M$ and >10 years until FDA approval. These numbers are frightening, but they reflect reality. If you want to compete in this category, you should be able to afford this order of magnitude of time and money or be ready to sell your ideas and business to some bigger organization.

The reasons of failures are more instructive than the reasons of successes. Understanding why a project failed at least warns you for not doing the same mistakes. Many projects failed because of over-ambition. This teaches us to be more modest or to adapt our ambitions to our resources (technical, intellectual, human, and financial). In the field of neuro-technologies, cheap and quick projects usually

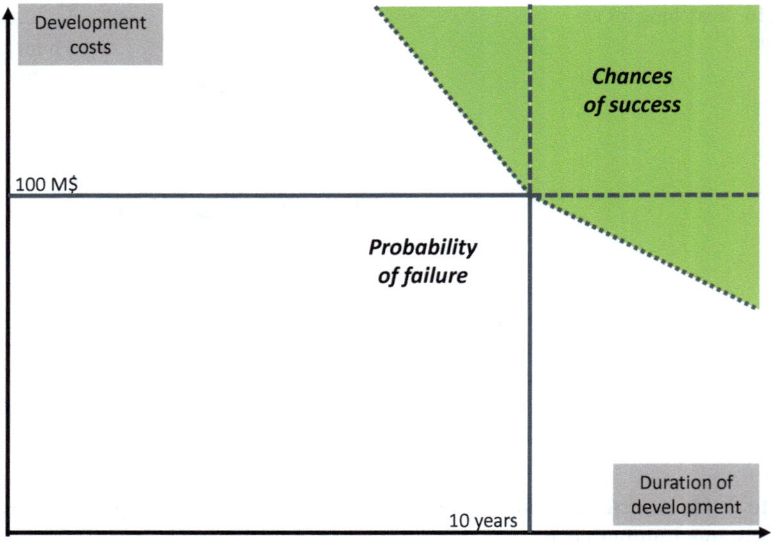

Fig. 2.17 The long and expensive route to success

fail. Why? Is it precisely because they are planned as "cheap and quick"? Often people challenge me over the expensive (>100 M$) and long (>10 years) route to approval (see Fig. 2.17). This is precisely the reality of those who got approval. Those who fail to get approval are not in our statistics.

Large companies with a lot of financial power may conduct parallel development steps in order to speed up the process. It might lead to an acceleration but will even increase the overall development costs. The opposite may not always be true: taking more time for the development may not automatically reduce costs, unless the additional time is allocated to guarantee that *"things are done right at the first time"* and to minimize development risks.

Section 3.2 will describe various neuro-devices which have reached the market. For each of them, we will review the reasons of their success and retain the lessons learned.

2.2.9 Cumulated Cash-Flow

As an illustration, I made a mini-business plan of a virtual company (NeuroVirtual) described in more details in Annex 2. It is not a real company, but it looks like several companies I have been involved with in my professional life. The best synthetic description of a start-up situation is the cumulated cash-flow over the years. To simplify, the annual cash-flow is the difference between the money flowing in the company (revenues of sales, licenses, etc.) and the expenses (salaries, taxes, patent

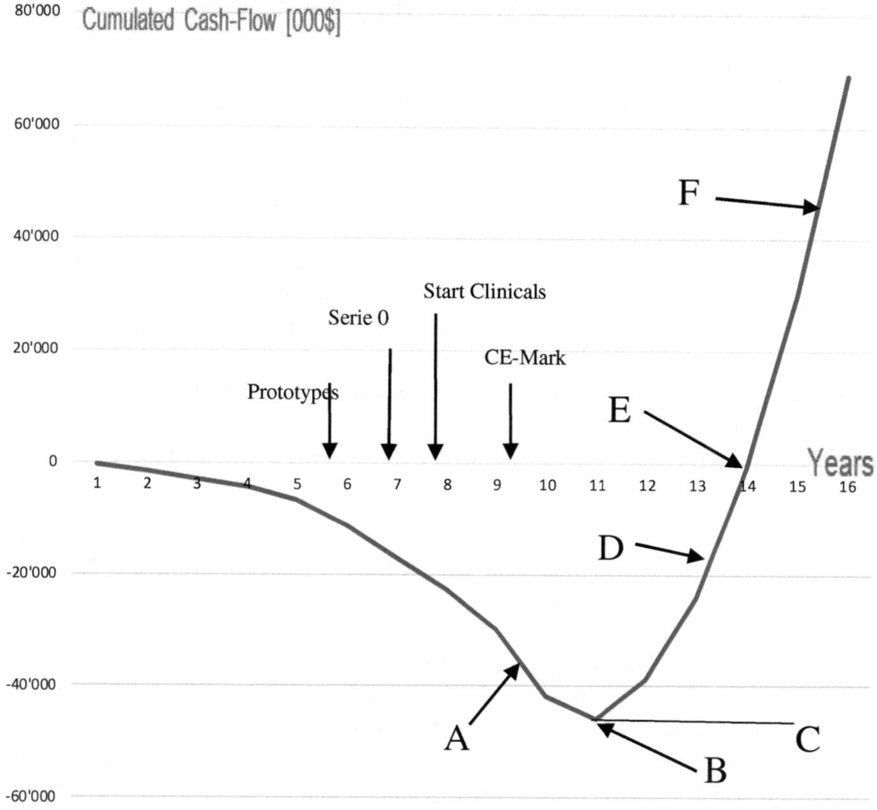

Fig. 2.18 NeuroVirtual, cumulated cash-flows

fees, services, production capacity, suppliers, etc.). In the early years, during development, validation, and clinicals, there is no revenue, so cash-flow is negative. Seed money, grants, and several rounds of financing are necessary to pay all the expenses. Cumulated cash-flow is the addition, year after year of the annual cash-flow. There are several points on the cumulated cash-flow curve (see Fig. 2.18) which are worth discussing:

A. *First sales:* They happen at the end of the development, validation, clinicals, and approval period. At that stage, revenues are small; expenses increase kicker than revenues because the company needs to ramp-up production, to build inventories, and to set distribution channels in place. First sales do not mean that investors get any returns. Often, they even need to add more cash to fuel ramp-up.
B. *First profits:* For the first time, cash-flow is positive. This is the inflection point of the cumulated cash-flow curve. From this stage, no additional cash is requested from the investors. First profits may be delayed if the company decides

to further invest in the ramp-up process, for example, by acquiring more production capabilities or expanding the sales forces.

C. *Exposure:* It is the total money absorbed by the company until first profits. The total of all investments, from seeds money to the last financing round, covers this accumulation of negative cash-flow. The deeper the curve goes and the longer time it takes for the first profit, the bigger the risks are (for the investors). A business plan (BP) showing a large exposure and a long return on investment is not likely to attract investors.
D. *Growth of revenues:* The slope of the recovery after point C depends on sales volumes and margins. If reimbursement occurs rapidly after product approval, sales may grow substantially. But, in case of a late decision for reimbursement, or if sales remain limited to a few countries, growth might be modest.
E. *Return on investment:* Accumulated profits have compensated investors for their various financing efforts.
F. *Profitable business:* From this point, the slope of the cumulated cash-flow indicates the real success of the company.

In this realistic example, we see that many years are necessary to get the first sales and many more to transform the venture in a real profitable business. Entrepreneurs must be fully aware of this before launching their project. Getting approval is far from the end!

Often, the original business-plan must be reviewed because of unexpected delays. Using the same virtual company model as above, I added 2 years of delays during the development phase, for example, due to a redesign or to changes required after a failed test. It postpones all the critical points of the cumulated cash-flow curve by 2 years, but it also worsens the exposure, meaning that investors need to contribute additional cash for covering the delays.

Some companies also get late during the clinical trials or to get approval. It is not rare that notified bodies (NB) require supplementary tests or clinical evidences. NB tends also to be overloaded and cannot grant approval on time. This is especially true in Europe since the introduction of the new Medical Device Regulation (MDR) and the reduction of the number of NB in 2017.

We also simulated the situation where the virtual company NeuroVirtual gets twice delayed in their plans, first by 2 years during development and a second time by 2 years in the clinical/approval phase (see Fig. 2.19). Details can be found in Annex 2. Such delays have a dramatic impact on the cumulative cash-flow curve.

2.2.10 Reimbursement

Somebody must pay for therapies and diagnostics issued from high technologies. The bill may be taken by insurances, social security systems, hospitals, associations, foundations, philanthropists, or by the patient itself. Whoever is paying, the price is an important factor in the success of a product.

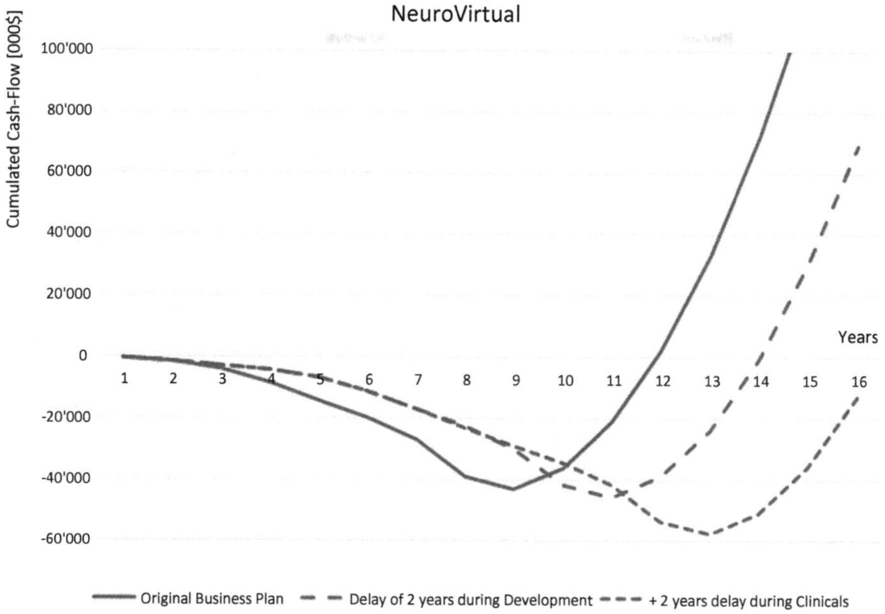

Fig. 2.19 Impact of delays on cumulative cash-flows

When a therapy is reimbursed, it may rapidly get a large acceptance and a fast growth. Getting reimbursement is somehow a recognition of the efficacy of the therapy. The criteria to get reimbursement are complex and vary widely from country to country. In this book, our aim is not to give readers any lesson about how to get reimbursement for neuro-devices, but we want to point out that it is an important parameter in our environment. Reimbursement strategy, expectations, and likelihood for the new product to be reimbursed should be part of the early discussions at project launch.

"Me-too" products are more likely to get reimbursed than very innovative products in specific therapeutic niches. Neuro-devices and BCIs are covering indications at the frontier of medicine, in areas where only few products have obtained reimbursement. There are not many predicate devices on which to base a sound reimbursement and pricing strategy.

Often, reimbursement is granted only after the product has proven its efficacy, superiority, and global economic/societal advantages. This may happen only after several years of post-market follow-up. Late reimbursement has an impact on the success of a product and the value of the company. These aspects must be well understood and included in the business plan. Even for projects where financial profitability is not the main driver, reimbursement is a critical factor of success.

2.2.11 Global Social Costs

Neuro-technologies and BCI projects are complex endeavors with challenging technical objectives added to long and expensive development efforts. The final sales price of these therapies and diagnostics is therefore high. In addition, most of these initiatives are aimed to treat unmet medical needs, meaning that the costs of these new treatments are not part of the current health expenditures. There is a conflict between demand of better treatments, increased quality of life, solutions for unmet needs, and high pressure on healthcare costs. In some countries, the financial burden of health in the national economy has become unbearable. The complexity of the therapies discussed in this book may mean that they might be available only in a few countries or for a few privileged patients.

Transfer of charge is also a matter difficult to apprehend. Payers for a new treatment may not be the same as the payers for the ancient solution. Take the example of urinary incontinence, a real unmet medical need. People suffering from incontinence pay from their own pocket for absorbing pads and diapers. If they get a sacral nerve stimulator, the costs are likely to be covered by insurances or social security. This transfer of charge may be perceived as an increase of healthcare costs. In fact, on the long term, sacral nerve stimulation has a lower cost than the recurrent costs for pads. So, seen globally, the implant saves money. Seen from national healthcare economics, the implant is an additional expenditure.

The burden of advanced healthcare technologies on global social costs should be taken seriously during the definition phase of a new project. To be successful, such a project must provide a solution that makes sense regarding the global costs for society. This includes analyzing the balance between the additional costs of the new therapy, with the global savings it induces. Savings may cover a large spectrum of expenditures, from reduction of personal assistance, homecare instead of hospitalization, reduction of medication, to facilitation of diagnostic. It is much more difficult to define costs or savings related to improved quality of life, longer life, increased mobility, or social interaction. These are unquantifiable matters which for sure have value which counterbalance costs.

2.2.12 Intellectual Property (IP)

Disregarding or underestimating IP owned by others is also a classical source of project failure and a lack of global vision on the environment. Any innovative project is at risk that somebody already had the same or a similar idea before. In this case, IP is owned by somebody else, either constitute prior art, limiting the possibility to file for a patent, or it is a barrier to FtO. In the latter situation, royalties may be requested by the owner of a previous patent. If, after several years of expensive development, your project is shown to be infringing the IP rights owned by another company, then your negotiating power is weak. If this other company is a large competitor, the only way to settle may be to give up your independency and let the project be absorbed by this competitor. Such cases were not rare in the last decades. In the early development stages of your project, large companies will not tell you anything about a potential infringement. They will let you continue the development and take all the risks, and they will get back to you at a later stage, when the project is ripe and de-risked. You then have no way to defend your project from a take-over.

I have frequently seen start-up companies or academic labs claiming their IP situation was good, because they had a patent pending. This is a naïve statement. A pending patent does not mean that all the fundamental claims will be granted. Sometime, all prior art does not show up during the examination of the patent. Patents are regularly granted, in good faith, but will not sustain a trial in court, because the prior rights of the opponent have not been respected. It is capital to get patents for good ideas, but these patents do not always provide the expected protection. Assessing the actual protection potential of a patent, therefore its value, is an art rarely mastered by start-ups or academic labs.

References

1. https://www.wysscenter.ch/
2. https://en.wikipedia.org/wiki/International_Telecommunication_Union
3. https://en.wikipedia.org/wiki/Federal_Communications_Commission
4. https://ec.europa.eu/growth/sectors/medical-devices/regulatory-framework_en
5. https://www.fda.gov/NewsEvents/Newsroom/PressAnnouncements/ucm631844.htm
6. https://www.fda.gov/ucm/groups/fdagov-public/@fdagov-meddev-gen/documents/document/ucm631786.pdf
7. https://2016.export.gov/cemark/eg_main_017269.asp

Chapter 3
Targets of Neuro-Technologies

3.1 Interfacing with the Nervous System

The enormous complexity of the brain has been described by numerous scientists. I'll not come back to the astronomic numbers quantifying the human brain. I will just retain some facts:

- Neurons do not function as any electronic component.
- Neurons are interconnected to thousands of neighbors.
- Complex electro-chemical reactions propagate though this incredible mesh.
- The brain cannot be compared to digital computers:
  - Quantum of information is not defined.
  - Not a binary representation.
  - Not linear behavior.
  - No clock.
  - No central processor.
  - No fixed architecture.
- Many interconnected sub-systems, more-or-less dependent, with no hierarchy, are competing and complementing each other.
- Scientists have identified areas of the cortex and volumes of the subcortical space corresponding to motor functions, vital functions, emotions, instincts, reflexes, memory, learning, reasoning, language, senses, and so on.
- Plasticity, or the unique capacity of the brain to reinvent itself, means that the structure and interconnections are in constant dynamic evolution.

Every day, scientists are progressing in the discovery and understanding of our nervous system. Advanced imaging systems, with various temporal and spatial resolutions, allow us to penetrate some of the mysteries of the brain. Considerable efforts are focused on identifying the communication mechanisms between neurons, brain sub-systems, and nerves. These scientific achievements open the door to

C. Clément, *Brain-Computer Interface Technologies*,
https://doi.org/10.1007/978-3-030-27852-6_3

the development of therapies to cure or alleviate neurological disorders and to better diagnose diseases. Without these fundamental research programs, the advancement of neuro-technologies would not be possible. Developers of BCI need to understand the intimate brain mechanism before entering in any development process.

We are only at the very beginning of the long journey of discovering the brain. A lot remains unknown, even mysterious sometimes. It can be compared to the discovery of the cosmos. Scientists, with the help of engineers, have managed to ship a man to walk on the moon and have sent various machines to probe our solar system, and Voyager has left our solar system. The frontiers of the brain, as of the cosmos, are still very far away.

In our ambitious goals to interact with our nervous system, we should follow the same philosophy we adopted for the conquest of the cosmos: do small steps which already will lead us to great successes.

Faced to the unlimited complexity of the brain, we should remain modest. Understanding all the details is out of reach. Understanding the globality is also utopia. We need to *use our brain to trick the brain*! If we have some understanding of tiny areas of this vast field, let's follow an exploratory path, like following a river in the jungle. We will get only a limited knowledge of what is happening along the river, even less about the global dynamics of the jungle, but we may find some treasures.

This is where the combination of cutting-edge sciences and advanced technologies may lead to medical progresses and solutions. Already some decades ago, scientists understood that applying electrical signals deep in the brain or at the roots of peripheral nerves on the spinal cord may block the symptoms of Parkinson's disease or chronic back pain signals, respectively. Technologies issued from the pacemaker industry made these first neuromodulation therapies available to a large population. It was the starting point of DBS and SCS, two pioneer approaches of interfacing with the brain, the equivalent of walking on the moon in our space analogy. The marriage of sciences and technologies made it possible, even if all the fundamental mechanisms were not totally scientifically understood and even if the technology was not fully adapted to interfacing with these not well-understood issues.

I will use another analogy to explain how we should "trick the brain": a football stadium. When being outside of a stadium, you hear the crowd shouting because a goal was scored. This is already valuable information: something did happen, and you noticed it. You'll be able to record several occurrences, maybe of different nature. But you will not be able to know the details. Was it your favorite team? Was it your favorite player? Was it a nice goal?

Using an EEG cap (see Fig. 3.1a) to collect brain signals is somehow comparable to being outside a football stadium: you get valuable information, but you miss the details. Sometimes, global information is enough, and, with the help of advanced signal processing, the EEG signals are adequate to steer a wheelchair, for example. In other cases, like getting a BCI able to decode a dozen of degrees of freedom, you need the details: you must enter the stadium.

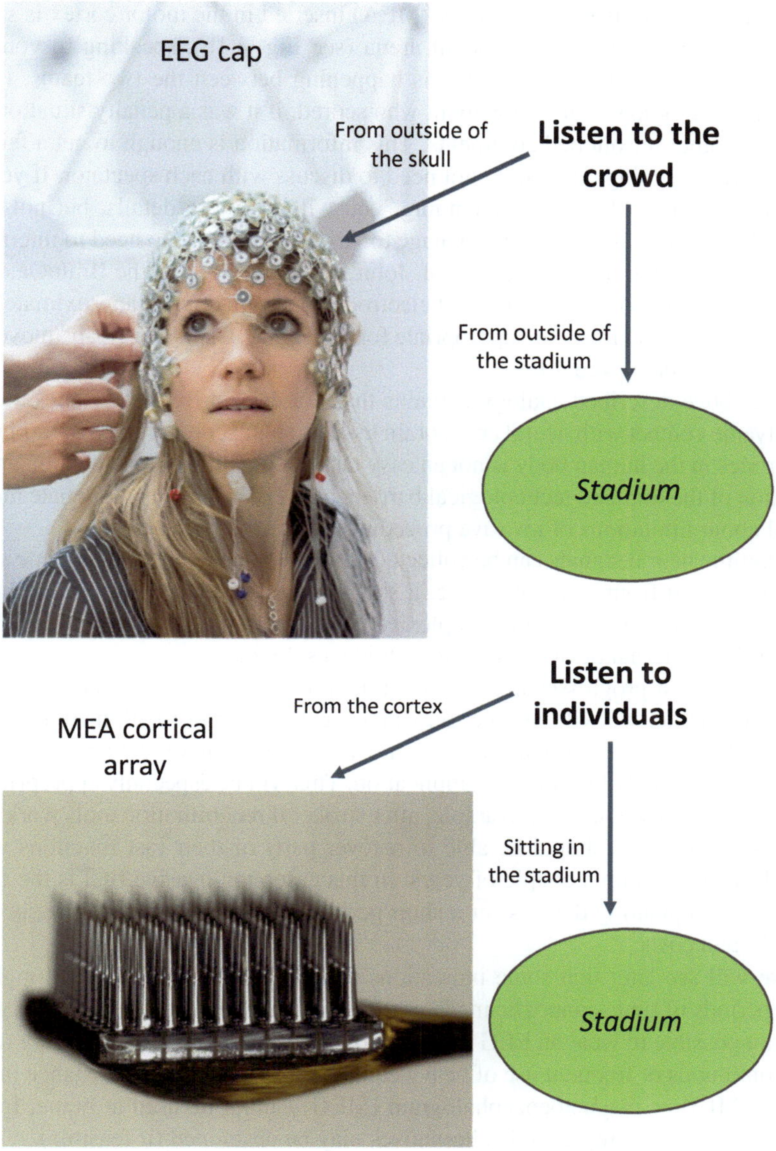

Fig. 3.1 (**a**) Analogy with a stadium: listen to the crowd. *(Credit picture: Wyss Center for Bio and Neuroengineering)*. (**b**) Analogy with a stadium: listen to individuals. *(Credit, picture: Blackrock Microsystems LLC)*

Having a multiple electrode array (MEA) inserted in the motor cortex is similar of having you sitting in the football arena (see Fig. 3.1b), speaking to your two neighbors. They will describe what is happening between the two teams, in real time. You will know everything about who scored, if it was a penalty situation, and plenty of other unsolicited comments. This information is enough to get a fair picture of what is going on. You do not need to discuss with each spectator. If you ask the opinion of all the people attending, you will get more details, but not much more. By remaining modest, we manage to "trick the brain." No need to interrogate each neuron of the motor cortex. Prof. John Donoghue [1] and the BrainGate consortium have demonstrated that 100 electrodes, in contact with approximately the same number of neurons, are appropriate for decoding "reach and grasp" movement intention of a paralyzed arm.

The above described analogy explains that, for certain BCI, electrodes must be in galvanic contact with neural cells, brain tissues, spinal cord, or nerves. Implanting electrodes in the human body is not an easy task. Depending on the location and the duration of the implant, technological barriers are very high. We will debate more in detail about limitations of invasive procedures in the following chapters.

Anytime neural signals can be collected from outside the body or anytime stimulus can be sent from external source, it should be the preferred option. Implanting devices has such an impact on complexity, costs, and risks that it should be considered only if external devices failed to provide a solution.

With the fast progresses in signal treatment, it is not excluded that, in a reasonable time frame, we will be able to extract much more information from external BCI. If this is the case, it will facilitate the use of noninvasive technologies to control more accurately brain-driven application. This will be especially appropriate for rehabilitation procedure, for example, after stroke. If rehabilitation tools work properly, stroke patients should be able to recover parts of their lost functions within months or maximum a couple of years. In this case, noninvasive BCI is the option of choice. Implanting devices for a short period of time is not reasonable, in terms of risks and costs.

We will see later that some indications or therapies, applied day and night for long periods of time, cannot be made user-friendly with external BCI. For example, it is not possible to wear an EEG cap 24/7 for a lifetime. Similarly, a therapy requiring continuous or frequent use of heavy equipment like magnetic resonance investigation (MRI) or magnetoencephalogram (MEG) cannot be used at home. In such cases, wearable or implantable alternatives may be developed (if feasible).

For being able to enter in contact with certain neural tissues, we have seen the necessity of implanting electrodes. For a limited time period (up to a few months), it may be acceptable to have a transdermal cable linking the electrodes to the external electronics, with risks of infections and necrosis around the opening of the skin. For long-term (several years) or lifelong applications, transdermal passages are not acceptable, and the entire device (body interface, cable, electronics) must be implanted. Describing how to build fully implantable long-term BCIs is the main purpose of this book (Fig. 3.2).

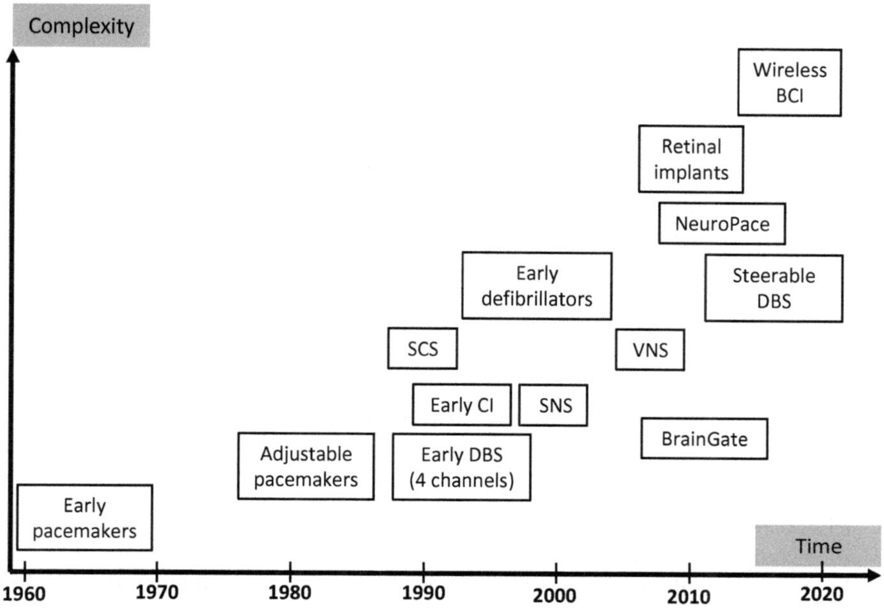

Fig. 3.2 Evolution of complexity over time

Selecting a fully implantable approach is attractive in terms of patient autonomy, aesthetics, and performance, but it is also the most difficult and regulated field of human medicine. Before entering this category, one must be sure to get enough resources, time, and competencies. The field of long-term implantable BCI is reserved to solid organizations.

In other words, be reasonable and target the nervous system following paths which are achievable by your team. If you cannot find an affordable solution: seek alliances and partnerships.

3.2 Invasiveness

Invasiveness is a fuzzy concept. Most people consider implants as invasive and external therapies as noninvasive. This black and white approach does not take in account patients' ergonomics, comfort, and quality of life. There is a scale of gray between these two extremes. Some external devices or actions from the outside might be perceived by patients as more invasive than their implants. For example, being introduced in a tunnel for MRI is a traumatic experience for some patients, certainly more "invasive" than the introduction of electrodes in their body. Invasiveness has also to be ranked in terms of severity. A subcutaneous tiny device placed under the skin in an ambulatory procedure under local anesthesia is less invasive than wearing an EEG cap several hours a day over long periods of time.

3.2.1 Noninvasive

In the conventional perception, a noninvasive device has the following characteristics:

- Not penetrating in the human body through the skin (devices placed in natural body orifices are considered as noninvasive).
- Usually short-term use (less than 30 days) or renewal at regular intervals (like skin patches for transdermal electrical nerve stimulation (TENS)).
- Patient comfort is a priority.
- Mainly wearable.
- Low cost.

In the field of neuro-therapies, the following devices are classified as noninvasive:

- Skin patches for TENS
- EEG caps
- Electromyogram (EMG) electrodes
- Devices for tibial nerve stimulation (TNS) for the treatment of mild incontinence
- External device for vagal nerve stimulation (VNS)
- Transcranial direct current stimulation (TDCS) devices
- Transcranial alternating current stimulation (TACS) devices
- Various types of skin electrodes placed on the head, face, or other parts of the body for various purposes

Diagnostic and imaging systems [2], like magnetic resonance imaging (MRI), computed tomography (CT), positron emission tomography (PET), or magnetoencephalography (MEG), are also considered as noninvasive as no physical hardware is introduced in the body. Nevertheless, these procedures may be perceived as quite invasive by patients.

3.2.2 Not So Invasive

Three categories of devices, which penetrate through the skin, are neither noninvasive nor invasive, as their impact on the body is minimal:

- Transdermal leads for intramuscular stimulation (IMS) or intramuscular myogram (IMMG). These wire electrodes are inserted in the muscles without major surgical intervention and remain in place for shorter durations. An example is FES of the arm of a paralyzed patient, activated from a BCI on the motor cortex. Several transdermal leads are inserted in muscles along the arm, from shoulder to hand, allowing "move and grasp" actions. In some cases, these transdermal

electrodes remained in place for months and even a couple of years without infection. A successful example of such work has been conducted by the team of Prof. Robert Kirsch from Cleveland FES Center [3] and Prof. Hunter Peckham from Case Western Reserve University in Cleveland [4], in the frame of the BrainGate initiative [5].
- Small-sized subcutaneous implants pushed under the skin through a small incision, under local anesthesia. In the field of cardiac monitoring, products like Reveal and Linq from Medtronic [6] have been marketed on a large scale since about two decades.
- Some other devices may belong to the category of "not so invasive;" in the sense they do not require opening the skull to access the brain. In a sense, VNS is "not so invasive," as surgery to access the vagal nerve is minor, most of the time done under local anesthesia. Other initiatives, like the Stentrode of Synchron Inc. [7], aim to read or even stimulate the brain from stent-like electrodes introduced in the jugular vein and push up to the brain, to areas proximal to the motor cortex. The objective is to collect cortical signals generated in the motor cortex from the Stentrode and to carry them to a subclavicular implant for further telemetry transmission connected to external decoders linked to actuators (2D screen, robot arm, or FES) to help paralyzed people to recover some movement capabilities. Synchron also works on using such electrodes introduced in the blood channels to add stimulating features and have close-loop systems for treating epilepsy, detecting the onset of a seizure and stimulating from the Stentrode to cancel the seizure. There is still a lot of fundamental work to be done to carry this system to a human application, but the concept is interesting. However, risks of lesions or clots induced by the presence of stent-like devices in brain veins are serious, and long-term feasibility in human beings has still to be demonstrated.

3.2.3 *Invasive*

Invasive is a term which is appropriate when a device is introduced deeper in the body and remains there for long periods of time. It implies a major surgical act under full anesthesia.

Deeper tissue interfaces are of different nature, size, contact impedance, rigidity, and materials. Some are designed for sensing, some for stimulation. The non-insulated surface of the interface is called electrode or contact. The electrodes are connected to insulated wires. These wires are gathered in a bundle or coiled or braided to form a cable. Alternative to cables are flat ribbons. The other end of the wires is either directly attached to the IPG or to a connector.

- Brain interfaces
  - Epidural (ECoG or paddle) (Fig. 3.3)
  - Subdural (ECoG or paddle) (Fig. 3.3)

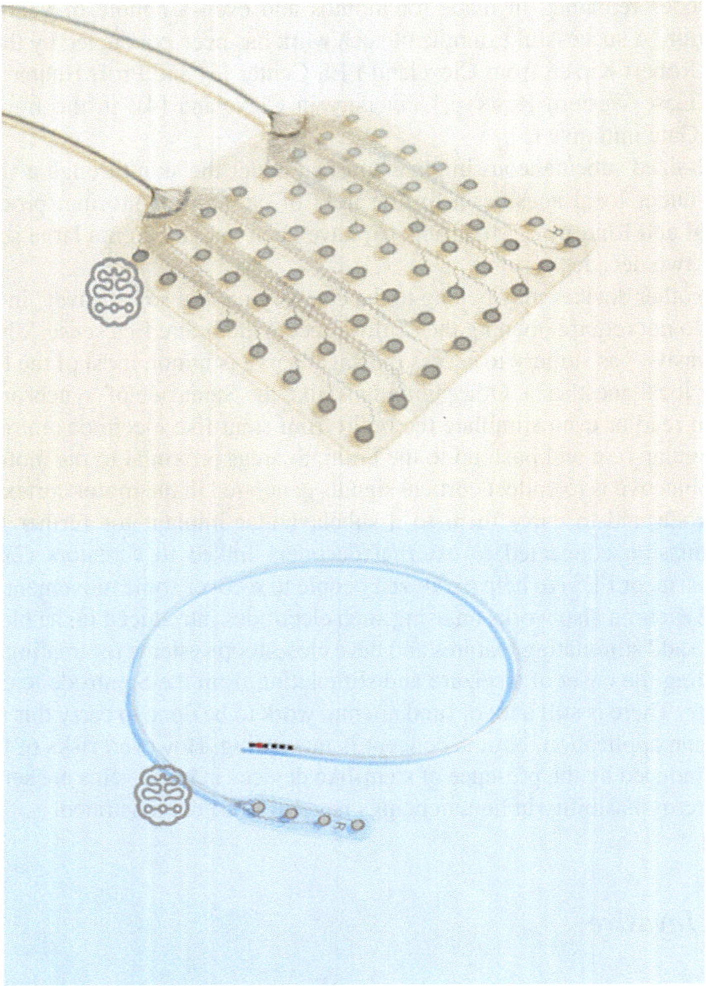

Fig. 3.3 Epidural ECoG grid and strip electrode. *(Courtesy of Cortec GmbH)*

  - Intracortical (MEA) (Fig. 3.4)
  - Penetrating (wire or multipolar in-line) (Fig. 3.5)
  - Deep (multipolar DBS) (Fig. 3.6)

- Spinal cord interfaces
  - Epidural (paddle or multipolar in-line) (Fig. 3.7)
  - Subdural (paddle or multipolar in-line) (Fig. 3.7)

- Nerve interfaces
  - Cuff electrodes (Fig. 3.8)
  - Fascicular electrodes (Fig. 3.9)

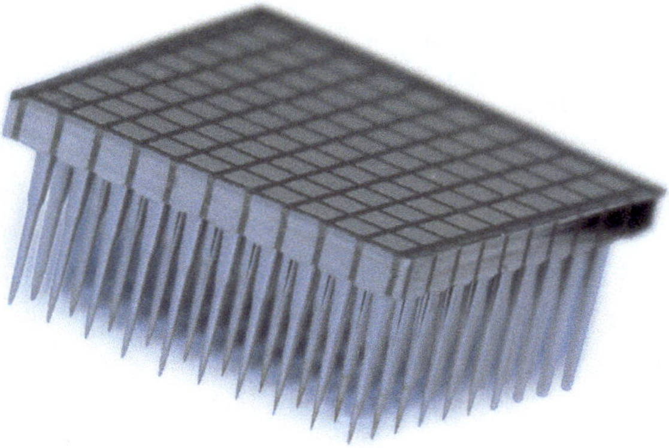

Fig. 3.4 Intracortical microelectrode array (Utah array). *(Courtesy of Blackrock Microsystems LLC)*

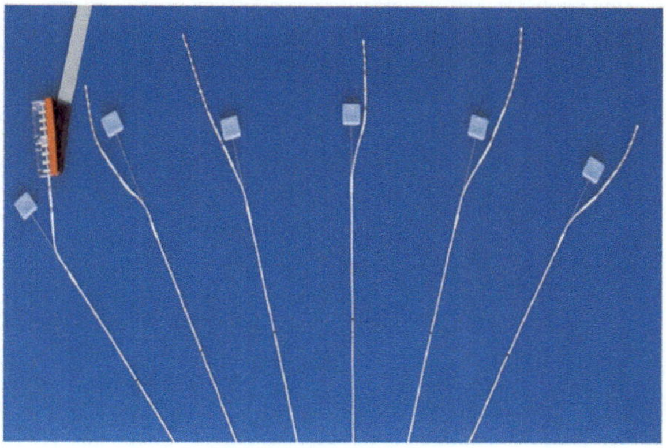

Fig. 3.5 Penetrating electrodes. *(Courtesy of Ad-Tech Inc.)*

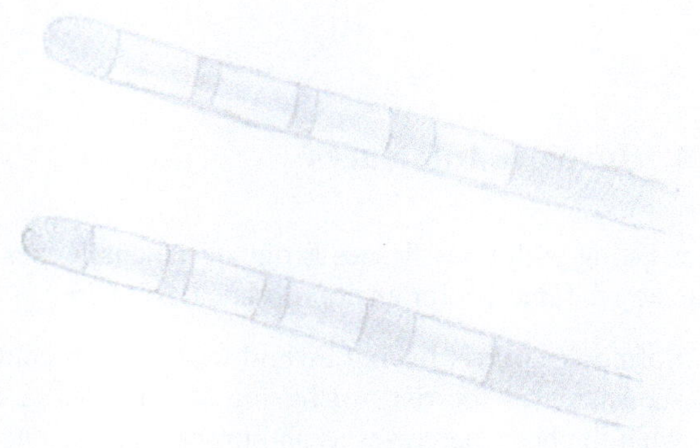

Fig. 3.6 DBS electrodes

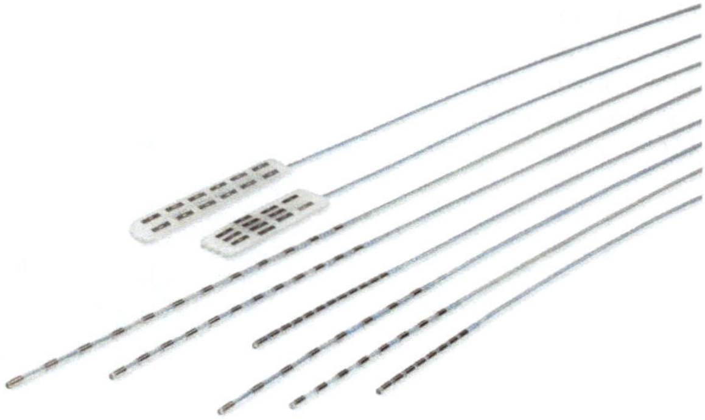

Fig. 3.7 Epidural and subdural linear or paddle electrodes. *(Courtesy of Nuvectra Inc.)*

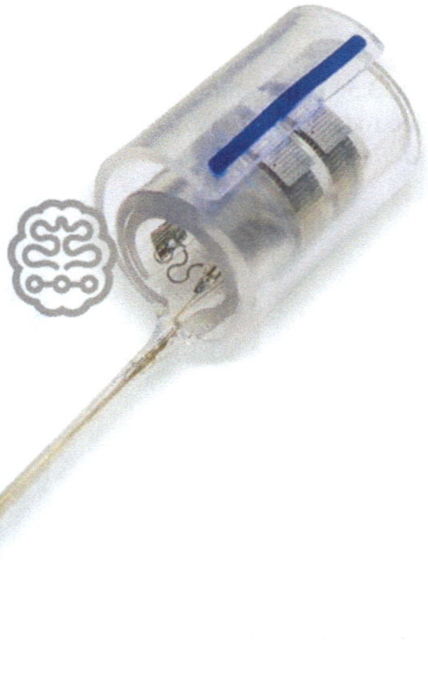

Fig. 3.8 Spiral cuff electrode. *(Courtesy of Cortec GmbH)*

3.3 Invasive Interfaces

Interfacing with body tissues is one of the challenges of active long-term implants. The key parameters for an appropriate design are:

- Softness/rigidity: having a rigid electrode in contact or introduced in soft tissue increases the risks of having scar tissues or fibrotic encapsulation around what the body perceives as an intruder. The science around the tissue growth on implants is not fully understood. We know it is dependent on the type of material, on the quality of the surface, on presence of residual contaminants, and on the blood perfusion. One cause of tissue irritation is due to the relative movement of the implant in the surrounding tissues. Hard implants are more susceptible to move than soft implants which will "follow" the displacement of tissues.

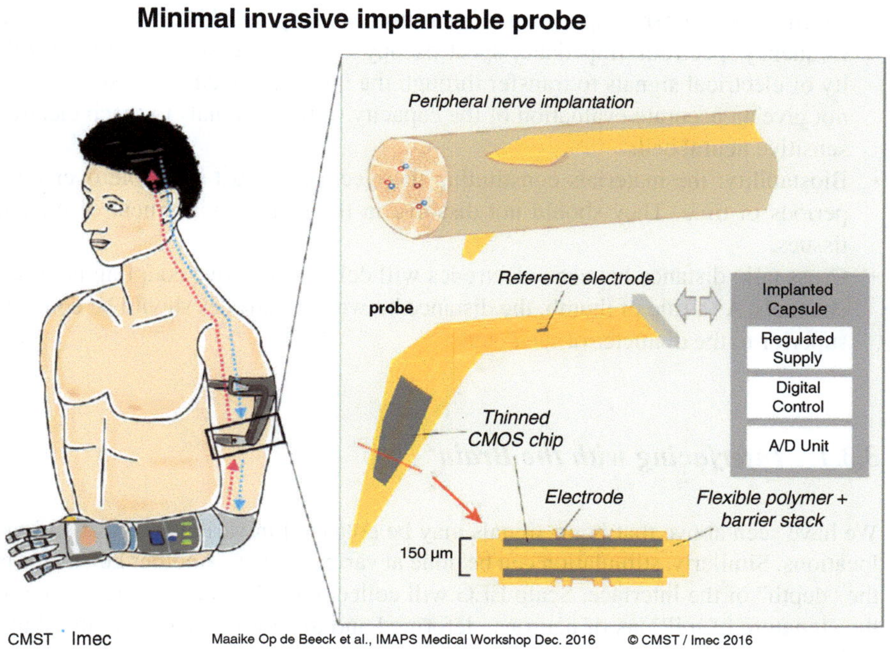

Fig. 3.9 Fascicular electrodes. *(Courtesy of IMEC)*

The origin of the movements of tissues depends on the location. In the brain, there are natural movements due to blood pulsing and to respiration. In addition, there are relative movements caused by moving the head. Larger displacements may be generated by shocks or traumatic impacts.

- Contact impedance: an important parameter, especially for sensing small electrical potentials. The contact impedance is a serial impedance between the source of electrical signals (tissues) and the amplifier. Contact impedance depends on the surface of the electrode, on the material, and on the structure of the surface. Some electrodes have fractal structures meant to increase the actual contact surface. When trying to sense single neurons, the surface is by nature very small (tip of a pin of the Utah array), so the contact impedance is rather high (above 100 kOhm). Spikes fired by neurons are in the range of 100 μV. It means that the amplifier must be designed for optimizing signal-to-noise-ratio (SNR) and gains taking in account the high serial impedance. Some devices feature the possibility to measure contact impedance (injection of a voltage and measurement of the current). In the field of neurology, impedance is usually measured at 1 kHz. Measuring impedance from time to time gives an image of how the physical contact between the electrode and the tissues is evolving. For example, an increased impedance may reflect building up of fibrotic tissues around the electrode. An even better assessment of the quality of contact is impedance

spectroscopy, where impedance is measured not only at 1 kHz but over a large frequency spectrum. Impedance spectroscopy is only a measurement of the ability of electrical signals to transfer through the interface electrode-tissue. It does not give an accurate evaluation of the capacity of these signals to reach electrosensitive neural cells.
- Biostability: the materials constituting the electrodes must be stable over long periods of time. They should not dissolve in the harsh environment of human tissues.
- Cross talk: distance between electrodes will define capacitive coupling between channels. As a rule of thumb, the distance between electrodes should be equal or superior to the diameter of the contact.

3.3.1 Interfacing with the Brain

We have seen above that brain signals may be collected in various ways at various locations. Similarly, stimulation can be done at various places. Another key factor is the "depth" of the interface. Scalp EEG will collect global signals, an aggregate of the signature of millions of neurons, deformed and diffused by the CSF, the skull, and the skin.

Global brain waves are categorized by their frequencies:

- Delta waves: 0–4 Hz. Delta waves are related to relaxation and restorative sleep. Delta waves seem to be linked to unconscious body functions like cardiovascular and digestive systems. Sleep may be linked to delta waves.
- Theta waves: 4–8 Hz. Theta waves are defined as suggestible waves related to hypnotic state, daydream, emotions, anxiety, and sleep.
- Alpha waves: 8–12 Hz. Alpha waves are described as frequency bridges between subconscious waves (theta) and conscious thinking (beta). They are related to obsessive-compulsive disorders (OCD), anxiety, and stress.
- Beta waves: 12–40 Hz. Beta waves are found in awake humans, related to cognitive reasoning calculation, speaking, reading, and thinking. Higher levels are signs of anxiety and stress. Lower levels indicate depression and lack of attention.
- Gamma waves: 40–100 Hz. Gamma waves indicate stress and anxiety at high levels and attention deficit hyperactivity disorder (ADHD) or depression at lower levels.

Going subcutaneous, below the scalp, improves a bit the quality of the measurements. Sub-glial placement partly eliminates artifacts caused by muscles, for example, originating in movements related to chewing or speaking. Placement of subcutaneous electrodes is easy, through a small incision and using introduction tools to push or pull the electrodes in place. Most of the time, subcutaneous scalp electrodes do not need to be fixed in place.

One further step deeper in the direction of the brain is penetrating through the skull and placing an ECoG grid over the surface of the brain. Placement of an ECoG grid requires a large craniotomy. As of today, the main regular use of ECoGs is limited to the evaluation of epileptic patients intended to get an ablation of the part of the brain being the focal source of seizures. This evaluation is done at the hospital over a period of a couple of weeks. The piece of bone removed for accessing the brain is put back in place after the insertion of the grid and cables are exiting through a transdermal passage. An external recorder will collect the signals picked by the ECoG days and nights over up to 2 weeks. Seizures during this period will be accurately measured. Collection of this large amount of data will facilitate an accurate location of the focal point, preparing for an optimal subsequent resection surgery. ECoG contacts are still quite big, a few millimeters in diameter, so they capture global signals of a large population of neurons.

The only way to enter in contact with a few or one single neuron is to insert tiny electrodes in the cortex. The necessity to measure spikes has been discussed earlier in this book. Future BCIs, with high spatial and temporal resolution, are going to require tiny penetrating electrodes in direct contact with neurons. In consequence, MEA is the interface of choice for advanced BCIs. Today, the only human grade MEA is the Utah array, supplied by Blackrock Microsystems LLC [8] (see Fig. 3.10).

A standard round burr hole (the same as for the introduction of DBS leads) is done in the skull. The MEA is introduced through this hole and localized on the top of the motor cortex corresponding, for example, to the right arm. With the help of a pneumatic inserter, the 100 pins of the MEA are punched through the arachnoid, down approximately 1.5 mm. In the current configuration, the MEA is connected to a transdermal connector, called pedestal, with a bundle of thin gold wires, wire-bonded at each end. The total length of the bundle is 13 cm. The pedestal is screwed on the skull. Even if constituted of thin wires, the bundle has a certain rigidity. The surgeon needs to carefully form the bundle in a way that the cable does not exercise pressure on the MEA.

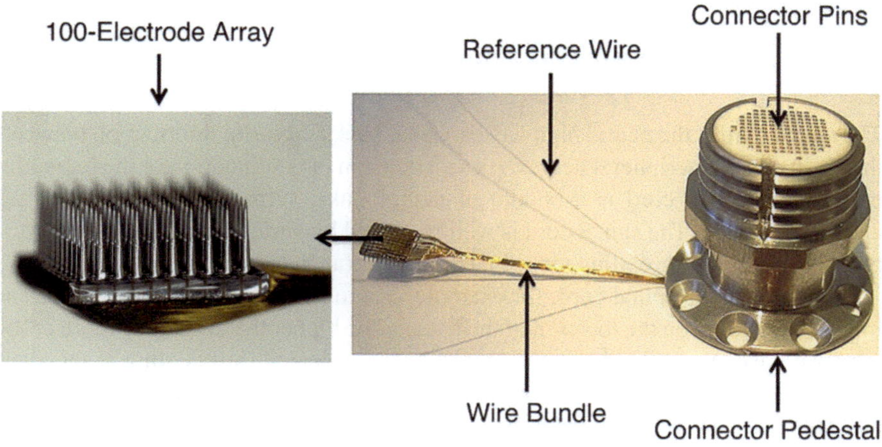

Fig. 3.10 Utah array connected to pedestal. *(Courtesy of Blackrock Microsystems LLC)*

In a near future, the pedestal will be replaced by a fully implanted wireless communication device but still connected to the Utah array. It will be further discussed in Sect. 7.3.

The Utah array (see Fig. 3.10) does a smart use of the mechanical performance of silicon while taking advantage of the semiconducting properties of this material to have a double diode insulation between channels. Its production is based on many years of accumulated experience and secret know-how. Mainly used for preclinical research projects, the Utah array has found its way in about two dozen human beings. Large variations have been seen in terms of contact impedance, evolution of impedance over time, and insulation between channels. Some Utah arrays are still functioning after years of implantation. Others degrade already after a few months. We have no solid explanation about this.

Only the tip of the pins is exposed to tissues. The root of the pin is insulated with Parylene [9], a material known since decades as being stable and reliable in other parts of the body. Some puzzling failures have been identified in the Parylene insulation of the Utah array when placed in human brains. One explanation could be unexpected high level of H_2O_2 around the array, which is a powerful destroyer of all plastic materials, Parylene included. The interactions between the materials of the array, the biofilms on the surface, the thicker fibrosis layer developing with time, and the exotic chemistry happening in this capsule are still to be understood. Parylene has shown excellent stability in cardiac applications. Surprisingly, this material may not be as well adapted for use in the brain, but these apparent and not confirmed weaknesses must be further studied before drawing any conclusions.

We are discovering that the brain is a very special environment. Technologies which were proven stable in other parts of the body seem to behave differently in the brain. Efforts are required now to develop reliable interfaces with brain tissues, which will last for decades without degradation of performance. Artificial aging test procedures, adapted to the specificities of brain tissues, are currently being developed and assessed (see Sect. 4.7).

3.3.2 Interfacing with the Spinal Cord

The spinal cord is the neural highway carrying, back and forth, information between the brain and the peripheral nerves. As an extension or continuation of the brain, the core of the spinal cord is a bundle of neural links, surrounded by a protecting envelop, the dura. The spinal cord is well protected by vertebras. At each intervertebral space, nerves are exiting the spinal cord and branch at various parts of the body.

Damages to the spinal cord, for example, due to accidents, will have a different impact depending on the location of the lesion. The higher the lesion, the more problems we have, as all the downstream nerves are impacted. Stimulating at the level of the lesion may improve recovery and even rebuilding synaptic connections, as it has been demonstrated by Prof. Grégoire Courtine from EPFL in Switzerland [10].

In a near future, we also believe it will be possible to stimulate the spinal cord at the appropriated level to command limb movements in paralyzed people, hopefully in direct interaction with a BCI decoding movement intentions in the motor cortex.

Another interaction with the spinal cord is SCS, already described earlier in this book. There, the action is to "freeze" pain signals at the root of the nerve and prevent them to be conducted by the spinal cord up to the brain.

In any of the above situations, electrodes must be placed at the right location of the spine, for stimulation purpose. Introduction of the flexible paddle electrodes is done through the interspace between two vertebras and pushed upwards.

Flexibility of the electrodes is required, as body movements will lead to bending the spinal cord. In addition, stretchability of the electrode would be a great improvement to avoid displacement of the contacts following bending of the spinal cord. Prof. Stéphanie Lacour, EPFL in Switzerland [11], is leading the way in the development of flexible and stretchable electrodes for spinal interfaces.

3.3.3 Interfacing with the Vagal Nerve

After the spinal cord, the vagal nerve (or vagus nerve) is the second highway of neurological communication in the body. It is the tenth cranial nerve and consists of afferent and efferent fibers. The afferent fibers connect to the central nervous system (CNS) in complex links through the nucleus of the solitary path. In short, afferent fibers carry information from the periphery to the CNS.

Efferent fibers carry information from the CNS to the periphery. For the treatment of epilepsy, the afferent fibers are involved. We do not yet understand the exact mechanisms of the effects on the brain resulting from stimulation of the vagal nerve.

Regarding the interface, vagal nerve stimulation is simple. The cuff electrode has only two contacts. It also makes the connection to the IPG simple and inexpensive. Insertion of the cuff electrode is done through a small incision. The cuff is wrapped around the left vagal nerve at the base of the neck. In a sense, the vagal nerve is an easy indirect access to the brain.

It is also possible to stimulate the vagal nerve from an external device (see Fig. 3.11). ElectroCore [12] got approval for a handheld device with two electrodes applied on the neck along the vagal nerve. The electric field between the external skin electrodes stimulates the nerve. It is indicated to help control migraines and cluster headaches.

3.3.4 Interfacing with Peripheral Nerves

There is a wide range of applications where it is necessary to interface with nerves. Here are some examples:

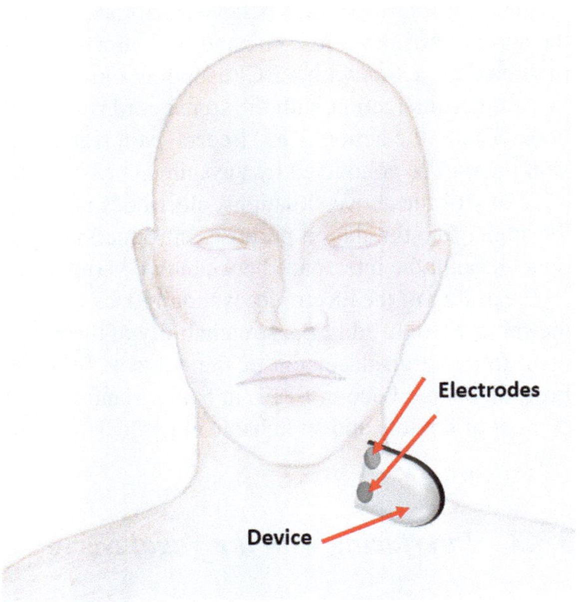

Fig. 3.11 External vagal nerve stimulation

- Amputees:
  - Controlling a hand prosthesis directly from the nerves of the arm with intrafascicular or cuff electrodes
  - Haptic feedback, transferring touch information collected by sensors at the tip of the prosthesis fingers to the nerves with intrafascicular electrodes
- FES: Intramuscular stimulation requires a lot of energy. Stimulating on the nerves instead provides the same movement capabilities but with much less current. The interface is preferably cuff electrodes.
- Erectile dysfunction after prostatectomy: an interface looking like an ECoG is placed on the pelvic floor to stimulate nerves which have been disrupted during the removal of the prostate. This work has been initiated by Prof. Nikos Stergiopulos at EPFL Switzerland [13] and is now developed in a start-up called Comphya [14].
- Hypoglossal nerve stimulation is used to treat obstructive sleep apnea (OSA) [15]. Stimulation of the hypoglossal nerve allows the muscles of the tongue to preserve normal tonus and avoids it to fall back in the throat and obstruct airways. A fully implantable device has been developed for this purpose by Inspire Inc. [16]. It consists in a chest IPG connected to a pressure-sensing lead and to a stimulation electrode.
- Facial mirror stimulation: for people suffering from unilateral facial paralysis, sensing electrodes are placed under the skin of the valid side, and mirror stimulation is sent to stimulating electrodes on the paralyzed side.

- Sacral nerve stimulation for urinary and fecal incontinence also belongs to peripheral nerve stimulation. It is described in other chapters of this book.
- Tibial nerve stimulation is, surprisingly, showing that acting on a certain nerve in the legs has a positive impact on bladder control.

Hypoglossal, sacral, and tibial nerve stimulations have already found successful commercial introduction and approval. Some other promising PNS therapies are still in the lab, but we will see a fast growth in the future. Each application requires adapted nerve interfaces, near the nerve, around it, or in it. The electrodes may vary widely in terms of shape, number of contacts, material, and flexibility. Alternatively, peripheral nerves may be stimulated by an external device in contact with the skin.

3.3.5 Interfacing with Organs

We have already discussed some applications which are not directly BCIs but having a natural technological link with the characteristics of brain and spinal cord interfaces. These developments are very instructive in the way they interface with organs. Let's do a quick review of these specific tissue interfaces.

- **Retina:** electrodes interfacing with the retina have a lot in common with interfacing with the cortex. For getting enough resolution, many channels are needed. Another similarity is the long-term biostability, as it will be very difficult to replace the electrodes if they get inefficient. Retinal electrodes must be thin enough to be placed on the retina (epi-retinal) or under it (subretinal). They also must be flexible in a way to conform to the curvature of the bottom of the eye. Most of the current retinal implants have thin film electrodes based on polyimide substrates. As polyimide absorbs moisture on the long run, special care should be done to avoid delamination of the contacts and the conductive traces (see Fig. 3.12).

- **Cochlea:** is a very small and fragile organ. Introducing electrodes in this spiral-shaped canal is a challenge. The total length of the cochlear spirals is about 2.5 turns. Current electrodes usually cannot cover more than 1.5 turns. The thin electrode is pre-shaped in the form of a spiral to facilitate introduction. The body of the electrode is made of soft silicone rubber, with up to 22 platinum contacts, attached to tiny platinum wires (about 10 μm in diameter) insulated with Parylene. The assembly of such a miniature electrode is a challenge (see Fig. 3.13).

- **Vestibular organs:** together with the cochlea, two other elements rule the vestibular function. The semicircular canals indicate rotation, and the otoliths detect linear acceleration. Companies active in cochlear implants are using similar technologies to access these organs in order to treat disorders in balance and equilibrium. No product is yet approved in this field. Like the cochlea, vestibular organs located near the inner ear are extremely small, rendering the introduction of electrodes very difficult.

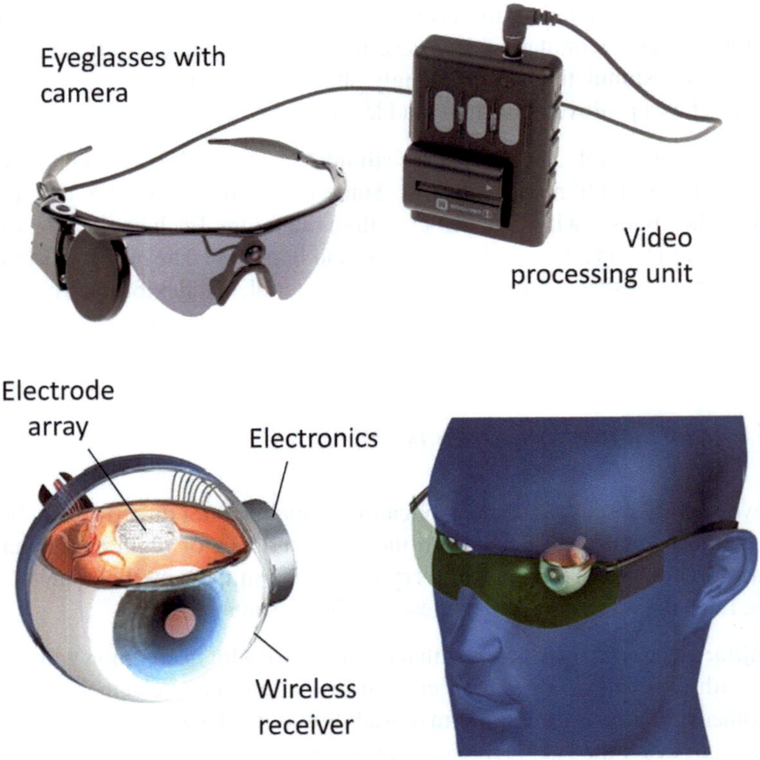

Fig. 3.12 Retinal implant. *(Courtesy of Second Sight Inc.)*

- **Digestive system:** electrodes are placed in the muscular walls of the stomach to reduce vomiting and nausea in cases of gastroparesis. The same gastric nerve stimulation (GNS) device is being evaluated for obesity. The Enterra (Medtronic) [17] (see Fig. 3.14a) is a device which got a Humanitarian Device Exemption (HDE). The interfaces with the stomach wall consist of two bipolar wire leads.
- **Heart:** since more than 60 years, many types of electrodes have been designed for cardiac stimulation and sensing. Early pacemakers and defibrillators have been using epicardial electrodes in the form of meshes. Current cardiac electrodes are introduced through the subclavian vein to reach the right side of the heart (see Fig. 3.14b). Ventricular leads are simply pushed down to the bottom of the ventricle with the help of a guide wire inserted in the hole in the center of the lead. Atrial leads are shaped in a form of a "J," so they can be attached on the top part of the atrium. A straight guide wire forces the "J"-shaped distal section of

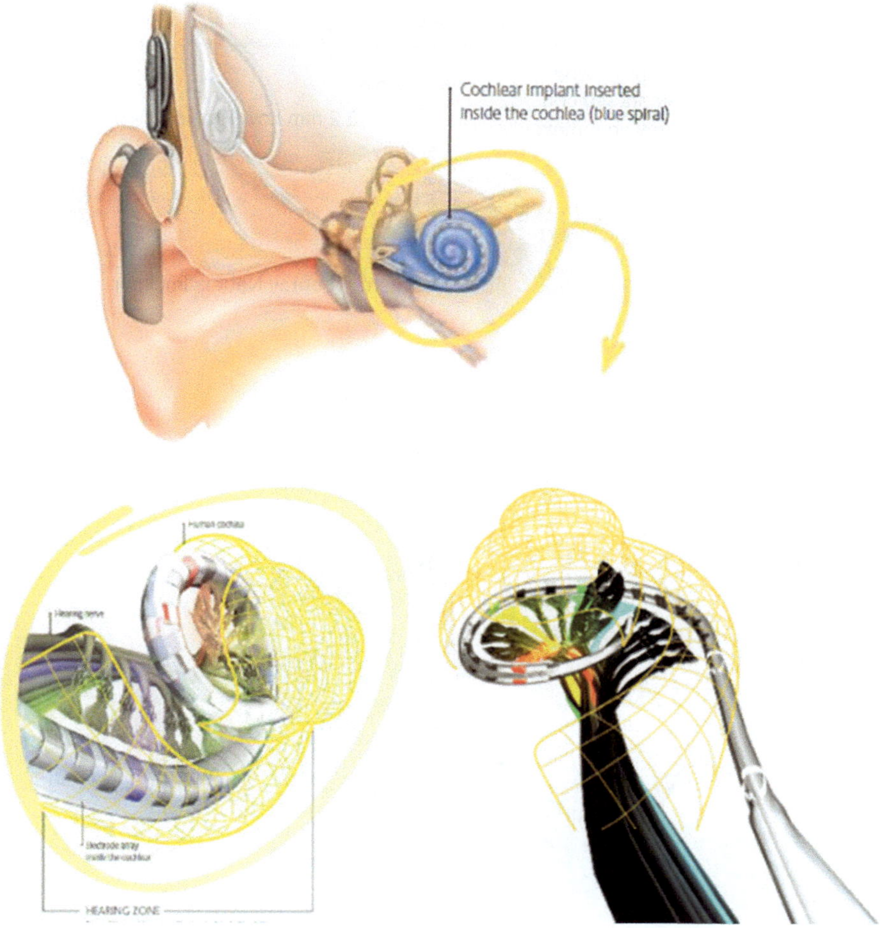

Fig. 3.13 Cochlear implant. *(Courtesy of Cochlear Inc.)*

the atrial lead to be straight and, when in place, slowly pulled out to allow the tip of the electrode to regain its "J" shape. This trick could be useful to reach certain parts of the brain which are not easily accessible, like in the convolutions. As an example, the area of the motor cortex corresponding to the lower limbs is down a convolution which cannot be easily reached by a surface electrode or a MEA. The tip of cardiac leads is attached to the fibrous surface of the inner wall of the heart by tines or with a screw mechanism rotated by the insertion guide wire. Currently, most cardiac leads are bipolar, with a ring proximal to the tip, for the return of the current.

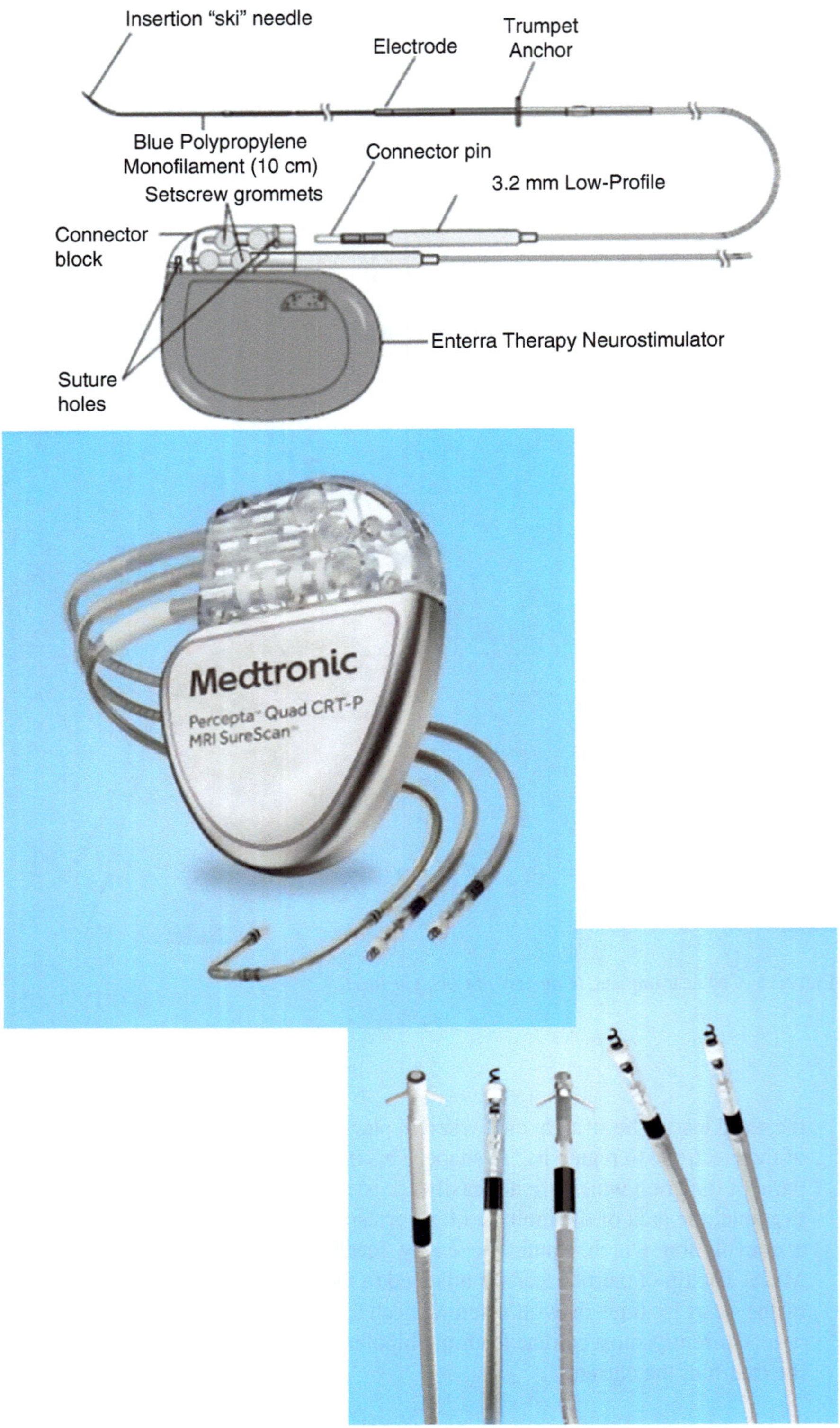

Fig. 3.14 (**a**) Gastric nerve stimulation. *(Courtesy of Medtronic plc)* (**b**) Pacemaker and pacing leads. *(Courtesy of Medtronic plc)*

3.4 Achievements

In this sub-chapter, we will review the existing neuro implants and their main technical features. We will retain only the products which did achieve their translation and are available for a large patient population. For each of them, we will extract the lessons to be learned in view of the realization of BCI systems.

A lot has been done so far and it is a valuable source of inspiration for future BCIs. It is capital to get an overview of this industry and learn from successes but also from mistakes (Fig. 3.15).

3.4.1 Cochlear Implants (CI)

Interfacing directly with neuroreceptors of the inner ear was the first commercial large-scale achievement of neuro-technologies. The main indication of CIs is children born with a nonfunctioning conduction of the sound waves from the eardrum to the cochlea. These children have a fully functioning cochlear, but no sound waves activate the liquid of the cochlea. The disorder itself (nonconduction) is not neurological, but the solution is. A tiny electrode (12–22 channels) is introduced in the cochlea and stimulates the natural neuroreceptors of the inner ear. The electrode is connected to an implanted electronic in a hermetic housing (see Figs. 3.16, 3.17, and 3.18), placed under the skin behind the ear, sometimes in a recess milled in the bone. Introducing the electrode in the cochlea is a difficult task. The cochlea is a small spiral-shaped channel. The neuroreceptors are delicate thin hair cells, which could easily be damaged by the electrode itself during the introduction. It is rare that an electrode could be removed and replaced by a new one, because it will make too much destruction. Therefore, the entire system has been designed for a unique insertion. For a young kid, it means that the same electrodes will remain in the cochlea for several decades, maybe for all his/her life, if no better technology is discovered in the future.

Children born deaf and eligible for a cochlear implant must be implanted as soon as possible; otherwise, lack of hearing may lead to severe speech disorders, for missing sound feedback. Some of them get their cochlear implant before the age of 1 and are likely to keep it until their death. With a life expectancy of 100 years, it creates exceptional constraints in terms of reliability, biostability, and robustness of the implant. Designers of CIs have included these tough requirements in the overall concept: the implanted electronics is as simple as possible, a simple transmitter of signals received from the external device to the electrodes. As the implant will not be upgradable for decades, the complexity of signal treatment is done in the external speech processor, which can be modified easily. The design strategy was therefore to keep the implanted part simple and reliable, the external part being upgradable and flexible.

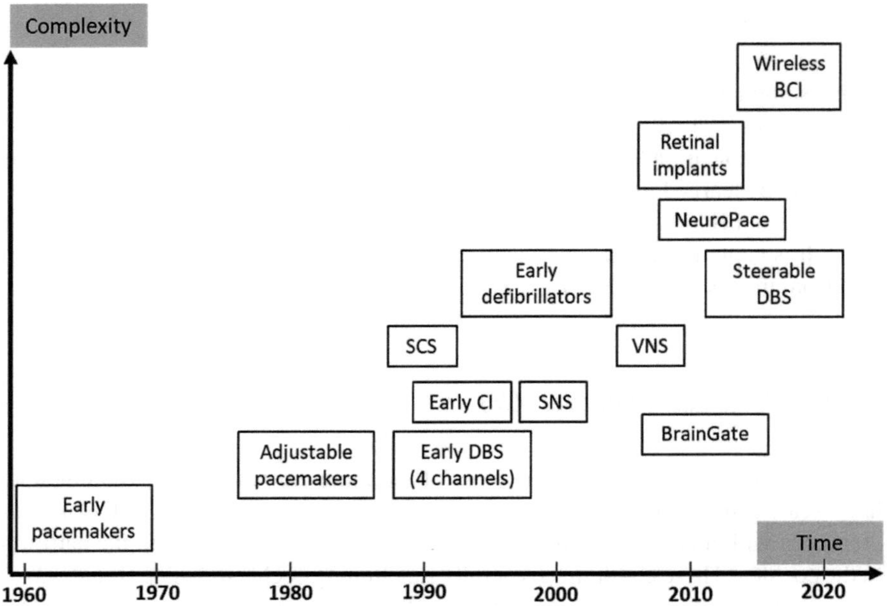

Fig. 3.15 Evolution of the complexity of active implants

Another difficulty arises due to the configuration of the bones around the cochlea. The entrance of the cochlea, the round window, cannot be accessed perpendicularly through the ear canal. In consequence, a hole must be drilled from behind the ear, where the titanium housing will be located, to the round window. This is a very delicate surgical act, especially when done on children. An additional difficulty comes from the proximity of the facial nerve. Great care must be taken to avoid any damage to the facial nerve during the drilling process. Touching the facial nerve may mean permanent facial paralysis. As children will have the implant for a very long time, it is worth milling a recess in the skull bone, to have a better inclusion of the titanium can and less aesthetical impact.

Cochlear implants are also indicated for older patients suffering from age-related hearing loss. Alternatives are behind-the-ear or in-the-canal external hearing aids, which have been available since a long time.

The implanted electronics receives signals from an external sound processor positioned behind the ear. Natural sounds are picked by a microphone and processed by the external unit. The external unit includes a source of energy, a small battery which must be changed frequently (every day or so, depending on the model). Part of this energy is passed to the implant by an inductive coupling consisting in an external coil aligned with an implanted coil, just separated by skin. As the distance between the two coils is small, inductive coupling is good. The alignment of the coils and the fixation of the external coil on the skin are assured by two magnets attracting each other through the scalp. One of the magnets is implanted in the center of the internal coil; the other one is in the center of the external coil. When

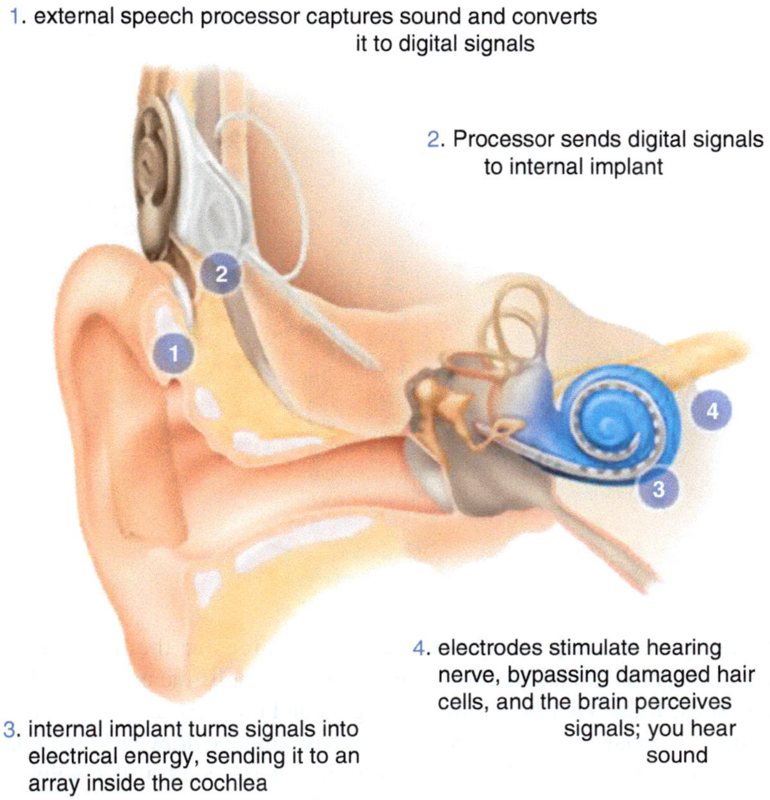

Fig. 3.16 Cochlear implant. *(Courtesy of Cochlear Inc.)*

Fig. 3.17 Cochlear implant system, N7 and Kanso Portfolio. *(Courtesy of Cochlear Inc.)*

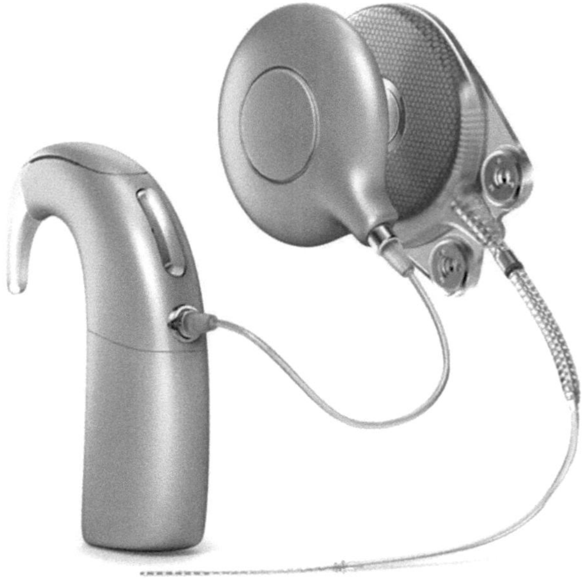

Fig. 3.18 Innovative cochlear implant in ceramic housing. *(Courtesy of Oticon Medical)*

the patient goes to sleep or when he/she wants to remove the device for convenience, for example, to take a shower, he/she simply pulls the external behind-the-ear unit away. It can later be put in place again and properly aligned by the simple attraction of the magnets. This clever system has only one drawback: the implanted magnet is not MRI compatible. There too, designers found a nice solution. The implanted magnet is trapped in the middle of the soft silicone patch encapsulating the antenna and can be pulled out of the patch due to the softness of the silicone rubber. Patients who must go to an MRI examination of their head simply go through a minor intervention under local anesthesia. A small incision is done in the skin above the patch antenna, the implanted magnet is removed, the MRI investigation is done, and then a new magnet is placed in the center of the patch.

As the space available is very limited, the titanium housing containing the implanted electronics must be as thin as possible, to avoid a visible protrusion under the skin and to minimize skin erosion. Modern cochlear implant housings are about 4 mm thick. For children, most surgeons mill a 2 mm deep cavity in the bone, to minimize protrusion and assure a good fixation. For adults, in some cases, surgeons decide not to mill and simply place the implant above the skull. The limitation of space also prevents the use of a detachable connector between the housing and the leads. In today's technology, miniature connectors with 22 channels do not exist for long-term implantation. Intensive work is currently done by companies like Bal Seal Engineering Inc. [18] to design miniature connectors adapted to cochlear implants. As of today, the lead cable is not detachable from the housing. It means that the wires are permanently directly attached to the housing, without possibility of disconnecting the housing from the leads. As the implant must remain in place over several decades, it implies that the implanted electronics must be very reliable. This is the reason why the internal electronics is kept as simple as possible. It only contains the circuits to convert the energy transferred by induction and 22 stimulation channels fed by externally preprocessed signals. All the complexity is in the external unit, which can be replaced or upgraded.

The market is dominated by Cochlear Inc. (~55%) [19], followed by Advanced Bionics (AB), Sonova group [20], MED-EL [21] (~20% each), and two smaller firms. Since about 30 years, an amazing total of 700–800,000 cochlear devices have been successfully implanted.

Historically, pioneer work has been done in Australia by Cochlear, followed by the Alfred Mann Foundation (AMF) [22] which later spin-off Advanced Bionics. AB was then acquired by Boston Scientific (BSc) [23] and finally ended up in the Sonova group, world leader of external hearing aids, known under the trade name of Phonak. The technology of cochlear implants has progressed over the years, especially at the level of the speech processors and patient's controls.

One of the weak points of CIs is the presence of an implanted coil-patch attached on the side of the titanium housing. Tiny displacements around the implant are induced in the movements of the chaw, creating potential fatigue rupture at the link between the rigid housing and the soft patch. As titanium is a shield for electromagnetic power transfer and for signals sent by the external device, the coil must be deported outside the hermetic can. Some manufacturers tried to replace titanium by ceramic housing, which, because of the electromagnetic transparency of ceramics, allows to place the induction communication coil inside the implant. Unfortunately, ceramics are fragile and breakable, especially if thin. Some of these ceramic encapsulations did crack when impacted externally through the skin. This has led to a new standard called "impact test," which has been a major restriction to the use of ceramic housings placed on the skull. Today, a wide majority of CIs are still made of titanium with a deported patch antenna.

An interesting case is the French company Neurelec [24]. France has been historically a pioneer player in the field of CIs. Year 2013, Neurelec merged with Oticon Medical [25], a Danish major firm in bone-anchored and external hearing aids. Oticon Medical is part of the William Demant Group [26]. Recently, Neurelec (Oticon Medical) reinvented ceramic encapsulation and launched a highly miniaturized CI, with a donut-shaped zirconium housing including a magnet in the center and the coil inside. This design combines electromagnetic transparency and impact test resistance.

This device (see Fig. 3.19) is a good example of a thin, robust, compact implant for above-the-neck applications. It may inspire designers of the BCI of the future.

What lessons did we learn from the CI industry in view of BCI devices?

- Long-term implants cannot have:
  - Primary batteries if the electrodes cannot be disconnected
  - Rechargeable batteries, as the number of recharging cycles is limited

Fig. 3.19 Ceramic housing, detail. *(Courtesy of Oticon Medical)*

- Inductive coupling for power transfer to the implant is:
  - The only way to energize long-term implants (maybe alternatives like energy harvesting will become available in the future)
  - Efficient only if the distance between the two coils is maximum a few millimeters.
  - Sensitive to the alignment between the coils, elegantly solved in CI by the addition of two magnets.
- Battery-less implants permanently require an external counterpart or headpiece.
- If the implant can be done thin enough to be almost invisible, the headpiece may be an aesthetic issue, unless carefully covered by hair.
- Using the same induction coils for both data transmission is limiting the communication bandwidth.
- If the electronics is encapsulated in titanium, the induction coil must be deported outside the housing.
- Electronics encapsulated in electromagnetic transparent materials, like ceramics, glasses, or sapphire, have limitations:
  - Poor resistance to impacts
  - Larger overall thickness compared to titanium can
- The Neurelec device has the advantage of including the coil inside the housing, improving long-term reliability.
- As no miniature connectors are available for long-term implants, the electrodes must be pre-attached to the housing. Consequences are:
  - Difficult surgery, as the housing might make the introduction of the electrode more cumbersome.

  - In case of trouble with the electronics, it is not possible to exchange it without also explanting the electrodes.
  - Reliability of the electronics must be assured for decades.
- Above-the-neck implants must be slim and not protruding more than 2–4 mm. Thicker implants must be totally or partially buried in cavities milled in the skull.

Some work is currently ongoing to add rechargeable batteries to cochlear implants, with the objective of getting rid of the headpiece. Patient's comfort will be greatly improved, allowing showering and swimming. The entire system becomes invisible. It is based on a concept of middle ear actuator which mechanically induces vibrations on the oval window of the cochlear. Therefore, there is no penetrating electrode in the cochlea. As such, this device is not a neurological implant, but it remains a great source of inspiration. The device is named Carina® [27] and has been developed by Cochlear Inc. (see Fig. 3.20). There are two major challenges:

- Due to the limited number of recharges of the battery, the titanium housing, unlike conventional CI, must be exchangeable when the battery reaches its end-of-life (EOL). As the mechanical transducer is not easily removable, designers have included a detachable connector between the titanium housing and the actuator. This was possible due to the fact there are only two wires, compared to the large number of channels of a CI. The lesson to learn from Carina® is that cables with a few channels can be made detachable. This remark also applies to simple neurostimulators like VNS or SNS.

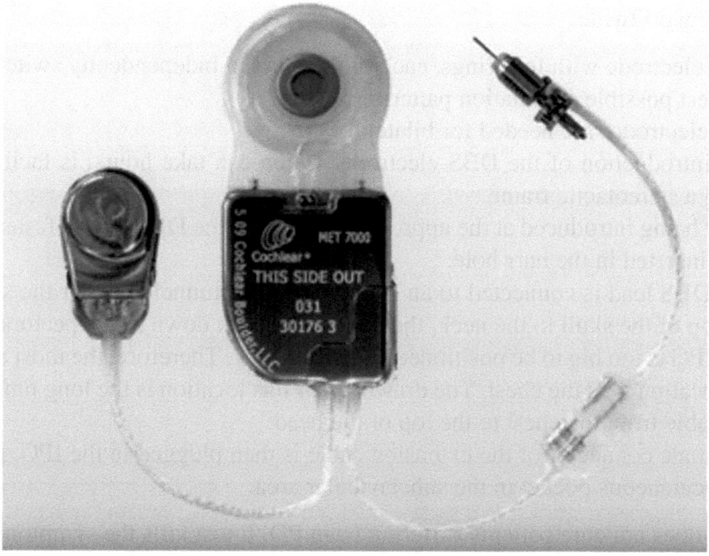

Fig. 3.20 Rechargeable middle ear implant. *(Courtesy of Cochlear Inc.)*

- An implantable microphone must be added to the system. The thin membrane, sensitive to sounds, wave must be able to deflect while keeping long-term hermeticity. Preserving hermeticity is a recurrent barrier to the implantation of sensors. In this embodiment, it has been achieved by laser welding a very thin titanium foil on the microphone casing.

The major weakness of CI is the lack of long-term implantable connectors. As of today, it is not possible to explant the titanium housing only, because it is permanently connected to the electrode. Exchanging just the electronic implant is not possible; the electrode must also be removed from the cochlea, which is a risky operation. Some surgeons can remove the original cochlear electrode and place a new one, but it remains an exception.

3.4.2 Deep Brain Stimulation (DBS), Parkinson's Disease (PD)

At the end of the 1980s, Prof. Alim Louis Benabid [28] was first to identify that stimulation, at a given frequency of the subthalamic nucleus, was blocking the uncontrolled shaking movement of patients suffering from PD. From a hole on the top of the skull, he introduced long electrodes deep in the brain to reach an area the size of a pea. This early work was made possible by using a modified pacemaker (from Medtronic) as source for the stimulation.

Then, Medtronic further developed the system to what is now commercially available worldwide:

- DBS electrode with four rings, each of them being independently switchable for the best possible stimulation pattern.
- Two electrodes are needed for bilateral PD.
- The introduction of the DBS electrode, which can take hours, is facilitated by using a stereotactic frame.
- After being introduced at the appropriate location, the DBS lead is fastened in its base inserted in the burr hole.
- The DBS lead is connected to an extension cable, tunneled under the skin from the top of the skull to the neck, then along the neck down to the pectoral area.
- The IPG is too big to be positioned above the neck. Therefore, the most appropriate location is in the chest. The drawback of this location is the long tunneling of the cable from the chest to the top of the head.
- The male connector of the extension cable is then plugged in the IPG, placed in a subcutaneous pocket in the subclavicular area.

DBS does not cure patients suffering from PD. It just kills the symptoms, which is anyways a tremendous improvement in quality of life. Patients dispose of a remote control enabling them to switch the device off or to adjust stimulation intensity.

Unlike pacemakers which only stimulate for a few milliseconds every second, DBS systems stimulate in a continuous fashion, consuming much more energy, resulting in the need of a large battery. In consequence, DBS IPG is two to four times larger than pacemakers and cannot be placed on the skull under the scalp. It explains why the IPG is in the pectoral area.

The system (see Figs. 3.21 and 3.22) having only four channels per side, it was possible to design four channel connectors, small enough to be placed under the scalp. In fact, there are two connectors: the first one in the header of the IPG, the second one to connect the extension to the DBS stimulating lead. Even if the neck is a very mobile part of the body, four channel coiled leads are flexible enough to resist fatigue.

In some cases, two leads, one for each side, are connected to one single eight channels IPG.

Early days DBS IPGs had primary batteries which last a couple of years. At the end-of-life of the battery, a small incision is done, the IPG is extracted, and the leads are unplugged and replugged in a fresh IPG, as it is done regularly with pacemakers. To avoid having a minor surgical intervention every second year, manufacturers have developed rechargeable systems, which would last up to 10 years. Recharging

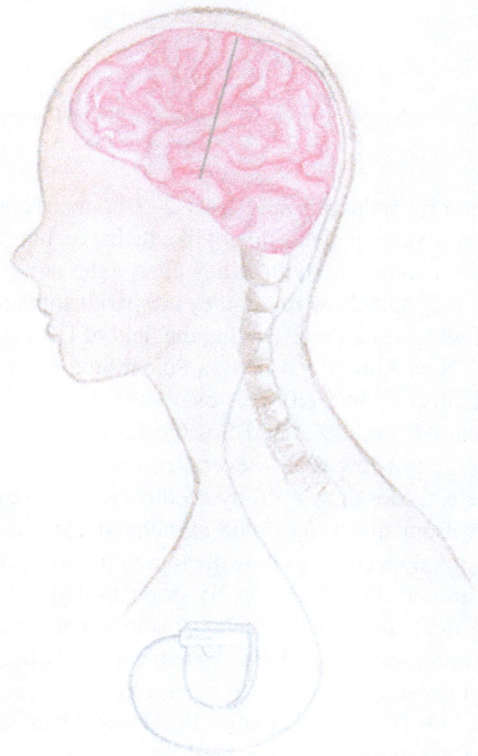

Fig. 3.21 Implantation of a DBS system

Fig. 3.22 DBS system

is done by induction and lasts 2–3 hours. Time between two recharges is 1 to 2 weeks. Some patients do not like rechargeable DBS systems, as they are reminded of their condition anytime they charge the device. With primary battery DBS systems, they somehow forget they are parkinsonian, at least until the next surgery.

Medtronic has been leading the field of DBS since the FDA approval of Activa in 2002. Now Abbott and Boston Scientific are gaining market shares to the expense of Medtronic. In 2020, it is estimated that Medtronic will have 51% of the DBS market, Abbott 22%, and BSc 20% [29].

In the last few years, several companies [30, 31] moved from the original four ring electrodes developed by Medtronic to more complex DBS leads with the goal to electronically "steer" the stimulated zone (see Fig. 3.23). These steerable or directional electrodes allow doctors to fine-tune the delivery of energy on a specific area around the lead. Properly oriented delivery increases efficacy of the therapy while decreasing side effects. Boston Scientific and Abbott [32] (former St. Jude) both commercialized DBS systems with, respectively, six and eight contacts per lead, exposing Medtronic to a fierce competition.

In Europe, two companies developed leads with large numbers of contacts for better and more accurate steerability:

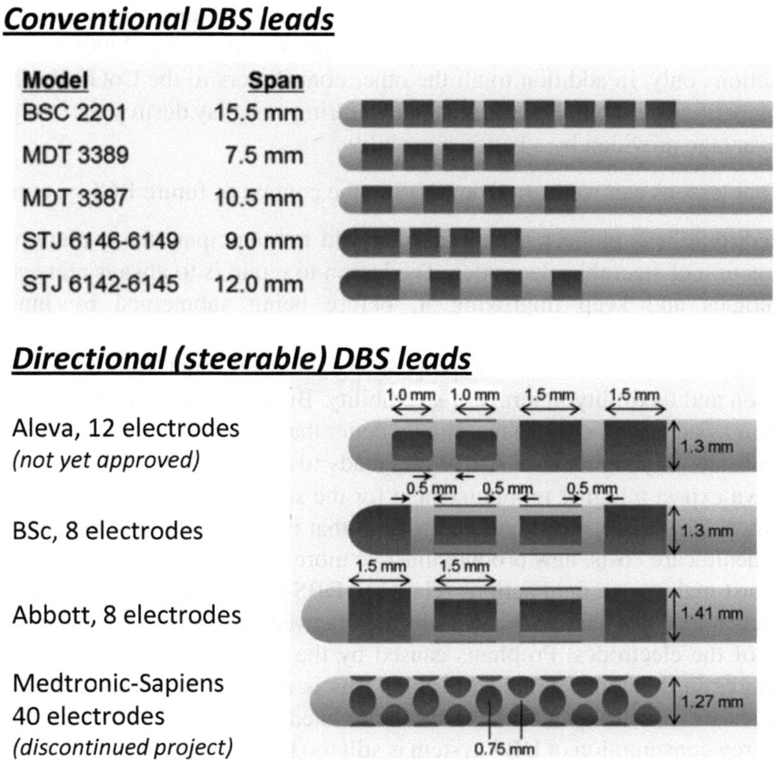

Fig. 3.23 Comparison of various DBS leads

- Sapiens Steering Brain Stimulation in Holland went up to 40 contacts per leads. As it is not feasible to tunnel a 40 wire flexible cable along the neck, Sapiens decided to add a demultiplexer at the root of the DBS lead, at the top of the skull. A flexible cable with only a few wires connects the IPG to the small titanium-encapsulated demultiplexing hub. This hub is a marvel of integration and miniaturization. Unfortunately, the cost of such a system is probably about three times higher than a conventional DBS, for a marginal superiority. Sapiens has been acquired by Medtronic in 2014. Medtronic abandoned the concept a couple of years later.
- Aleva Neurotherapeutics [33] in Switzerland settled on 12 contacts per lead. Aleva only developed the DBS leads and was therefore forced to find a partner for a 24 channel IPG. First, Medtronic was the partner of choice, but the deal was cancelled when Medtronic acquired Sapiens. Now Aleva is partnering with Nuvectra [34, 35], using their Algovita® IPG, already approved for SCS. As for Sapiens, the major problem for Aleva will be the cost of the device. Their electrode is more expensive to produce than the conventional DBS electrode of Medtronic. Even the six or eight channel electrodes from Abbott and BSc are cheaper than Aleva's proprietary lead technology. In addition, 24 channels IPGs are very expensive, because of the number of channels. I estimate the manufac-

turing and assembly cost of one hermetic channel (FT and connector) at about 100$ per channel. Therefore, 24 channels represent more than 2000$ for the connections only, in addition to all the other contributors to the CoGS. In the very competitive field of DBS, a high manufacturing cost may destroy the competitive advantage provided by a better steerability.

What lessons can we learn from DBS in the context of future BCI systems?

- Medtronic was pioneer in this field but did not anticipate the evolution in the direction of steerable electrodes. The lesson to retain is to always reassess technologies and keep improving it, before being submerged by innovative competitors.
- Sapiens and Aleva were first to propose high channel count, offering a great precision and flexibility in terms of steerability. But they maybe went too far away. Their systems will only be marginally better than the three big competitors but at a substantially higher cost. Are payers ready to cover the additional costs? Could Aleva expect a higher reimbursement for the superior therapy? It remains to be shown. The lesson we should learn from that is in a world of constant pressure on healthcare costs, new product must be more performant *and* cheaper.
- A vast majority of field actions related to DBS systems reported by the surveillance authorities have root causes located between the hermetic housing and the tip of the electrodes. Problems caused by the IPG itself are rare compared to failures in the connections and cable. This is an important lesson: mechanical issues are more frequent than electronic-related issues.
- Energy consumption of DBS system is still too high, requiring large batteries and consequently large IPGs, which cannot be located above the neck. Implantation in the pectoral area requires long leads, extensions, connectors, and difficult tunneling procedures. The lesson to retain is that the priority is to design stimulation technics which are less greedy in energy. In other words, we should work at the level of body interfaces to manage to get the same effect with less current.

3.4.3 Spinal Cord Stimulation (SCS) for Chronic Back Pain (CBP)

The cost of chronic pain for the society is enormous, in the range of 100 B$ per year only in the USA. About a quarter of this burden is related to CBP. Chronic pain is usually treated by pain reliever drugs, especially opioids. Due to the constantly increasing number of cases of overdoses, authorities are imposing severe restriction to the use of opioids for pain relief. SCS is an alternative to drugs, with the advantage of having no secondary effect. The use of SCS for failed back surgery syndrome (FBSS) has been approved by the FDA in 2014.

SCS is the largest market of neuro-technologies and one of the most mature. Early clinical trials were done in the late 1960s. First commercial steps were done in the early 1980s by Cordis (now Johnson and Johnson [36]) and by Medtronic with a product called Itrel® and Advanced Neuromodulation Systems (now Abbott). As for DBS, Medtronic dominated this field, for two decades. Today, the other big firms are gaining market shares, and new companies are entering with innovative products. Projected market shares for 2020 are Abbott 24%, BSc 23%, Medtronic 21%, Nevro 20%, and Nuvectra 4% [29]. The emergence of new players, like Nevro [37] using high-frequency stimulation, Saluda [38] with a close-loop approach, and Nuvectra (see Fig. 3.24b), seriously challenges the position of the big three.

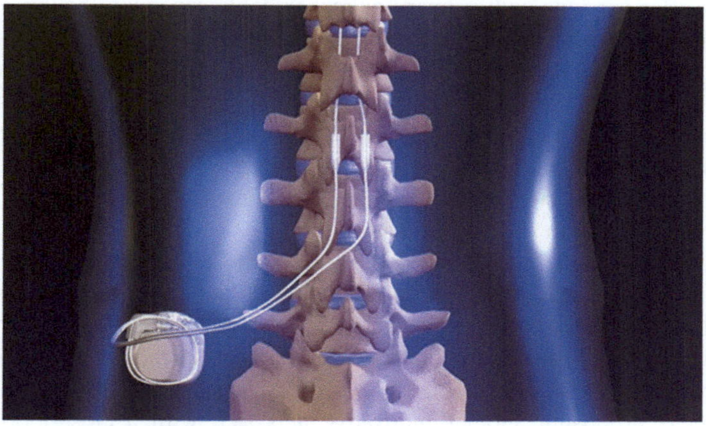

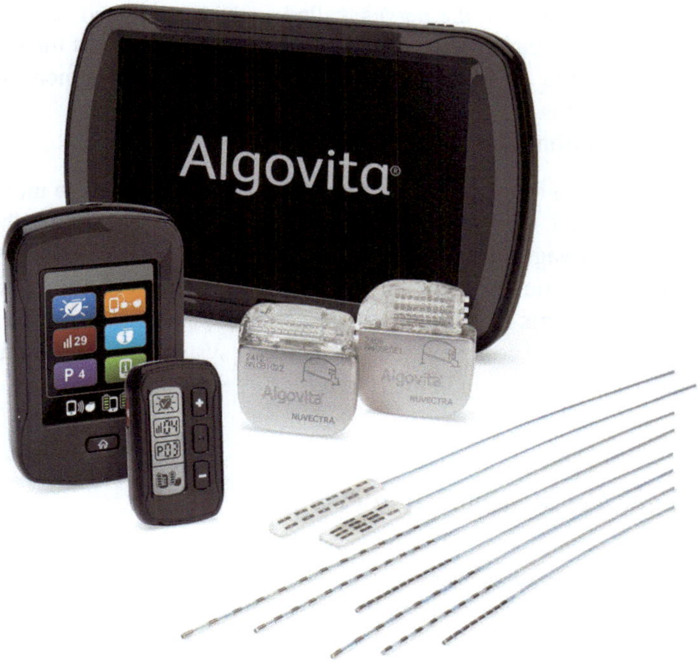

Fig. 3.24 (**a**) Spinal cord stimulation. *(Courtesy of Medtronic plc)*. (**b**) Spinal cord stimulation system Algovita. *(Courtesy of Nuvectra Inc.)*

Epidural patch electrodes, with 4–16 channels, are inserted along the spinal cord (see Fig. 3.24a). A flexible coiled cable with an in-line connector is leading to an IPG implanted in the back. Like pacemakers and DBS stimulators, SCS IPGs are hermetically encapsulated in a titanium housing. Insertion of the leads and connection to the IPG are easier compared to DBS and require shorter operation room time. There is a bidirectional low-frequency RF communication between the implant and external devices (remote control for the patient, programming unit for the physician). Stimulation being continuous, energy is also the major concern. Most manufacturers offer versions with primary or rechargeable secondary batteries. Patient acceptance of the latter is better than for rechargeable DBS systems.

The main technical barrier is linked to the movement of the spine, which may displace the paddle electrode. The paddle electrode must be flexible enough to accommodate bending of the back. Ideally, the electrodes should also be stretchable to follow relative longitudinal movement. No commercial electrodes are fully conformal, flexible, and stretchable. Several developments are ongoing to improve leads' compliance with body movements.

Stimulating at high frequency (HF) (in the range of 10 kHz) is an interesting new trend. Efficacy has been demonstrated, even if the scientific community cannot explain the effect. Nevro is a fast-growing company selling high-frequency SCS devices. Nevro has already taken 20% market shares from traditional SCS manufacturers.

Saluda is proposing another innovative approach with a close-loop system, where electrical signals from the nerves, called evoked compound action potential signals (ECAPS), are measured and used for real-time adjustment of the stimulation pulses. This close-loop system is claimed to provide a better treatment of chronic pain. Saluda's device is not yet approved.

Lessons learned from SCS:

- A better adaptation of the body interface (paddle electrode) to the movement of the spine is capital. This remark is applicable to other parts of the body, like peripheral nerve, vagal nerve, and the brain itself.
- High-frequency stimulation has shown its efficacy without us fully understanding how it works. This is a reminder that we are very far away from having a full comprehension of the complexity of the nervous system. Theoretical models of today may be proven wrong in the future.

3.4.4 *Vagal Nerve Stimulation (VNS)*

Stimulation of the left vagal nerve has been shown to influence area affected by epileptic seizure on both sides of the brain. A second indication is treatment-resistant depression. Cyberonics has been pioneer in the development of VNS and is today world leader in this field. Since its first commercialization in the USA in 1997, the Cyberonics system has been implanted almost 100,000 times for epilepsy and 25,000 times for depression.

In the year 2015, Cyberonics merged with Sorin Medical to form LivaNova [39]. A few other companies are active in the field of VNS for other indications. EnteroMedics got approval in 2015 for a device to treat obesity based on vagal nerve blocking. Low-energy high-frequency signals applied to the vagal nerve have an impact on several metabolic functions, resulting on weight loss and better glycemic management. EnteroMedics changed name to ReShape Lifesciences Inc. [40].

Cyberonics' first generation system included a patient magnet for activation of a pre-programmed pulse sequence. The recently approved second generation, AspireSR, is an automated close-loop stimulator based on sensing heart rate as an indicator of an incoming seizure. Stimulation of the vagal nerve is done with a bipolar helix cuff electrode wrapped around the nerve (see Fig. 3.25). Placement of the electrode is easy compared to DBS or SCS. The IPG has only two channels and is therefore not more complex than a pacemaker. Simplicity is the main characteristic of VNS. Even if efficacy may be limited compared to more sophisticated devices, the cost/benefit ratio VNS may be attractive.

Besides approved indications (epilepsy and depression), several clinical researches are conducted with the objective of treating other disorders by stimulating the vagal nerve. Among others, clinical work is being done on chronic heart failure, arrhythmia, chronic pain, addiction to alcohol, anxiety, and autoimmune disorders [41].

Are there lessons to be learned from VNS with regard to BCI?

- There are routes, like the vagal nerve, permitting an easy access to the brain, as it has been demonstrated by Cyberonics for the treatment of epilepsy. Accessing directly to the brain requires either opening the skull or having clumsy external electrodes. Using the alternative route of the vagal nerve is appealing.
- A low number of channels keep the cost of the implant at low level.
- Cuff electrodes are easy to manufacture and to implant.

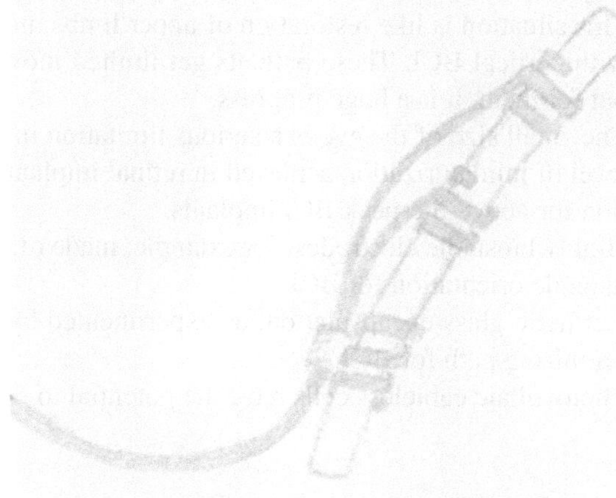

Fig. 3.25 Vagal nerve stimulation leads

3.4.5 *Retinal Implants (RI)*

We will not make a detailed description of retinal implants, as this indication is rather specific. Nevertheless, some technical aspects may be of interest with regard to BCI.

Since three decades, many research institutes have worked on restoring some vision capability for totally blind people. Various approaches have been explored, in epi-retinal and subretinal configurations. Second Sight [42] was the first company to get approval (2011 CE-mark, 2013 FDA) for a retinal implant. A second generation has also been launched. So far, only a few hundreds of patients have benefited from the system. Other companies have made significant progress and are reaching the market: Bionic Vision (Australia) [43], Pixium (France) [44], Retina Implant (Germany) [45], and Nano Retina (Israel) [46]. They use a variety of technologies for the electrodes, electronics, image transfer and powering. They all encounter the same hurdles: the small size of the eye, its mobility, and the limited resolution achieved by retinal electrode arrays. Long-term biostability is an issue, especially if electronic chips are implanted in the eye. Nano Retina has designed a fully hermetic implant, encapsulated in glass. Bionic Vision has developed an electrode array made of diamond, with exceptional longevity. Several groups are developing photovoltaic retinal implants which have the capital advantage to be cableless. Each cell of the implant, placed on the retina, consists in an independent photovoltaic sensor connected to a stimulation electrode. The simplicity of this system is promising, but much work is still needed to make the implant biocompatible and biostable on the long term.

What can we learn from all this work in the perspective of building BCI?

- Early retinal implants are limited to 64, 128, or 256 pixels. This means a very limited resolution compared to natural vision. For blind people, this gross vision is a fabulous change. Being able to see the frame of the door, a coming car when they cross the road, or some fuzzy image of their relative is a major achievement. This situation is like restoration of upper limbs movement of paralyzed patients with cortical BCI. These patients get limited movements like grasp and move, but for them, it is a huge progress.
- The small size of the eye is a serious limitation in terms of implant design. The level of miniaturization achieved in retinal implants is a good source of inspiration for above-the-neck BCI implants.
- Highly biostable electrodes, for example, made of diamond, could also be a sustainable orientation for BCI.
- Hermetic glass encapsulation, as experimented by Nano Retina, may also be a promising path for BCI.
- Photovoltaic cableless cells have the potential to replace more complex devices.

3.4.6 *Urinary Incontinence (UI)*

Medtronic was a pioneer in applying pacemaker-inspired technologies to stimulate the sacral nerve for treating mild to moderate forms of urinary incontinence (see Figs. 3.26 and 3.27). Already in 1997, Medtronic got FDA approval for InterStim® [47], which was approved in Europe in 1994. In 2011, the device got approved for fecal incontinence. So far, about 200,000 patients benefited from the InterStim therapy.

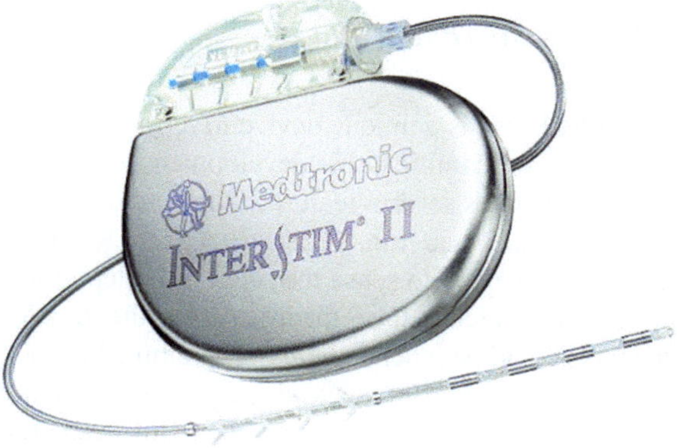

Fig. 3.26 Sacral nerve stimulation system. *(Courtesy of Medtronic plc)*

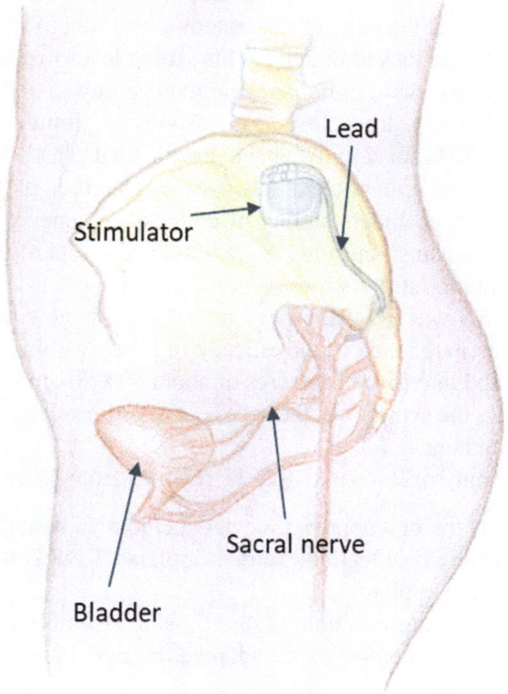

Fig. 3.27 Implantation of a SNS system

The IPG looks like a pacemaker, with four in-line connectors. The device has a primary battery with a longevity of more than 5 years. The four-contact electrode is placed at the root of the sacral nerve.

Other companies are addressing the SNS market like Nuvectra's Virtis® device pending approval [48] and Axonics's miniaturized device [49].

If not directly related to BCI, SNS shows a long-term success of peripheral nerve stimulation.

3.4.7 NeuroPace

NeuroPace [50] is a unique device in terms of technologies. Even if it had not been sold in large quantities, this development is of high interest for the readers of this book. It includes several features which are of high interest with regard of BCI.

The device (see Fig. 3.28a–d), called Responsive Neurostimulator System® (RNS), is intended to sense the onset of epileptic seizures and to stimulate the brain in a manner which reduces the intensity of the seizure and then assess the effect. Eight channels could be independently programmed for being stimulation or sensing channels. Different types of electrodes can be connected to the titanium can, either penetrating leads for stimulation or paddle ECoG leads for sensing. The housing is curved and formed in three segments, one hosting the primary battery, one for the electronics, and one for connections. This shape follows approximately the curvature of the skull. A craniotomy of the shape of the contour of the device hosts a ferrule which overlaps the skull at the edge of the craniotomy. Then, the titanium housing is inserted in the ferrule and locked in place. This setting leads to a minimal protrusion above the surface of the skull, minimizing aesthetical impact and risks of scalp erosion.

RNS is a true close-loop responsive stimulator. Through wireless telemetry, recorded brain activity can be downloaded. This offers a unique way to record brain activity on epileptic patients over long periods of time. It is a breakthrough feature for a better understanding of epilepsy. Whenever an abnormal cortical activity is detected, small currents are automatically sent to the stimulation leads. There is no continuous stimulation.

RNS was FDA approved in 2013 and is now implanted in about 1600 patients. Experts argue about the efficacy of RNS. Various studies report reduction of number and intensity of seizures of about 50%. Even if stimulation is not reducing drastically the symptoms, the access of real-time electrocortical activity is a great source of fundamental data.

Plenty of lessons should be retained from NeuroPace RNS:

- It is the first approved device including a battery located in the head.
- Beside cochlear and retinal implants, RNS is the first approved above-the-neck active implant.
- RNS features a unique side-by-side assembly (three segments: battery, electronics, connectors) better adapted to cranial implantation than conventional stack-up concepts.

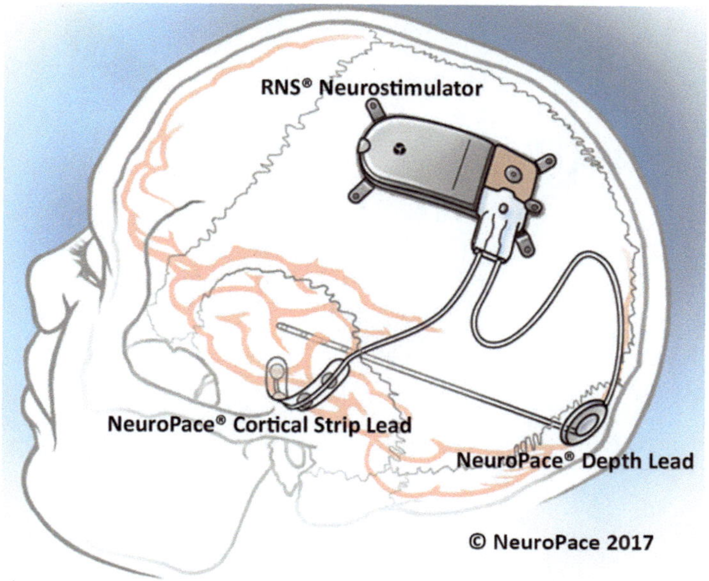

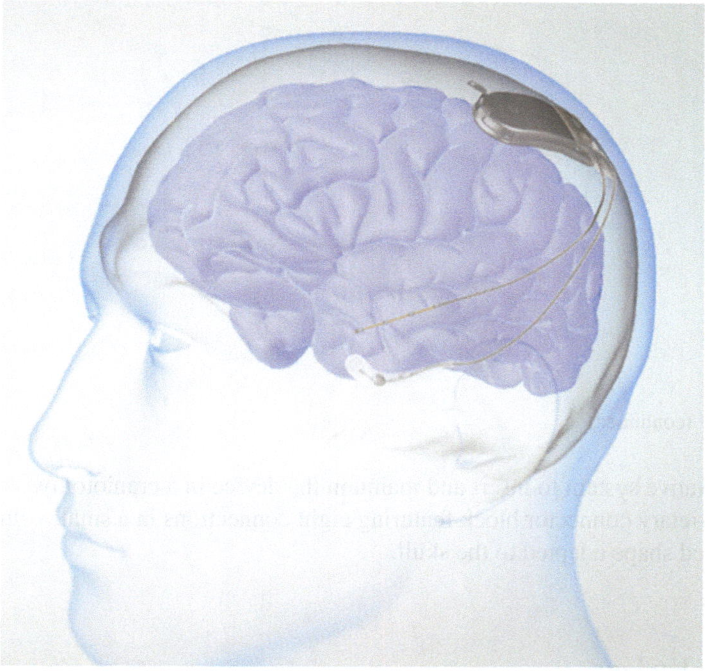

Fig. 3.28 (**a**) Implantation of the RNS® System. *(Image courtesy of NeuroPace, Inc.)*. (**b**) NeuroPace RNS® System. *(© 2017 NeuroPace, Image courtesy of NeuroPace, Inc.)*. (**c**) NeuroPace RNS® Neurostimulator. *(Image courtesy of NeuroPace, Inc.)*. (**d**) NeuroPace RNS® System. *(Image courtesy of NeuroPace, Inc.)*

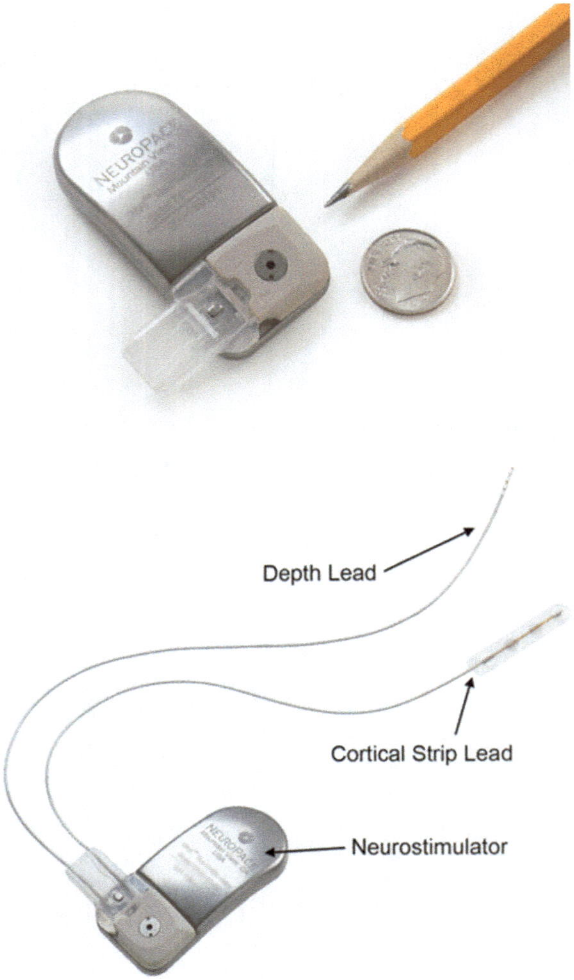

Fig. 2.28 (continued)

- Innovative system to insert and maintain the device in a craniotomy.
- Proprietary connector block featuring eight connections in a small volume.
- Curved shape adapted to the skull.

3.4.8 Various

Other approved devices are inspirational for designers of the future BCI. A few examples are mentioned below:

Programmable implantable pumps, like SynchroMed-II from Medtronic, are currently used to treat neurological disorders by direct injection of drugs in the intrathecal space: baclofen for essential tremors and spasticity [51] and morphine for intractable pain [52]. Another application of programmable implantable pumps is patient-controlled anesthesia (PCA), where the patient, via a remote control, can activate and control the delivery of drug. For future BCI systems, we can envision closed-loop systems where the pump is automatically activated by signal collected in the brain. For example, early warning of an incoming epileptic seizure could trigger the local delivery of an appropriate drug in a focal area of the brain or in the cerebrospinal fluid (CSF).

Functional electrical stimulation by fully implantable means, networked neuroprosthetic (NNP) system [53] (see Fig. 3.29), was manufactured by Synapse Biomedical [54] and developed by Case Western Reserve University, Institute for Functional Restoration, all in Cleveland, under the leadership of Prof. Hunter Peckham and Prof. Robert Kirsch. The FES Center [55] in Cleveland is also involved in this initiative. The indications are movement restoration of paralyzed patients or movement assistance for ALS patients.

The modular system consists in a master implant connected to an expendable chain of slave units, also in hermetic titanium small housings. Each unit is linked to the next one by four-wire addressable bus. Satellites are having various functions, like EMG stimulation, nerve stimulation with cuff electrode, EMG sensing, movement sensor, accelerometer, temperature sensor, etc. The bus, connectors, and cables are proprietary concepts.

The NNP system was one of the first devices receiving approval from the FDA through the Expedited Access Pathway (EAP), an accelerated procedure launched in 2015 [56].

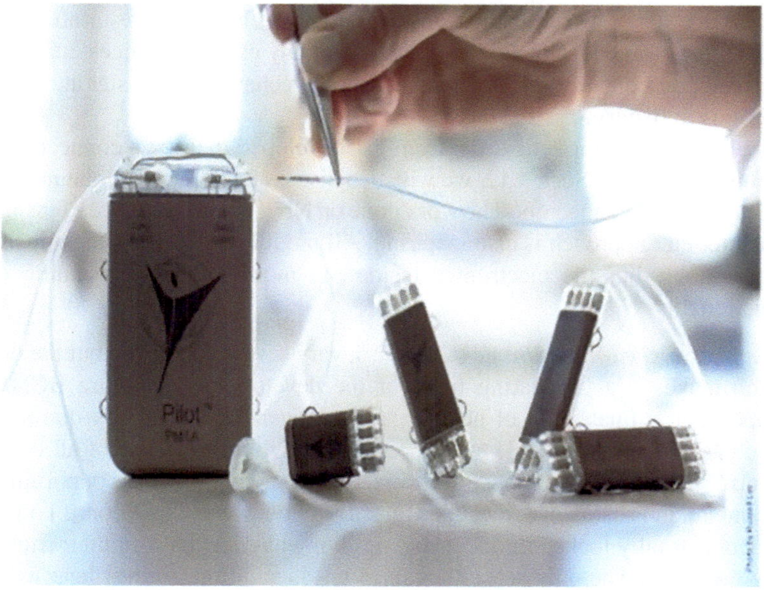

Fig. 3.29 Networked neuroprosthetic (NNP) system for FES. *(Courtesy for FES Center)*

There are many interesting points to retain from NNP in the scope of BCI:

- The master-satellite concept is a precursor of implants with distributed intelligence. Wisely, the team remains in a wired configuration. It makes the surgery quite complex, but the system works. This conventional approach has led to a solid validation of the modular concept. Having a wireless communication between the modules will come later (see Sect. 7.4.3). In the current state-of-the-art wireless technologies, we are not yet able to assure secured RF communication between many modules in the human body.
- The master contains rechargeable batteries and an induction coil inside the titanium can. Heating of the implant because of eddy currents generated by the magnetic field during recharge has been managed in a way that the temperature does not rise by more than 2 °C on the surface of the implant. However, in the current design of NNP, the external coil becomes too hot and must be cooled by water circulation. We will see later, in Sect. 4.11, that providing energy for the functioning of implanted electronics is a major barrier for further integration of BCI systems.
- Only the master has a battery; satellites are battery-less. So, power must be transferred from the master to the electronics embedded in the modules. It is known that one should avoid DC power transfer over a cable between two implants, as DC voltage will induce circulation of ions in surrounding body fluids and will corrode contacts and connectors. To avoid this, NNP includes an AC power transfer. This issue is covered in detail in Sect. 4.4.
- The day we will be able to have wireless communication between the various modules, the issue of powering the satellites will become a serious one. If modules must remain as small as possible, they cannot incorporate a battery and must therefore be energized by induction. This requires to permanently wear an external device for power transmission, which maybe bulky for restoration of limb movement. The Cleveland team did not want any external devices during operation and consequently selected the option of a rechargeable master and wire connections to battery-less satellites.
- For a better miniaturization, the team has developed their own connectors. Later in this book, under Sect. 4.12, we will also discuss the topic of miniature implantable long-term connectors. The work done on NNP is a great source of inspiration.
- The modular approach is also inspiring. All modules have the same housing and connections, but the electronics inside the modules vary depending on the functionality.

Complete locked-in syndrome (CLIS) is a very severe condition, but the ongoing research makes it a fascinating field for the development of future BCIs. CLIS patients are at a final stage of neuromuscular degeneration, mainly due to amyotrophic lateral sclerosis (ALS), where patients slowly lose control of all their muscles, also finally including eye movements. Usually, the last opportunity for communication is using eye trackers. But some patients even lose control on the eye and become totally locked in, incapable of giving any feedback to their friends and family members. The only sense they have left is audition, so they hear us but are not able to answer our questions.

Pioneer work is currently being done to find a way to interact directly with the brain of CLIS patients by using BCI. As it is not yet a commercially available product and even might never be a fully commercial initiative, it should not belong to this chapter. I decided to include it here because it is a source of inspiration. A thorough discussion on the future evolution in this field will be addressed under Sect. 7.3.6.

Early experiments with CLIS have used heavy tools, like EEG or fMRI to read brain activity when patients where asked simple questions with yes/no answers. With a modest but promising level of confidence, researchers have been able to classify the answers. Some communication, very limited and slow, has been established. Yes/no answers are far from satisfactory to regain some capability for the patient to express feelings.

The next step is to place a cortical interface on the brain surface or a penetrating electrode array, allowing CLIS patients to operate a speller with an auditory feedback. In the first stage, the use of a Utah array connected to a pedestal is being tested.

The inspiration we get for this experimental work is that people in extremely severe conditions may not only benefit from an open window of communication but also will greatly contribute to the progress of future BCI and accelerate the development of translational devices addressing less severe needs.

3.5 Long-Term Clinical Perspectives

This chapter is dedicated to the targets of neuro-technologies. Let's briefly conclude this chapter by discussing the next decade evolution of the existing targets and of the perspectives of having new development scopes in the field.

Existing targets for implantable neuro-technologies have shown that electrical stimulation and sensing have a substantial potential to improve human health. We also understand that we are at the very beginning of the industry of neuro-technologies.

The foreseeable evolution of therapies described in Sects. 3.4.1, 3.4.2, 3.4.3, 3.4.4, 3.4.5, 3.4.6, 3.4.7, and 3.4.8 over the next decade might take the following directions:

- *Cochlear Implants*
  - Addition of a detachable connector allowing the exchange of the implanted electronics if needed
  - Multiplication of the number of channels
  - Electromagnetically transparent housing and integration of the coil inside the housing (already achieved by Neurelec)
  - Integration of the speech processor, an implantable microphone, and a rechargeable battery in the implant, for having a fully implanted CI without external headpiece
  - Phase-synchronized bilateral CI for source of sound orientation

- *Deep Brain Stimulation Implants*
  - Addition of new therapies and targets
  - Penetration of steering technologies
  - Introduction of close-loop DBS
  - Improvements of the lead introduction procedures
  - Miniaturization
  - More channels for better steerability
  - Placement of the IPG above the neck
- *Spinal Cord Stimulation Implants*
  - Introduction of flexible stretchable paddle electrodes
  - Penetration of HF stimulation
  - Extension of close-loop SCS
  - Miniaturization of the IPG
- *Vagal Nerve Stimulation Implants*
  - Addition of new therapies and targets
  - Improvement of efficacy
  - Addition of sensors for close-loop VNS
  - Miniaturization of the IPG
- *Retinal Implants*
  - Multiplication of pixels for better resolution
  - Better long-term stability
  - "Passive" photovoltaic RI with no cable between the sensing and stimulation function
  - Migration of the stimulation electrodes to the optic nerve and visual cortex
- *Urinary Incontinence Implants*
  - Better focus on the control of the bladder and sphincters
  - Extension to more severe incontinence cases
  - Extension to sexual dysfunctions
- *NeuroPace Implant*
  - Extension from 8 to 16 channels (ongoing)
  - Extension of battery life
  - Improvement of fixation mechanism in the craniotomy, for example, with patient-specific bone plate replacement
  - Dissemination to a larger population
  - Improvement of seizure prediction algorithms
- *Various Implants*
  - Programmable implantable drug delivery pumps: miniaturization for above-the-neck implantation and direct delivery of drug in the brain

  - FES implants: extension and dissemination of miniaturized NNP systems and RF communication with wireless BCI, possibly through an external hub
  - CLIS implants: improvement of control algorithms and demonstration of superiority over conventional external BCI

In addition to the applications which have been successfully explored so far, there are many unmet or poorly met medical needs for which neuro-technologies may provide a solution in a reasonable time frame:

- Migraine and chronic headaches
- Phantom pain
- End-of-life pain
- Haptic feedback for amputees
- Full control of prosthesis for amputee by direct connection with peripheral nerves
- Epilepsy seizure forecast, prediction, and inhibition
- Control of schizophrenia, phobias, anxiety, depression, and other psychiatric disorders
- Better vision restoration
- Fully implantable hearing restoration
- Neurofeedback technologies for tinnitus and pain
- Re-synchronization of brain wave to treat dyslexia
- Restoration of vestibular disorders
- Mirror stimulation for unilateral facial paralysis
- Control of severe incontinence
- Implantable stroke rehabilitation systems
- Better cortical decoding for more degrees of freedom in movement restoration
- Extension of cortical decoding for the control of and restoration of walking (paralyzed lower limbs)

As described earlier, the pharmaceutical industry and biotechnologies have reached some limits regarding the treatment of neurological disease or disorders. More and more, we will see the conjunction of drugs and technologies in combination products, focused on unmet neuro medical needs. By joining forces, biotechs and med-techs should be able to push the frontiers of neurological disorders. New systems should soon become reality, like automated local drug delivery controlled by a BCI, optical stimulation based on optogenetics, and combination of stimulation and injection of stem cells for nerve, spinal cord, and brain reconstruction. This may happen within the next two decades.

Personalized medicine is another future evolution which will impact neurology. Adapting treatment to the individual biological and genetic signature of a person is not limited to drugs. The concept of "à la carte treatment" can be extended to neurotechnologies. Already today, each patient connected to a BCI goes through a personal adaptation of the algorithms and calibration procedures. BCI interfaces are, by nature, personalized.

Neuro-technologies also are a fertile ground for a fast application of the strong trends of tomorrow: big data, machine learning, and artificial intelligence. In the field of BCI, these are not buzz words but the tools we will use to leverage BCI concepts.

References

1. https://www.braingate.org/team/john-donoghue-ph-d/
2. https://en.wikipedia.org/wiki/Neuroimaging
3. http://fescenter.org/about-fes-center/leadership/kirsch-robert-phd/
4. https://case.edu/universityprofessor/past-recipients/p-hunter-peckham-phd
5. https://www.braingate.org/
6. https://www.medtronic.com/us-en/healthcare-professionals/products/cardiac-rhythm/cardiac-monitors/reveal-linq-icm.html
7. https://www.synchronmed.com/
8. https://blackrockmicro.com/
9. https://en.wikipedia.org/wiki/Parylene
10. https://people.epfl.ch/gregoire.courtine?lang=en
11. https://people.epfl.ch/stephanie.lacour?lang=en
12. https://www.electrocore.com/
13. https://people.epfl.ch/nikolaos.stergiopulos?lang=en
14. https://www.comphya.com/
15. https://en.wikipedia.org/wiki/Obstructive_sleep_apnea
16. https://www.inspiresleep.com/for-healthcare-professionals/our-technology/
17. https://www.medtronic.com/us-en/patients/treatments-therapies/neurostimulator-gastroparesis/enterra-2-neurostimulator.html
18. http://www.balseal.com/
19. https://www.cochlear.com/us/en/home
20. https://www.sonova.com/en
21. https://www.medel.com/
22. http://aemf.org/
23. http://www.bostonscientific.com/en-US/Home.html
24. https://www.oticonmedical.com/cochlear-implants
25. https://www.oticonmedical.com/about-oticon-medical
26. https://www.demant.com/
27. https://www.cochlear.com/uk/home/discover/carina-middle-ear-implants
28. https://en.wikipedia.org/wiki/Alim_Louis_Benabid
29. Cavuoto J. Neurotech reports 2018: The market for neurotechnology: 2018-2022, Updated Sep 2018, http://www.neurotechreports.com/
30. Rossi J et al (2016) Proceedings of the third annual deep brain stimulation think tank: a review of emerging issues and technologies. Front Neurosci 10:119, Edited by Paolo Massobrio
31. Anderson DN et al (2018) Optimized programming algorithm for cylindrical and directional deep brain stimulation electrodes. J Neural Eng 15:026005
32. https://www.abbott.com/
33. https://www.aleva-neuro.com/
34. https://nuvectramedical.com/
35. https://globenewswire.com/news-release/2016/02/01/806602/0/en/Aleva-Neurotherapeutics-and-Greatbatch-Collaborate-on-Next-Generation-Device-for-Deep-Brain-Stimulation-DBS.html
36. https://www.jnj.com/

37. https://www.nevro.com/English/Home/default.aspx
38. http://www.saludamedical.com/home/
39. https://www.livanova.com/en-US
40. https://www.reshapelifesciences.com/
41. https://en.wikipedia.org/wiki/Vagus_nerve_stimulation
42. http://www.secondsight.com/
43. https://bionicvision.org.au/
44. https://www.pixium-vision.com/en
45. https://www.retina-implant.de/en/
46. https://www.nano-retina.com/
47. https://www.medtronic.com/us-en/healthcare-professionals/products/urology/sacral-neuromodulation-systems/interstim-ii.html
48. https://neuronewsinternational.com/nuvectra-issues-update-on-us-fda-and-ce-mark-applications-for-virtis-device/
49. http://www.axonicsmodulation.com/
50. https://www.neuropace.com/
51. https://www.medtronic.com/us-en/patients/treatments-therapies/drug-pump-severe-spasticity/about-itb-therapy/synchromed-ii-pump.html
52. https://www.medtronic.com/us-en/healthcare-professionals/products/neurological/drug-infusion-systems/synchromed-ii.html
53. http://fescenter.org/research/technology-programs/networked-neuroprosthetic-system-nnps/
54. https://www.synapsebiomedical.com/
55. http://fescenter.org/about-fes-center/
56. http://fescenter.org/wp-content/uploads/2015/10/Press-Moynahan-and-Peckham.pdf

Chapter 4
The Human Body: A Special Environment

4.1 Active Implantable Medical Devices (AIMDs)

In most cases, new technologies have been developed for a purpose which is not related to human health. High integration of fast electronics was induced by the need of faster and smaller computers. Low-consumption circuits found their origins in the space industry, watchmakers, and wearable consumer products. Advanced radio-frequency (RF) components and subassemblies were fostered by space, militaries, and mobile phone and by the recent push for wireless communication in our everyday life. The process for laser welding of titanium casings under controlled ultradry atmosphere (see Sect. 4.9.2), so widely used in the AIMD industry, has been originally developed for the NASA Apollo program.

In other cases, some fundamental progresses have been triggered by the medical industry itself. For example, new biocompatible and biostable materials have been developed and improved to meet the specific requirements of AIMDs. This chapter will lead us through this unique playground for engineers: the human body. We will also see what is recommended to do if you want to implant electronic devices and let them in the body for a while. At this stage, we start to cover the important notion of *How to Build*.

AIMDs share the following characteristics:

- *Active*: the device has an electrical or mechanical activity, which means the presence of a source of energy, like a battery (rechargeable or not) or and induction coil and/or an electronic circuitry and/or sensors and actuators and/or energy harvesters.
- *Implantable*: describes devices which remain inserted in the human body for more than 30 days.
- *Medical*: defines devices which provide a therapy or a diagnostic or collect data in the body.
- *Device*: electromechanical entity which is encapsulated in a way that ensure the following:

C. Clément, *Brain-Computer Interface Technologies*,
https://doi.org/10.1007/978-3-030-27852-6_4

- Hermetical protection:
  - Prevents any fluid or gas to penetrate in the device
  - Prevents any fluid or gas to escape from the device
- Biocompatibility: all the materials exposed to the body tissues or fluids must have no toxic or adverse effects on them.
- Biostability: all the materials exposed to the body tissues or fluids must remain unaltered for the entire expected time of implantation.
- Sterility: the final assembly intended to be implanted must be sterilized (meaning biological contaminants must be killed before implantation).
- Cleanliness: non-biological contaminants, such as dirt, oil, dust, or other inert materials, must be reduced to a minimum to avoid tissue irritation or pyrogenic effects.

In addition to these five basic functions, the recent trends in the industry are to integrate in the encapsulation the function of providing a protection against or compatibility with electromagnetic disturbances, especially coming from mobile phones, anti-theft security systems, and MRI (magnetic resonance imaging) equipment (see Fig. 4.1).

4.2 A Special Environment

The human body was here before technologies! It is the fruit of million years of evolution and adaptation to its surrounding. Some of our most precious organs, like the heart, the brain, or the eyes, are amazingly sophisticated constructions, far superior to anything scientists and engineers have created in the last centuries and predictably also superior to what our genius will build in the next centuries. Our little technology tricks are shamefully modest compared to the immense complexity of the human body.

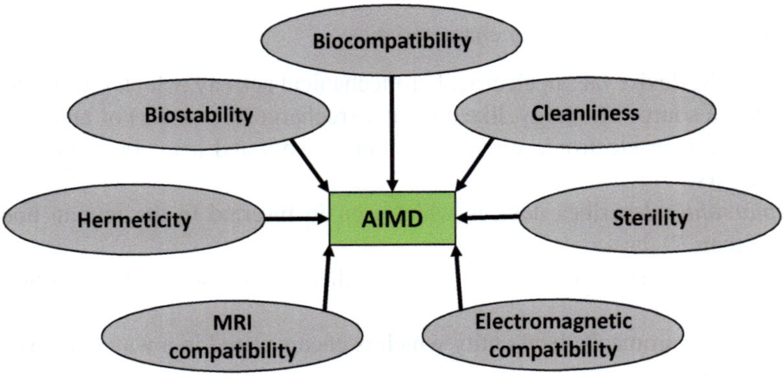

Fig. 4.1 Main characteristics of an AIMD

The chemistry, biology, and physics of the tissues surrounding an implant are interrelated worlds in constant evolution. We have not yet fully understood all the reactions happening in the body when we decide to insert an implant. There are phenomena which we are not yet able to predict or control. Mastering material sciences is a cornerstone of the art of implantology.

4.2.1 Not Negotiable

If we want to interface, integrate, repair, and even improve the human body, we must take it as it is. Our body and organs have a given size, cellular constitution, chemistry, physics, and behavior. This is the frame of our playground. We cannot change the way the human body is constituted; it is not negotiable.

We may be able to do minor changes to the human body, for example, by taking a contrast agent for the improvement of an imaging diagnostic or by replacing a defective function by an implant. But these are not fundamental changes. The age of the bionic man/woman, having several of his/her major organs replaced by technology, is still very far away. I even doubt it will ever happen.

There are some obvious constrains within the human body. Not mentioning that we need some air, food, water, and fun to survive, there are constitutive elements which cannot be bypassed. As an example, the guest star of this book, the brain, consists in billions of cells and molecules, interconnected in an amazing network, all gathered in the form of very soft tissues. As this gel-like mass has no internal rigid structure and no skeleton, it must be contained in a hard box: the skull. This outer protection also makes the brain difficultly accessible from outside. We will discuss further about this limitation in Chap. 5. The nonnegotiable presence of the skull means that either we cannot access the brain directly (e.g., so we do limited measurements from outside, like EEG) or we will need to break the barrier to get a direct physical access to the soft brain tissues (e.g., a craniotomy for the placement of an electrocortical grid (ECoG)).

Another example is the limitations of the propagation of radio-frequency waves in the body. At the very high frequencies required for the transmission of large flows of information, the successive layers of tissues attenuate, diffuse, and scatter radio waves. It means that either we restrain the volume of information to be communicated through tissues or we minimize the thickness of tissues above the device, for example, placing the antenna just below the skin.

4.2.2 Laws of Physics

Body tissues have specific biological, physiological, chemical, and electrical characteristics. They nevertheless also obey the laws of physics. We mentioned above some limitations related to the transmission of RF electromagnetic waves in the body. Similar difficulties are also encountered by other types of waves, for example, light or ultrasounds. Thick layers of tissues are opaque to visible light. Only thin layers of tissues will let infrared (IR) or near-infrared (NIR) light go through. If we want to stimulate neurons with light, we must conduct light to the spot where we want to act, for example, by using an optic fiber.

When submitted to various electrical and magnetic fields, body tissues show some reactions. These effects are used, for example, to do volume imaging, like magnetic resonance imaging (MRI). Unfortunately, such fields also interact with implants, electrodes, and other foreign materials introduced in the body for therapeutic purpose.

Electromagnetic fields in the body also heat up the tissues [1]. Excess of heat in tissues may create permanent damages. A standard (ISO-14708-1) sets the limit of increased temperature to +2 °C. Temperature slightly above this limit is acceptable in transitory situations. Inductive energy transfer or high-frequency RF communication can easily increase the temperature by 2 °C in the bulk of the tissues or at the surface of an implant. The laws of thermodynamics will govern the removal of this excess of heat by radiation, by conduction, or by blood circulation. Heat dissipation from an implant will vary widely depending on the surrounding environment, from the bone to CSF, blood perfused tissues, or subcutaneous locations.

These limitations related to electromagnetic fields in body tissues must be carefully understood at the beginning of disruptive projects. For example, several teams are trying to spread, in the entire volume of the brain, tiny electronic modules (neurograins or neural dust). These submillimetric battery-less modules have tiny antennas which should be able to collect enough electromagnetic power to make the modules work and communicate wirelessly. The laws of physics teach us that this will only work if the field reaching the microscopic antennas is large and powerful. This will require enormous external coils to be placed on the head, fed with large currents. Consequences are going to be excessive tissue heating. We may also wonder if it makes sense to miniaturize implants to a level which require enormous external devices.

Easier to understand: the laws of mechanics. An implant will be submitted to acceleration forces in case, for example, of a car accident. It must therefore be attached in the body in a way which will prevent displacements prone to induce damaging to surrounding tissues. Other limitations are related to the soft nature of tissues constituting the nervous system. The electrodes or tissue interfaces of BCI systems are usually mechanically stiff and rigid. As tissues are constantly moving, due to body movements, blood circulation, or respiration, there are risks of relative movements at the interface between tissues and electrodes.

The design of neuro-devices must take in account the limitations induced by the law of physics. We cannot just deny them.

4.2.3 *Surgical Aspects*

The brain, spinal cord, and nerves are soft and fragile tissues. The surgical procedures to place electrodes in or on nervous tissues are difficult. It requires not only skills but also adapted surgical instruments, tools, imaging systems, or surgical robots. The brain is meshed with a dense network of blood vessels. Introducing penetrating electrodes in the brain with minimal damage to blood vessels is a challenge.

As mentioned earlier, direct access to the brain requires making an opening in the skull. The surgical technics to do craniotomies are mastered by neurosurgeons, but opening the skull is always a big step, especially for patients. Opening the skull and penetrating the brain are perceived as the highest possible level of invasiveness.

Surgical subdural interventions in the brain require great care, as the natural barrier of the dura has been opened. Infections may propagate along the cables linking the brain interface to the electronics.

Another difficulty for neurosurgeons is the variations of the brain environment at the time of implantation, depending on age, weight, and other physical conditions. Post-surgery evolution is also a challenge. For example, the placement of the electrodes in the cochlear of a 1-year-old child must take into account the growth of the head in the following years.

4.2.4 *Reaction of Body Tissues*

Implants are not well accepted by the body. Surrounding tissues react to the arrival of an intruder. First, biological, chemical, and immune mechanisms try to destroy the foreign body. As implants are too big and too biostable to be killed like mere bacteria, the body builds an insulation barrier around the implant, a capsule of fibroid tissues.

For implants or parts of implants which are not in electrical contact with the body (e.g., the IPG of a DBS system), this fibrotic capsule is not a major issue. In some cases, the body does such a good job that the capsule becomes really a barrier, preventing or refraining natural chemical and biological exchanges. It may result in exotic chemical reactions inside the capsule, at the surface of the implant. Atypical pH values and formation of acids have been reported.

For electrodes, the formation of a fibrotic capsule may degrade the quality of the electrical contact with tissues and increase the contact impedance. It must be noted that the mechanisms of fibrosis around brain and nerve electrodes are far from being understood. Substantial differences have been seen between patients, but we lack scientific explanations. The geometry, roughness, cleanliness, activation, and materials of the surface of the electrodes seem to have a large impact on the electrical characteristics of the fibrotic capsule and its evolution with time.

4.3 Biocompatibility

In the following sections, we will discuss on biocompatibility, biostability, corrosion, cleanliness, and sterility (see Fig. 4.2). All these notions are related and must be considered as critical factors when designing an implant. Sometimes, people mix up these terms. For example, cleanliness and sterility are not synonymous. If you take a dirty implant and put it in a sterilization chamber, at the end of the process, the dirt has been sterilized, but the implant is still dirty. Assembly of medical devices is done in so-called clean room, in order of remaining as clean as possible. But a clean room is far from a sterile room. Biocompatibility refers to chemical acceptance of the surrounding tissue. Sterility characterizes the absence of biological contaminants. A sample could be biocompatible and not sterile or the opposite.

Biocompatibility may be defined as the ability for an implant to be chemically tolerated by the human body for a long period of time. There are several other definitions of biocompatibility, meaning it is a complex concept. For the purpose of this book, my favorite definition is: "Biocompatibility is the capability of a prosthesis implanted in the body to exist in harmony with tissue without causing deleterious changes" [2].

An implant or an assembly is considered as biocompatible if it does not intoxicate nor substantially modify the surrounding cells and tissues. Biocompatibility tests are regulated by ISO-10993, and the FDA has issued a related guidance "Use of International Standard ISO-10993, 'Biological Evaluation of Medical Devices Part 1: Evaluation and testing'" [70].

The nature and severity of the tests to be conducted prior a first implantation in humans depend on the duration of implantation. For short-term implantations (<30 days), biocompatibility tests are less severe than for long-term devices.

It is very important to note that biocompatibility must be demonstrated for an assembled, clean, and sterile implant. Even if all the materials in contact with body tissues or fluids are independently and intrinsically biocompatible, it must be proven that the processes used to assemble and join the individual parts have not impacted

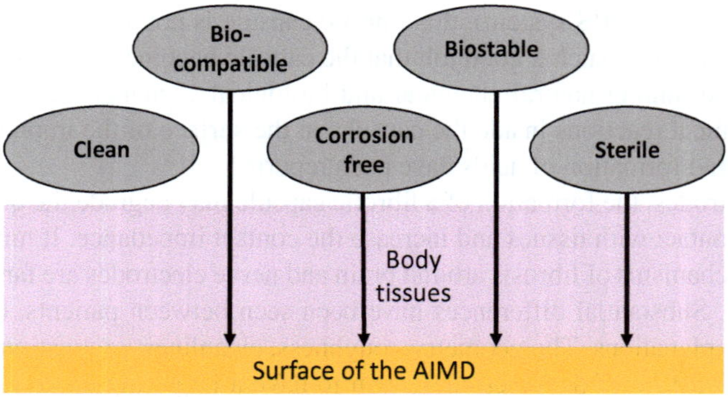

Fig. 4.2 Properties of the surface of an AIMD

the overall biocompatibility. It has been reported that some processes do modify the physical and chemical properties of material traditionally considered as biocompatible. For example, laser welding of two different metallic alloys, both individually biocompatible, may create intermetallic compounds which are not biocompatible.

Only a very limited number of materials are biocompatible. Out of this restricted choice, an even more limited subset is also biostable (see below). It means that designers of long-term implants do not have a wide spectrum of materials available.

Biocompatibility may be classified in categories:

- Inert biocompatibility: materials accepted by the body with minimal tissue reaction. This is the type of biocompatibility of interest for most AIMDs applications.
- Resorbable biocompatibility: materials which are reabsorbed by the body and replaced by natural tissues. Materials in this category may find their way in some BCI applications, for example, to protect or reinforce electrodes during surgery.
- Bioactive biocompatibility: materials which react strongly with surrounding tissues and build strong links with them. Could be of interest for bone inserted BCI.

Biocompatible materials of interest in the scope of this book are:

- Metals and alloys:
  - Pure titanium (grades 1–4)
  - Titanium alloys (Ti6Al4V)
  - Alloys of cobalt, chromium, molybdenum
  - Surgical stainless steel (Fe, Cr, Ni), MP35, 316L
  - Nitinol (Ti, Ni)
  - Precious metals: gold, platinum, platinum-iridium, palladium
  - Niobium
- Ceramics:
  - Inert:
    - Aluminum oxides, alumina, sapphire, ruby
    - Silicon oxides
    - Zirconium oxides, zirconia
    - Titanium oxides, titania
    - Glass ceramics
    - Vitreous or glassy carbon (C)
    - Carbon silicon (C-Si)
    - Diamond
  - Bioactive:
    - Hydroxyapatite (HA)
    - Tricalcium phosphate (TCP)

- Polymers:
  - Polymethyl methacrylate (PMMA)
  - Polytetrafluoroethylene (PTFE), commercial name: Teflon™
  - Polyethylene terephthalate (PET), commercial name for textile: Dacron™
  - Liquid crystal polymers, LCP
  - Dimethylpolysiloxane (silicone rubber), PDMS
  - Ultrahigh molecular weight polyethylene (UMWPE)
  - Polyetheretherketone (PEEK)
  - Polyurethane (PUR)
  - Parylene™ (used only for coating, not bulk)
  - Polysulfone (PS)
  - Poly(3,4-ethylenedyoxythiophene) polystyrene sulfonate (PEDOT: PSS) conductive polymer used for low impedance electrode coating
- Composite:
  - Carbon-PTFE
  - Carbon-PMMA
  - Alumina-PTFE
- Bio-absorbable:
  - Polyethylene glycol (PEG)
  - Hydrogels
- Protein-based biomaterials:
  - Collagen
  - Fibrin
  - Silk

Materials are usually evaluated for their bulk properties. Biocompatibility is mainly a surface characteristic. Surface properties may be widely modified by mechanical or chemical treatments. With regard to biocompatibility, surfaces are influenced by their:

- Wettability
- Cleanliness
- Surface energy
- Corrosion resistance

Reaction and adhesion of tissues on intrinsically biocompatible materials can be widely influenced by surface modifications and coatings, like:

- Passivation: stabilizes the oxide layer on metals by immersion in nitric acid
- Acid etching: removes the superficial layer to increase surface roughness and promote tissue adhesion
- Plasma etching: acts as acid etching and promote adhesion of coatings
- Sand blasting: increases surface roughness and hardness

- Brushing: increases roughness
- Atomic layer deposition (ALD): protects the bulk material with very stable atomic layers
- Physical vapor deposition (PVD): thicker protection
- Sputtering
- Spray coating
- Parylene coating

Even if a bulk material is biocompatible, surface reactions may generate non-biocompatible compounds. Similarly, corrosion products must be tested for cytotoxicity.

Joining two biocompatible metals or alloys by fusion (laser, spot, resistance welding) may liberate non-biocompatible intermetallic sub-products. For this reason, biocompatibility tests must be done on fully assembled systems.

Biocompatibility is requested for all materials and joined materials which are in contact with tissues or body internal fluids. In the conventional configuration of hermetically encapsulated implants, there are two domains: the inside and the outside of the hermetic housing.

Materials and components which are inside a hermetic housing do not need to be biocompatible. As hermetic sealing is done before sterilization, materials inside the housing are not sterile (see Fig. 4.3). One of the objectives of hermetic encapsulation is protecting non-biocompatible materials (e.g., an electronic board) from exposure to body fluids and tissues. Hermetic encapsulation is adequate to protect from body fluids to sip through or to diffuse and reach the electronics.

The second role of the hermetic encapsulation is to avoid the migration of nasty chemicals from inside the can to the outside. Due to the presence of highly reactive lithium, batteries present a high risk if exposed to moisture. The safe practice for implanted batteries is to have a double hermetic barrier: hermetic encapsulation of the battery itself and hermetic encapsulation of the housing surrounding the electronics/battery assembly (Fig. 4.4).

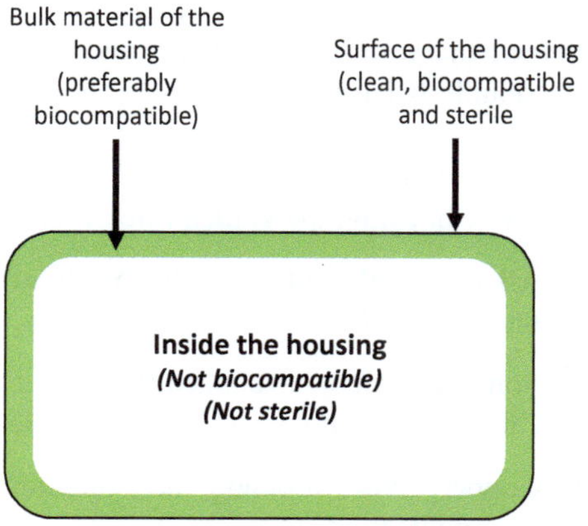

Fig. 4.3 Hermetic housing

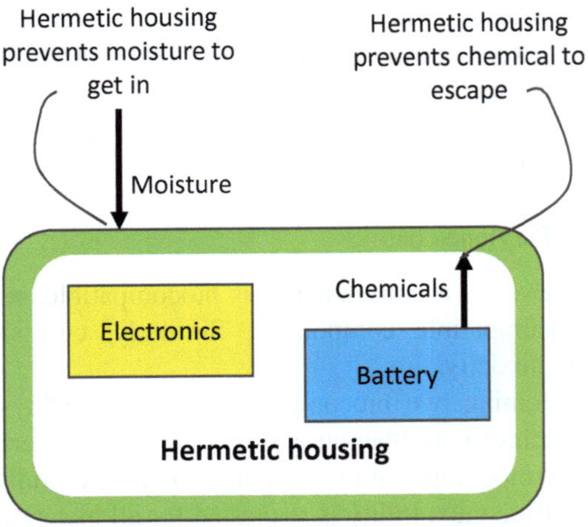

Fig. 4.4 Barrier properties of hermetic housings

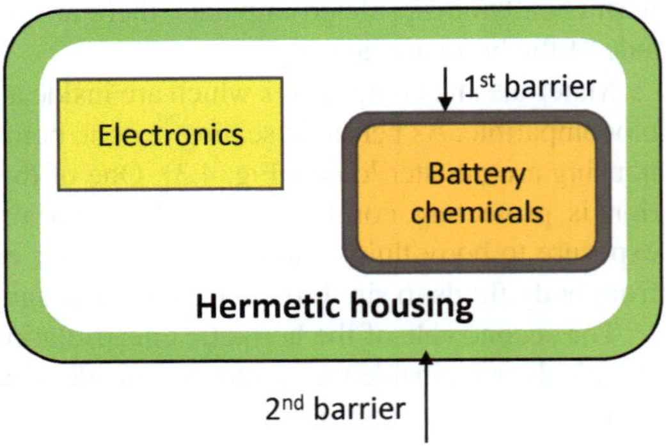

Fig. 4.5 Double barrier for batteries

The principle of double protection (hermetic battery in a hermetic implant) is of major importance when dealing with patient's safety (see Fig. 4.5). To my knowledge, with exception of two specific midterm products, all active implants on the market follow the rules of the double barrier for the battery. In AIMDs including a battery, especially if rechargeable, several if not a majority critical risks are related to the battery. Leaking of chemicals and lithium exposure risks are best mitigated by the double-barrier concept.

The above describes the importance of the hermetic barrier to keep non-biocompatible materials insulated from body contact. Regarding the outside of the hermetic housing, the golden rule of long-term AIMDs is:

All materials outside a hermetic encapsulation *must* be biocompatible and biostable.

Some designers have taken the risk of having non-biocompatible materials outside the hermetic housing. Some examples:

- Copper coil around the titanium housing for energy transfer and communication

- Ceramic hermetic housing sealed by a brazing process using non-biocompatible low fusion temperature alloys
- Soldering lead wires on feedthroughs with non-biocompatible brazing alloys
- Use of silver loaded conductive epoxy outside of the hermetic housing

In all these cases, the rational to not following the golden rule is the addition of a protective layer on the top of the non-biocompatible materials, in theory avoiding direct contact of the non-biocompatible materials with the body. The copper coil is potted in a thick over-molding of epoxy. The non-biocompatible brazing/soldering materials are covered with silicone rubber. The key of this discussion is risks taking. Potted or not, a non-biocompatible material remains non-biocompatible. The risk is that, after a while, the protection cover cracks, breaks, delaminates, or dissolves, finally exposing the non-biocompatible material to body fluids. For very long-term implants (several decades, like cochlear implants), it is strongly recommended to avoid such risks. For shorter-term implants, from a few months to a couple of years, potting non-biocompatible materials might be acceptable (see Fig. 4.6), as long as processes (cleaning, priming, curing) are well under control and fully validated through an artificial aging process (see Sect. 4.7).

Recently, exotic nonconventional materials have been considered for inclusion in innovative implants (see Sect. 7.4.2). For example, exciting research is going on in the field of implantable electronics, with major applications for BCI. The idea is to place electronics (amplifiers, multiplexers) directly on the electrode arrays. This would allow a substantial reduction of the number of connecting wires and/or an improvement of the signal-to-noise ratio (SNR). The concept is to realize these electronics not by conventional complementary metal oxide semiconductor (CMOS) process (doping silicon wafers) but by electrodeposition or printing organic semiconductor materials directly on the substrate of the electrodes. Such electronic circuits are not hermetically encapsulated. In consequence, the exotic materials used (e.g., gallium or compounds issued from nanotechnologies) must

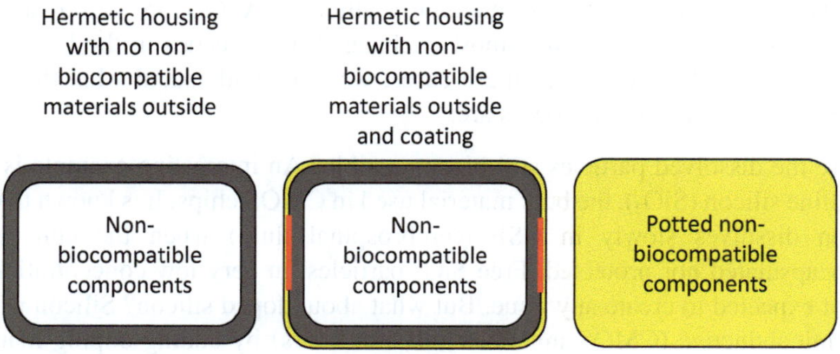

Fig. 4.6 Biocompatibility of housing materials

be biocompatible. Carbon nanotubes and graphene have also shown high potential for long-term implants. Today, not much is known regarding the biocompatibility of these materials on the long term. It is likely that new biocompatible materials will be discovered in a near future.

Miniaturization of active implants raises an interesting debate. When tiny grains of electronics are inserted in body tissues (see Sect. 7.4.3), would it be acceptable to have some non-biocompatible materials exposed? What is the maximum amount of toxic material accepted locally by tissues? Paracelse (Swiss alchemist, died in 1541) said, in French, "*le poison est dans la dose*," meaning that toxicity is related to the quantity of toxic material. Where is the limit? Are we, one day, able to do such small devices that we do not need to bother about biocompatibility? Basic research in this field is still to be done.

4.4 Biostability

Biocompatibility has been described as the characteristic of a material to be tolerated by the body. Somehow, biostability is the counterpart of biocompatibility (see Fig. 4.7a). Biostable materials have the property of being resistant to the aggressions of the body.

There is a solid base of knowledge related to biocompatibility of materials used in implants. On the contrary, biostability remains a field of uncertainties and doubts. Scientific evidences and rational explanations are still missing in many situations for long-term implants. Around 2 million AIMDs are implanted every year, and we do not know much about their long-term biostability. A given material may be biocompatible and preliminary selected for a long-term application, but it may be found as not biostable for the anticipated implant duration. One example is silicon used for CMOS chips. The bulk material is biocompatible, but not biostable.

Biostability is tightly related to time scale. Materials dissolve more or less quickly in the body. Some, like titanium or alumina (Al_2O_3), are very resistant to body aggression and will last, almost unchanged, for decades in the body. Some others will dissolve rapidly. In these cases, we must understand what the consequences of the dissolution process are:

- Are the dissolved particles still biocompatible? An interesting example is crystalline silicon (SiO_2), the bulk material used in CMOS chips. It is known that silicon dissolves slowly in CSF (cerebrospinal fluid) when the chip is not encapsulated nor protected. Free SiO_2 particles, in very low concentration, are not expected to create any issue. But what about doped silicon? Silicon is made semiconductive (CMOS junctions and transistors) by adding doping materials like boron (B) in the crystalline structure. Biocompatibility of B is not clear. In small concentration, B seems to have a positive impact on various diseases. What happens to free floating atoms of B circulating in CSF? Not much has been documented on this.

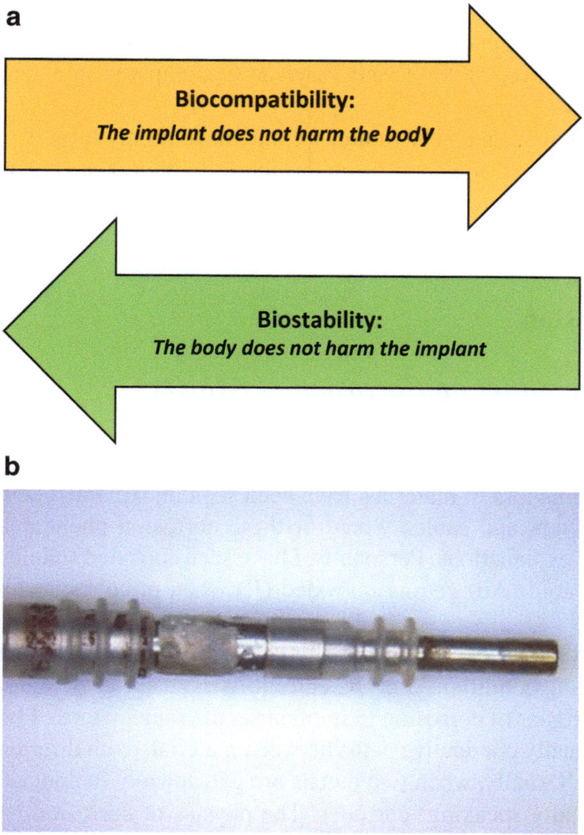

Fig. 4.7 (**a**) Biocompatibility/biostability. (**b**) Corroded IS-1 connector. ((**b**) Courtesy of Yttermed SA)

- Where are the dissolved particles going? Do they follow the blood stream, do they migrate through tissues, and do they circulate in CSF or in the lymphatic system? Do they accumulate anywhere?
- In cases where the dissolving biocompatible material is protecting a non-biocompatible material, how long does it take to have it exposed to body fluids?

New materials and processes are currently being developed, presenting significant improvement in terms of biostability and moisture penetration (see Sect. 4.9.7). Alternating very thin layers of various biocompatible materials show amazing performance as protective barriers. Atomic layer deposition (ALD) is one of these breakthrough developments [88]. Best results have been reached by alternating organic and inorganic layers. Compared to the same thickness of Parylene, ALD multilayer sandwiches have been reported to improve moisture resistance by a factor of 4000. The leading work on hermetic or near-hermetic encapsulation of AIMDs using multilayer ALD is done at Ghent University, Belgium, in the group of Maaike Op de Beeck [3–5].

Another example of the potential of multilayer coatings is an improvement of the long-term stability of the Utah electrode array (UEA) (see Sect. 3.3.1) by insulating it with a double layer of Al_2O_3 and Parylene C, compared to a single layer of the latter [6].

It is possible that, in a near future, we may have highly biostable and moisture-resistant protective layers, opening opportunities to break the golden rule, and have a proper long-term protection of non-biocompatible materials.

4.5 Corrosion

4.5.1 Generalities on Corrosion of AIMDs

Corrosion of materials implanted in the body is another area of uncertainties. Several odd behaviors of materials have been seen in explanted devices, for example, on connectors and cables. Some of these corrosion phenomena do not have clear scientific explanations. Personally, I have seen corroded stainless steel connectors and even gold (Au) getting corroded (!!) when in contact with platinum (Pt) wires.

Figure 4.7b shows an example of stainless steel IS-1 connectors exposed to rash accelerated aging conditions and DC current. We can see clear signs of corrosion.

The main trigger of corrosion in implants is the ionic nature of body fluids. This induces electrically conductive paths between materials with different electrochemical potentials. Usually, when two metals are galvanically in contact, the less noble (electrochemically speaking) corrodes. The physics of corrosion is very complex. Before going more in the details, we will simply warn designers of active implants that corrosion is an issue to be taken seriously. Here are a few guidelines to consider:

- Avoid having different metals in contact when the intermetallic junction is susceptible to enter in contact with ionic fluids. For example, implanted connectors will sooner or later become soaked with body fluids. If two metals with substantial electrochemical potential differences get exposed to this ionic environment, some corrosion may happen at the interface, to the expense of the less noble metal. This may happen in connectors but also in soldered joints or welds. A safe choice could be to use the same metal in the entire subassembly, for example, welding the Pt feedthrough to a Pt wire leading to a Pt electrode. The absence of intermetallic junctions is a safe way to prevent corrosion.
- Minimize ionic contamination by adding protective layers. Potting, coatings, and insulation layers are applied to protect junctions, welds, and connections. Moisture will diffuse through these protective layers, in the form of pure water (not conductive). If the surface under the coating is perfectly clean, then the diffused pure water will not trigger any corrosion. Perfectly clean surfaces do not exist in a manufacturing environment. Reality shows surfaces ionically contami-

nated before the coating was applied, then pure water diffusing through the coating will combine with ionic contaminants and become conductive, opening the door to corrosion. A thorough initial cleaning process is the secret of an efficient protective coating.

- Never carry DC electrical power along cables in the human body. As explained above, the slightest ionic contamination anywhere along the conductive path will induce corrosion. Stimulation signals should be "charge balanced," meaning that the average current has no DC component. Transfer of electrical power along a cable between two implants (e.g., from an implanted battery pack to an IPG) should be done in a switched polarity mode, resulting in a zero DC component.
- Avoid gaps and cavities where body fluids could enter and stagnate. All such gaps should be carefully cleaned before filling, underfill, or potting. It is also critical to avoid bubbles in the filling material or on the surface of the device, below coating or potting layers. Sooner or later, bubbles will get invaded by moisture.
- Do not use conductive epoxy, neither outside the hermetic housing nor inside. These epoxies are made conductive by the addition of particles of silver (which is not biocompatible), which tend to migrate and escape the bulk of the epoxy. In presence of even moderate electric fields, silver very quickly (within minutes) builds dendrites (see Fig. 4.8), susceptible to create short circuits between channels.

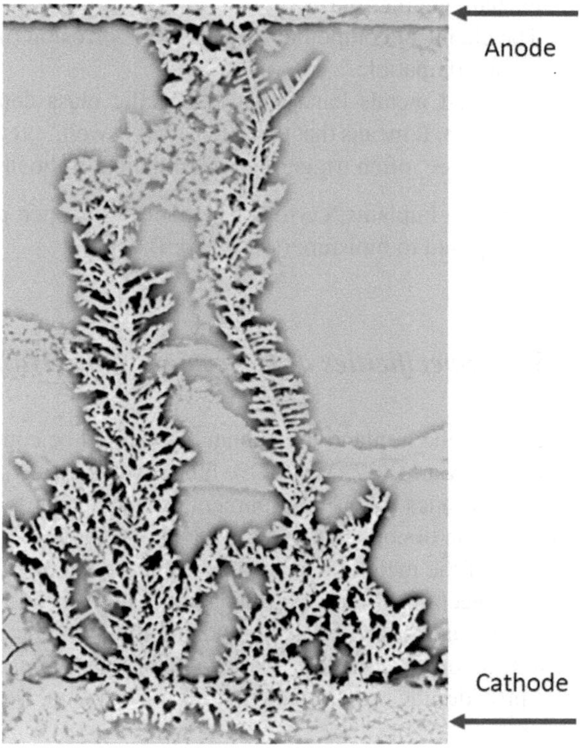

Fig. 4.8 Example of dendrites growth

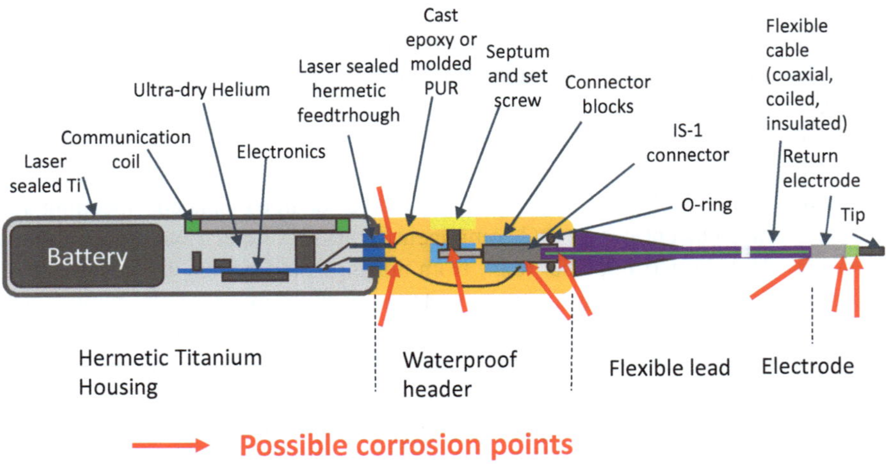

Fig. 4.9 Possible corrosion points on a pacemaker

The occurrence of corrosion in an active implant may have various impacts:

- The quality of the contact degrades, increasing the contact impedance and deteriorating the quality of the transmitted signal.
- It may become difficult to disconnect or remove the connecting pin when exchanging the IPG.
- The chemical compounds issued from corrosion (usually metal oxides) may not be biocompatible.
- Corroded metals have a lower specific mass density than non-corroded bulk materials. It means that corroded areas "swell," creating mechanical constrains at the surface, often triggering delamination of the insulation layer.

In active implants, corrosion is likely to happen at bimetallic junctions potentially exposed to moisture (see Fig. 4.9).

4.5.2 *Specificities of Corrosion in the Human Body*

Seen from the implant, the human body could be compared to the jungle along the sea: quite warm (37 °C), 100% humidity, and salty. Some implants will stay for several decades in this environment. Some standards have given guidelines related to corrosion of connectors in the human body [56].

One of the main objectives of the encapsulation of AIMDs is to provide a long-term protection of the electronics and battery inside the casing. This can only be assured if all the materials composing the hermetic encapsulation do not degrade with time or get corroded. The choice of titanium cases, seam welded with a laser ray, provides the best possible barrier to corrosion. Nevertheless, the titanium shield is only 99% of the total surface of the encapsulation. The 1% left, FTs, are part of the outer surface potentially exposed to moisture, gas, body fluids, and tissues. From time to time, some concerns are expressed regarding potential corrosion issues of FTs or around them.

Corrosion of metals in the human body is a very complex matter. Conventional electrochemistry cannot explain all the phenomena. The dynamics of the simple electrochemical cell model (two plates of metals in an aqueous solution) are already very complicated and only partly explained by scientists. Going to non-flat geometries, thin gaps, porous, or cracked materials surrounded by human tissues, make the picture totally fuzzy. If corrosion is more magic than science, I would say that corrosion of implanted devices relates to sorcery.

This lack of scientific data on corrosion in the human body is counterbalanced by decades of experience in manufacturing and implanting cardiac devices and dental and bone implants. Every year, about 1.5 million of implantable pulse generators (pacemakers, defibrillators, neurostimulators) are implanted in patients around the world, or one active device inserted every 20 s, 24/7. The total number of people living with an IPG today is in excess to 7–8 million (0.5–1% of the population of the USA, Europe, and Japan). All these devices have 1–32 FTs. I estimate the number of FTs currently implanted in human bodies to above 25 million. Adverse events related to failures of FTs due to corrosion are extremely rare. It is a proof that the AIMD industry has done a proper job so far. But we need to be careful, as BCI and neuro-devices will be more demanding regarding FTs. BCI implants will require many more channels in high-density configurations. The robust rather large pacemaker FTs with one, two, or four wires do not fit the requirements and specifications of neuro-devices. High-density miniaturized FTs will face new technical challenges in terms of corrosion.

Theories of corrosion identify many sorts of natural phenomena leading to corrosion. The physics and chemistries governing those various changes in materials are overlapping and interacting. Static reactions do not reflect reality. One must consider the evolution of corrosion with time. Corrosion is intrinsically a dynamic process or a combination of interrelated dynamic and complex processes.

In the literature we find descriptions of single mode corrosion phenomena. The most common ones are:

- Chemical corrosion
- Electrochemical corrosion
- High-temperature corrosion (not applicable to AIMDs, with exception of the impact of laser welding)
- Biological corrosion
- Atmospheric corrosion (drops of water on the surface; may be applicable to the cleaning processes of AIMDs)
- Corrosion triggered by fluid flows and cavitation (has been identified as a trigger of corrosion in heart valves)
- Corrosion in extreme pH situations (pH < 3 and pH > 10, might occur at the bottom of deep cavities and cracks in the implanted devices)

In real systems, like AIMDs, one may expect to have a combination of most of those effects, happening at various intensities and at various periods of the life cycle of the product.

Some chemical corrosion is happening early in the life cycle, even before implantation, and during assembly, cleaning, sterilization, and storage of the device. Chemical corrosion is mainly related to changes on the surface of metals, due to exposure of oxidative agents. In AIMDs, we may expect modification of the titanium oxide layer on the case during cleaning and sterilization.

Formation of biologically active layers may induce severe changes on the surface of metals. In our case, we assume that the device is perfectly sterile prior implantation. Nevertheless, biological corrosion happens in sterile situation, because devices are exposed to biological tissues as soon as they are implanted. The appearance of hard granulated poorly drained fibrosis tissues around titanium implants creates risks of accumulations of hydrogen, oxygen, or acids, which all may trigger some forms of nasty reactions with metals. We will come back to this later.

Electrochemical Corrosion

The classical model of corrosion is a "cell" composed of two plates of metal with different electrochemical potentials, immersed in an electrolyte and electrically connected through a resistor.

Immersed in water, any metal tends to dissolve at the surface, forming a double-charged layer (electrons at the surface inside the metal, metallic positive ions dissolved in the water). If the metal plate is not connected to another immersed electrode, the double layer remains stable, the dissolved ions being attracted on the surface by the electrons.

Having a flow of ionically charged body fluid (blood, CSF) on the surface or connecting to another electrode immersed in the same electrolyte will show a circulation of electrons from one plate to the other through the connecting cable, compensated by a circulation of metallic positive ions in the electrolyte. Material with the lowest electrochemical potential will corrode (loss of metallic atoms on the surface), and the electrode with higher potential will compensate the flow of electrons by retaining the metal ions. This process is related to many physical rules (diffusion, conductivity of the electrolyte, shape and relative surface of the electrodes, barriers to the circulation of ions, etc.) which place this simple model far from the actual situation seen by an implant.

The configuration changes somehow when the two metals are in contact with each other, the overall assembly being exposed to electrolyte or water-based solutions. In this case, we cannot consider two ideal homogenous surfaces related with a homogenous electric field. At the interface between both metals, the electronic current will flow, depending on the quality of the contact, on the oxide layer, and on the materials trapped at the interface. In the electrolyte over the interface between the two metals, the ionic current will depend on the geometry and on the electrolyte itself (freedom to flow, pH, presence of free oxygen or hydrogen, etc.).

In the ideal model of two metals perfectly placed side by side in a perfect electrolyte, the metal with the lowest potential will corrode (dissolve) along the interface and form a crevice, penetrating deeper and deeper along the interface. The speed of the dissolution will depend, not only on ionic concentration and materials but also on the flow of interstitial fluid.

As far as electrochemistry is concerned, body fluids are basically NaCl solutions in water, at 0.9% concentration (9 g/l) for the ideal situation. The body fluids surrounding the implant are dissociated in Na^+ cations and Cl^- anions, free to circulate and recombine depending on electric fields and potentials applied on conductive surfaces. Migration of Cl^- ions to areas saturated in hydrogen H^+ will create HCl acids prone to accelerate corrosion. On the other hand, Na^+ may recombine with OH^- complexes, leading to basic NaOH compounds also known corrosion triggers.

In a majority of IPGs, stimulation pulses are generated as soon as the battery is connected to the electronics. Therefore, even before electrodes get attached to the IPG, connectors might be exposed to electrical potentials.

Presence of a DC voltage between two poles, electrodes and connections, may induce transfers of ions in the body fluids, with creation of acids or bases, which might have some corrosive effects.

In most of AIMDs, voltages applied to parts outside the hermetic casing (connectors, leads, wires, or antennas) are pulses or of AC nature. The DC component is often small, and therefore DC-induced corrosion is minimal.

Corrosion in Deep Gaps and Cracks

Human fluids have rather neutral pH in the range of 7–7.5. Blood has a pH of 7.35–7.45, and the CSP is usually around 7.33. Much more acidity is found in the urinary tracts, inside some organs, and the most acid is the stomach (pH < 3.5). In brain tissues, in or around the spinal cord and nerve, pH is neutral or slightly alkaline, which should limit chemical corrosion to a minimum, if we assume that the circulation of the electrolyte maintains the pH in that range. There too, reality is more complex.

Body fluids trapped by capillarity in deep gaps or along thin cracks will undergo chemical changes due to time, oxidation, reduction, release of trapped gases, diffusion, or any other physicochemical reactions. Local density of free ions H^+ (acid) or OH^- (basic) will modify the pH locally and induce exotic dynamic phenomena of local corrosion.

Literature mentions a lot of situations where corrosion propagates in cracks, due to stresses on the material. Depending on the type of metal and the fluid penetrating in the crack, various parameters may speed up corrosion. For example, in a deep crack of bulk titanium, the newly created surface might not be able to oxidize as rapidly as in the open air. The newly exposed titanium will be deprived from its usual protective layer of titanium oxide. The body fluids propagating along the crack maybe have induced a reduction of oxygen, forming HCl. Such acidic solutions may corrode Ti.

These special situations are not likely to happen in gold and platinum, because of their natural inertness.

Effect of Human Tissues on Corrosion

After the implantation of an IPG or an electrode, the device gets "encapsulated" by a fibrotic layer of hard tissues. It is the natural mechanism used by the body to "reject" or "isolate" the foreign intruder. The implant is rapidly surrounded by various layers of tissues, most of them poorly irrigated. Circulation of body fluids through these tissues is impaired, so we cannot speak any longer of an ideal electrolyte, with freedom for anions and cations to move around. The mobility of Na^+, K^+, Cl^-, OH^-, H^+, and metallic ions issued from corrosion is limited, but some chemical exchanges remain (by diffusion, by Ca channels, etc.) through tissues.

We may expect to have fluids on the surface of the device with pH outside the range of "normal" body fluids. As mentioned earlier, buildup of HCl might impact Ti, but should not be strong enough to dissolve Au and Pt.

Recent researches also showed a high content of hydrogen peroxide (H_2O_2) in vicinity of brain-implanted electrodes. It may come from the immune defense mechanisms in the fibrotic capsule building around the implant. It could also simply come from the high oxygenation level of the brain. Whatever the causes of this excess of H_2O_2 on the surface of nervous implants are, consequences may be serious. H_2O_2 is actively degrading most polymers. Cases have been reported of degradation of the Parylene™ (chemical vapor deposited poly(p-xylylene) polymer, tradename by Union Carbide (SCS) Specialty Coating Solution division, sold to Cooksam Electronics in 1994) insulating layers of Utah arrays (Blackrock, Salt Lake City) placed in the human cortex for BCI projects [7].

Parylene is widely used, since decades, in long-term implanted applications. It is known as a highly biostable conforming coating for moisture protection and as a dielectric. In cardiac applications, Parylene is known as very stable over long periods of implantation time. Why does Parylene do not last as well in the human brain? We don't know but it is likely due to the higher content of H_2O_2. Another potential root cause might be related to impurities on the surface prior coating. This is one more element which preaches in favor of being careful when designing BCI. The environment of the brain and nervous system is still mainly unknown to us.

Explanted pacemakers sometimes show tissues with tight adhesion to Ti. Sometimes, explanting is done without seeing any adhesion. We do not explain such variations.

4.6 Cleanliness

We had seen above how capital cleanliness was regarding corrosion. Reducing corrosion sources is only one of the goals of setting appropriate cleaning processes at various stages of the manufacturing cycle. In this chapter, we will see why, when, and how we clean an active implant. First, we will review what are the types of contaminants which must be removed.

Contaminants

No surface is ever 100% clean. We can define cleanliness as the absence of contamination on a given surface at a certain time. Contamination is an evolutive phenomenon. For example, an exposed surface may accumulate dust, a part handled in an assembly plant gets contaminated by fingerprints, or even the surface may get modified by what is happening below the surface, like sweating of plasticizers in injection molded components.

The cleaning process will be specified and validated to reach an acceptable level of cleanliness, measured by the residual contamination and the nature of the residual contaminants on the surface after cleaning. Cleaning agents might be specific to certain types of contaminants. For example, removing oil on the surface of a machined metal part requires an appropriate liquid able to dissolve greasy layers.

Contamination on parts and subassemblies has different origins:

- Parts are received contaminated from the suppliers.
- Handling during assembly adds contamination (e.g., fingerprints or spits of saliva).
- Specific assembly processes may contaminate further some surfaces (e.g., deposit of flux vapors during soldering).
- Dust and particles coming from the environment (assembly of medical devices is done in clean rooms, to reduce this form of contamination).

Contaminants are of different nature:

- Biological: all forms of bacteria, spores, viruses, germs, or other living cells. Most of them will be removed during the cleaning process. Living organisms remaining on the surface and in cavities after cleaning are called bioburden. Sterilization will kill these organisms, but it must be remembered that the "dead bodies" remain after the sterilization process. After sterilization, these residuals are biologically inert but may create tissue reactions due to pyrogens and inflammation when implanted.
- Inert: chemicals, oils, surfactants, residues, oxides, particles, and dust.

Measuring cleanliness is difficult. In a lab, surface examination by scanning electron microscope (SEM) will help to identify residual contamination and possibly its nature. But surface observation is local and does not provide a global picture of the overall residual contamination. In production, a gross evaluation of contamination is done by measuring conductivity of the cleaning bath. It only gives an estimate of the residual ionic load but misses nonionic traces, like pyrogens.

Why Do We Need to Clean?

Proper cleaning is fundamental in the industry of AIMDs. A clean surface is a warranty of success for subsequent processes. For example, gluing and coating processes may fail if the surfaces are not perfectly clean. Absence of ionic contamination is especially important when non-biocompatible materials are coated to prevent exposure to body fluid.

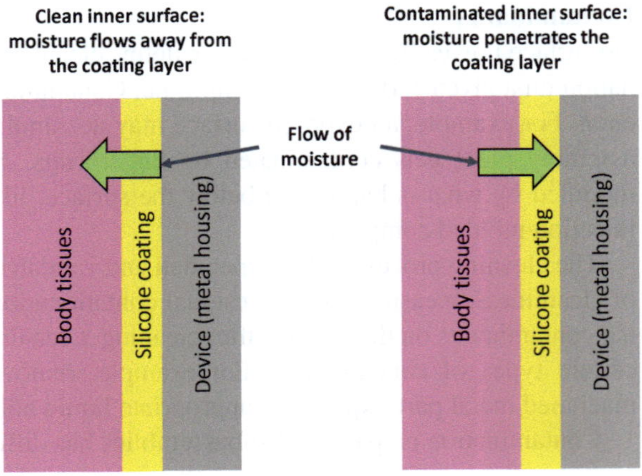

Fig. 4.10 Ionic pump

The importance of cleaning is best illustrated in the dynamic behavior of silicone coatings. The sustainability of the principle of covering non-biocompatible materials with silicone rubber is based on the so-called "ionic pump" principle (see Fig. 4.10). When a silicone-coated device is exposed to body fluids, the outer surface of the silicone layer is bathing in the natural high ionic concentration of human tissues. It creates an osmotic flow of moisture from the inside to the outside. Somehow, silicone is "dried" by this flow of water molecules. Therefore, at least for the few first months, there is no absorption of water in the silicone coating, and the non-biocompatible materials are perfectly protected from moisture ingress. With time, the inner interface of the silicone layer may evolve chemically. The non-biocompatible metal may start to release ions by intermetallic interfaces and natural dissolution or due to the release of adsorbed gas. In this case, if the ionic concentration becomes high under the coating layer, the osmotic pump may reverse and facilitate moisture flow from the outside to the inside. The same phenomenon may happen if the pre-coating cleaning was not properly done. Usually, cleaning is efficient on flat surfaces, but residual contamination often remains trapped at discontinuities, small cavities, glue joints, and brazing materials. Ions will be released from these tiny areas and initiate delamination of the coating. In consequence, cleaning is highly critical in coating processes.

When Do We Need to Clean?

There are several steps of cleaning when assembling an implantable device:

- Cleaning components before assembling the electronics boards. This will assure proper soldering of the components on the printed circuit boards (PCBs).
- Cleaning populated PCB. It will remove ionic contamination generated by the soldering process.
- Cleaning the inner assembly (PCBs + battery + coil + nests + spacers + other parts) after interconnection and connectivity test. Then, the inner assembly is placed in the housing and hermetic sealing is done. Residual contamination on the inner assembly will be trapped in the seal housing, which may lead to potential risk of building dendrites on the long term [53]. The impact of trapped contamination is strongly linked to residual moisture content (see Sect. 4.8).

- Cleaning the sealed housing prior welding connectors or lead wires.
- Clean again after welding external component to remove flux residuals or metal soot from laser spot welding or resistance welding.
- Cleaning of the sealed mechanical assembly prior header attachment and surface coating.
- Final cleaning after functional test and visual inspection. Then proceed to packaging. Final cleaning must remove most of the bioburden on the surface and open cavities of the finish device, to facilitate sterilization and minimize pyrogens.

How Do We Clean?
The above-described cleaning steps may be based on specific cleaning processes.

- Prior to sealing the housing, one usually immerges the subassemblies in aqueous solution with agitation (circulation of the fluid or ultrasonic agitation), followed by several rinsing steps in deionized (DI) water. A final rinse in isopropyl alcohol (IPA) facilitates the drying process, as IPA is a repellant of water. Proper drying in a vacuum oven is necessary to remove the moisture which might have been absorbed by plastic materials (PCB, epoxy encapsulated chips, insulators, etc.) during the immersion in the cleaning bath and rinsing processes. If subassemblies do no proceed immediately to hermetic encapsulation, they should be stored in cabinet with dry N_2 to minimize moisture absorption during storage. An alternative is to seal batches of components in plastic bags to preserve them from humid storage conditions.
- Mechanical parts, especially the ones manufactured by machining, like connectors or shields, often have a greasy surface, which must be cleaned by solvents strong enough to dissolve oil residues. Proper rinses are required to get rid of solvent traces.
- When the device is hermetically sealed, cleaning steps must take in consideration the fact that the inside of the housing (electronics, battery) may be sensitive to heat and vibration. Therefore, post-sealing cleaning processes are usually done at temperatures below 55 °C and without ultrasonic agitation. After attachment of connectors or soldering wires, cleaning processes are mainly aqueous solutions, followed by DI rinses and finally IPA.
- Final cleaning is done at the same time as final visual inspection. It consists in a gentle manual cloth wipe and Q-tip introduction in connector cavities. IPA or heptane is recommended for this final wipe. Final cleaning is done under laminar flow (class 100) to minimize risks of particles deposit. Then, the device is placed in the inner blister for thermo-sealing. From this point, the device is protected from any further contamination.

4.7 Sterility

The purpose of the sterilization process is to kill most of the biological contaminants (bioburden) left after the various cleaning steps. Sterilization is done when the device is already packaged. Implantable devices are usually packaged in double blisters or pouches:

- Inner blister: thermoformed foil of polyethylene terephthalate (PET), in which the device is placed right after final wipe cleaning. The central part of the blister is shaped to fit and maintain the device in place. Sometimes, accessories are added, like screwdrivers or introduction tools. Then, a semipermeable lid, of type Tyvek® [8], is thermo-sealed on the PET blister. Tyvek is a brand of DuPont. This paperlike material is made from polyethylene fibers and has the property to be permeable to vapor and small gas molecules but impermeable to water. The Tyvek lid is thermo-sealed on the blister by an appropriate thermo-sealing machine applying heat and pressure on the edge of the blister. From this point, the device is fully protected from any contamination coming from outside, biological or inert.
- Outer blister: the inner blister is inserted in another larger blister, which is also thermo-sealed with a Tyvek lid.
- The goal of the double blister concept is to warranty a maximum protection of the device when it is introduced in the operation room (OR). The double-packaged device is introduced in the antechamber of the OR, where the outer blister is peeled opened by assistant wearing sterile gloves. Then, the device, still protected by the inner blister, is introduced in the OR. When it is time for implantation, the inner blister is opened by an assistant wearing sterile glove; the device is extracted from the blister and put in the hand of the surgeon over the sterile theater. This procedure minimizes the risks of contaminating the sterile device.
- Blister or pouch sealing is assessed during the validation phase for integrity. Burst tests are conducted to verify that the blister is properly sealed. Such tests are regularly done in production to assure that the thermo-sealing process does not deviate from specified values.

There are several methods to sterilize a device. Sterilization is commonly used in hospitals for surgical tools. The most common method for reusable tools is autoclaving. After cleaning, reusable tools are placed in metal trays, introduced in the autoclave and exposed to high-temperature (121 °C) saturated pressurized steam during 15–20 min. This method applies to reusable insertion tools used in brain surgery. Disposable tools can also be autoclaved before use. This is usually done in the hospital, right before introduction in the OR. Such disposable tools are often packaged in single Tyvek pouches, which are permeable to steam.

For implantable devices, packaged in double blisters or pouches, sterilization is done by various methods described below:

Autoclave: The process has been described above for reusable or disposable surgical tools. It is also used for metallic bone implants (hip and knee prosthesis, bone plates, screws, etc.) as these devices are resistant to heat. AIMDs cannot be sterilized in autoclaves as batteries do not stand more than 55 °C.

Gamma Radiations: Many devices, in cardboard boxes, placed on pallets, are introduced in a large tunnel and exposed to gamma radiations produced by radioactive cobalt 60 emitting gamma rays. Gamma radiations penetrate through the entire mass of the load of the pallets and destroy all living cells by breaking covalent bonds of their DNA. The advantage is clearly the batch approach, where several devices can be sterilized at the same time. There is also no substantial heating of the sterilization load. Unfortunately, gamma radiations are so powerful that they damage electronic components. Therefore, gamma radiation is not an appropriate process for AIMDs. It can nevertheless be used for separately packaged electrodes. Great care must be taken regarding the plastic materials constitutive of electrodes or accessories sterilized by gamma radiation, as this process may liberate free radicals and make the plastic material become brittle. Silicone rubber is rather tolerant to gamma radiation. Gamma radiation sterilization plants are enormous installations requiring extreme safety protection, as the source is radioactive.

Electron Beam (E-Beam): Electrons are accelerated at high energy levels, so the resulting beta radiations penetrate through most materials. The sterilization effect is, like gamma radiation, based on the destruction of the DNA of microorganisms. As for gamma radiation, the applied dose is measured in kGray or Mrad (1 gray = 100 rad). Unlike gamma radiation, E-beam installations do not use radioactive sources. E-beam cannot be used in devices including electronic components and consequently is not an option for AIMDs.

Ethylene Oxide (EtO): EtO is a highly penetrating gas which effectively kills microorganisms, including resistant spores. EtO rips away cell membranes causing the death of microorganisms. It is a flammable and explosive gas, toxic for humans if inhaled or in contact with the skin. EtO sterilization is the process (see Fig. 4.11) of choice for AIMDs because of:

- Its low temperature, usually kept below 55 °C (max. temperature for Li-ion batteries)
- Its good penetration in deep cavities and hollow tubes
- Its good penetration in silicone rubber
- Its minimal impact on plastic material (does not liberate free radicals)
- Its standardization (small size chambers, single use EtO cartridges, parametric control, validated cycles usable with most implants)

There also drawbacks in the use of EtO:

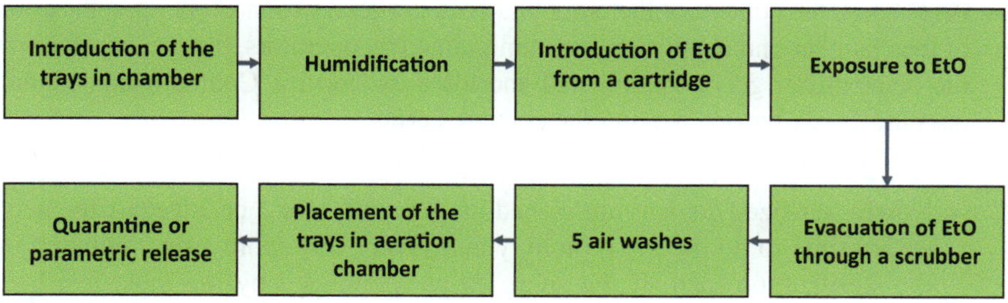

Fig. 4.11 Phases of an EtO sterilization process

- Dangerous for humans:
  - Safety measures: detectors in the room and worn by operators.
  - At the end of the cycle, EtO cannot be released in the atmosphere but must be eliminated through a scrubber.
- During exposure to gas, plastic materials, especially silicone rubber, absorb some EtO, which requires post sterilization aeration.
- EtO is only efficient in presence of moisture, requiring a pre-sterilization humidification phase.

The EtO process must follow a strict cycle for being efficient:

- Blisters are placed in baskets and introduced in the sterilization chamber. As the cycle is validated for full loads, in case there are not enough products to fill a basket, the load is completed by dummy blisters.
- Water vapor is injected in the chamber. The Tyvek lid is permeable to moisture, so the devices get humidified. The structure of Tyvek let only small molecules penetrate through the lid. Tyvek lets vapor and EtO molecules enter the blister, because they are small enough. Bacteria, spores, or other microorganisms cannot pass the Tyvek barrier.
- A single dose EtO cartridge is introduced and EtO invades the chamber. The Tyvek lid also let EtO through.
- The devices remain exposed to EtO during a few hours, with a continuous control of relative humidity, pressure, and temperature. In most cycles, pressure is maintained slightly under atmospheric pressure to avoid any leakage of EtO in case of default in the closing gaskets of the chamber.
- Then, the gas content of chamber (EtO and vapor) is pumped out and pushed through a catalyst bed (scrubber) which breaks the molecules of EtO in H_2O and CO_2.
- The chamber is vacuum pumped five times and fresh air is introduced. These five "air washes" remove more than 90% of the EtO residuals absorbed by plastic materials during exposure to EtO. The air washes also remove humidity and dry the content of the chamber and of the blisters.
- The baskets are moved from the sterilization chamber to an aeration chamber for final removal of the last residues of EtO. Depending on the weight, thickness, and type of plastic materials in the device, aeration will take longer.
- As a result, the content of the outer blister is sterile. As soon as we open the door of the chamber and the blisters are manipulated by operators, the outer surface of the outer blister gets contaminated and loses its sterility. Contaminants cannot pass the Tyvek barrier, leaving the content sterile.
- There are two ways to release the sterile product and authorize its shipment:
  - Newly sterilized devices are placed in quarantine for 1 or 2 weeks, the necessary time to do the microbiology analysis of test spore strips included in the sterilization load. If the spore strips (populated by the most resistant spores), placed in incubators during the quarantine, are shown to be sterile, then the batch is released. The disadvantage of this method is the long immobilization of batches in quarantine.

- A more modern method called "parametric release" avoids quarantine and allows products to be release immediately after aeration. All the relevant parameters (temperature, humidity, pressure, and concentration of EtO) are recorded during the entire sterilization process and compared to tight specifications set during validation. If all the parameters are within specifications, the batch is immediately released. Validation of parametric release is a long and heavy task, as parameters measurement and microbiology results with spore strips are compared. Revalidation is required every year.

Suppliers of EtO sterilizers recommend specific cycles. Selecting the right cycle and validating it for a given device are part of the design phase. Some medical device manufacturers have decided to use the same EtO cycle for all their products, even if some of them (e.g., electrodes) could be sterilized by gamma radiation. For the sake of standardization, they also use identical equipment in all their factories. EtO sterilization is used since decades and is well accepted by health authorities worldwide.

Hydrogen Peroxide Plasma (H_2O_2): The sterilization chamber is filled with gaseous hydrogen peroxide (H_2O_2) and activated by microwave radio frequencies in the GHz range to form a plasma. Activated H_2O_2 penetrates in the blister through the Tyvek lid and kills microorganisms. Compared to EtO, the advantage of this method is the nontoxic nature of hydrogen peroxide. When dissociated by the plasma, molecules of peroxide are transformed in water and oxygen. The weakness of plasma sterilization is a poor penetration in deep and narrow cavities, like long obturated tubes. The method is more recent and has not been fully accepted around the world. As some countries do not yet approve plasma sterilization, large manufacturers keep using EtO instead.

4.8 Accelerated Aging

It is fundamental to evaluate how implants will last over long term. How do we assess functioning over 10 or 20 years? Nobody can afford to test, in vitro or in vivo, a device during such a long time before putting it on the market.

To speed up the aging process, developers have conceived methods of acceleration, with the objective of assessing functionalities in 1 year, but simulating up to 10 years. The electronic industry has since a long time introduced an artificial aging process of electronic boards called "burn-in" (see Fig. 4.40, Sect. 4.9.4). Populated PCBs are put under voltage and let in ovens for several days. It is known that most of the failures in electronic circuits appear early in the life cycle, so-called childhood failure. Acceleration follows the law of Arrhenius [9] stating that every increase of the temperature by 10 °C accelerates aging by a factor of two. Therefore, increasing temperature by 40 °C will speed up the aging of electronics by a factor of 16.

Accelerating aging of implants, in the situation of surrounding human tissues, is much more complex than electronic boards in dry air. Increasing the temperature of saline solution in a beaker is far from a simulation of the human body in accelerated mode. The Arrhenius' law might be applicable with electronic boards in dry air up too rather high temperatures (e.g., 160–170 °C). Is it still valid in saline? Beyond a certain point, materials behave in a different way, not following Arrhenius' law. For example, polymers may start to transition. In the same way, polymerization, liberation of free radicals, diffusion, and degassing are physical processes which do not automatically follow Arrhenius's law.

Saline solutions [10] and phosphate-buffered saline (PBS) [11] are gross approximations of the human body. Increasing the temperature of these test liquids introduces further deviation from a reliable model. For example, the absorbed oxygen content in water decreases with temperature increase, meaning that rising the temperature makes the solution less aggressive. In consequence, the Arrhenius' law of accelerating aging by a factor of two for an elevation of 10 °C is not valid any longer. The relative decrease of oxygen must be compensated by adding reactive agent in the saline solution, like H_2O_2. The group of Cristin Welle at the FDA has issued a guidance for accelerated aging of neurological implants [12]. This in vitro method, if applied properly, allows conducting accelerated aging test with a better level of confidence. It presents the advantage to avoid unnecessary in vivo tests and save time. During the validation phase of a human implant, it is of highest importance to demonstrate that the device will last for as long as it has been designed for. The method, called reactive accelerated aging (RAA), is using elevated temperature (87 °C) and reactive oxygen species (ROS). The team of Cristin Welle has tested the evolution of commercially available intracortical penetrating electrodes. By careful measurement of the evolution of the performances of the test samples, it was possible to assess and quantify the impact of aging. Impedance spectroscopy (see Sect. 3.3) is the method of choice for the assessment of electrodes.

Some people judge the RAA procedure as too harsh. It is preferable to have a test a bit too aggressive rather than a gross underestimation induced by the 10 °C → 2× rule. Accelerated aging is one of the most important steps in the validation process of a human implant. Having a safety margin is part of the risk management. Personally, I recommend a safety factor of two: if the implant is intended to be implanted for 1 year, the accelerated aging should simulate 2 years of full functionality with a statistically relevant level of confidence. Such a high safety margin results from the very nature of the occurrence of a dysfunction due to aging (corrosion, loss of hermeticity, etc.). The severity of the damage is catastrophic and will occur many years after implantation. Identifying such a failure a long time after a product has been put on the market may lead to a massive recall with serious consequence for the product and its manufacturer.

Artificial aging tests described above are static. The reality of an implant is the body is far from static. There are movement of fluids, body movements, growth of fibrotic tissues, accumulation of ions, interference with external magnetic and electric fields, and other dynamic phenomena which may have a substantial impact on aging. Developing a test suite which reflects dynamic artificial aging is still to be done. Pioneer work in this direction has to be done by Dr. Pierre Fridez in Lausanne, Switzerland, in the scope of the EndoArt project [13], an adjustable gastric band remotely controllable. The electronics and the step motor controlling the adjustable diameter of the band were encapsulated in plastic material. To demonstrate long-term resistance to moisture and absence of corrosion, engineers conceived, more than 10 years ago, a unique artificial aging test equipment which was a precursor of the RAA method. This work introduced, for the first time, an artificial aging procedure which combines elevated temperature saline test with addition of oxygen and movement of the test samples. EndoArt has been acquired by Allergan in 2007 [14].

4.9 Hermeticity and Moisture Control

Hermeticity is the cornerstone of the industry of active implants. The ones who master hermeticity have a chance to succeed in such a special field. The best possible electronic board, if poorly encapsulated, will fail on the long term. In consequence, this chapter might be considered as the central pivot of the book.

As a general concept, encapsulation could be defined as the protective barrier around the active components of AIMDs: electronic circuits and battery. It also englobes the insulating and protective barrier around cables, connectors, electrodes, coils, and antenna. This book covers only encapsulations for implants intended to remain more than 30 days in the body. Encapsulations may be sorted in three categories:

- Hermetic encapsulation
- Near-hermetic encapsulation (see Sect. 4.9.6)
- Insulators, coatings, pottings (see Sect. 4.9.7).

The first active implantable medical devices to be commercialized were pacemakers in the late 1950s. Early-day pacemakers were very simple pulse generators, encapsulated in silicone rubber or epoxy. No feedthrough nor hermetic metal casing was then needed as the electronics (a few transistors, resistors, and capacitors) were assembled with large distances between conductive parts, so insulation was not a major issue.

When electronics became more complex, more integration and high density of components led to short insulating distances. Then rubber and epoxy encapsulation turned out to be not a good enough protection any longer. Leakage currents, corrosion, or short circuits between key parts became an issue. In the 1970s, the pacemaker industry started to encapsulate pacemakers in metal casing (mainly titanium casing), providing a much better reliability. At that time, batteries were the main concern, with some mercury compounds or even nuclear batteries. Protecting patients from leaks of nasty chemicals was then the highest priority. Later, battery chemistries moved to less toxic material, and implantable batteries got their own primary hermetic encapsulation. The main goal of hermetic encapsulation has then moved slowly to the protection of the implanted electronic circuits.

Another important driver of hermetic encapsulation is the highly integrated electronic circuits. When the density of the integrated circuits (ICs) increased, the risks of building dendrites on the surface of unpackaged chip increased accordingly. For the sake of miniaturization, some manufacturers used unpackaged chips, naked dyes directly mounted on the printed circuit board (PCB). The second reason to choose unpackaged chip is to avoid accumulation of moisture in the plastic materials used in standard chip packaging.

Hermetic metal encapsulation became a standard, especially for batteries. It was then recommended to have a double barrier for batteries: a hermetic encapsulation of the battery itself (metal can with FTs), plus an overall hermetic encapsulation of the device (Ti casing with FTs around the electronics and battery assembly).

There are no clear standards regarding the level of hermetic encapsulation an AIMD must meet. There are several reasons to this "no man's land":

- Test methods to find extremely small leaks are difficult to implement, and their results are far from obvious in terms of quality assurance.
- Consequences of poorly sealed device may take many years to show up.
- Regarding residual level of moisture trapped in the device, nobody has ever conducted any serious scientific evaluation of corrosion and its effects in function of time and moisture.
- The size of medical devices and the small volume of gas entrapped in them put the challenge of measuring small leaks to counting molecules escaping per seconds. There is no equipment with enough sensitivity to detect leaks in very small devices, at least not available on the manufacturing floor.

In consequence, the medical industry follows vague recommendations, extrapolations of standards used in other industries, and "rules of the thumbs" coming from the early days of the pacemaker industry.

Officially, there is no formal limit set for moisture in implants. The general guidelines of the FDA states:

> "In the absence of a standardized widely accepted test method, it is important that device manufacturers:
>
> a) know the scientific capabilities of the method being used
> b) conform to their own stated procedures and specifications
> c) properly calibrate and maintain their equipment."

Regarding hermetic sealing, leak test, moisture control, and residual gas analysis (RGA), the FDA refers to military standards (MIL-STD) [40] 883 [41, 42], 750 [43], and 202 [44] and to the American Society for Testing and Materials (ASTM) [45] F/34-72T [46].

For a manufacturer of AIMDs, the main difficulties are to select the right components and the right assembly processes leading to a highly hermetical encapsulation closed over a dry content. FTs and casing will be part of the characterization, qualification, and validation phases.

When a hermetic product is approved and commercialized, components, encapsulated device, equipment, and processes will be "screened" at several levels with an objective of assuring hermeticity and low moisture content:

- Receiving inspection of parts at the assembly plant
- Periodical assessment of quality of the parts
- Periodical audits of the supplier of the parts
- Helium leak test (100%) of the sealed devices
- Cross sections and metallographic analysis of sealed devices at least twice a day (beginning and closing of the production day)
- Periodical RGA measurements
- Annual revalidation of sealing processes
- Regular maintenance and calibration of equipment
- Periodical audits (internal and/or by notified bodies) of the assembly plant

It must be pointed out that no plastic material provides a 100% barrier against moisture. Moisture (water in vapor form) will sooner or later diffuse through all plastic materials. Moisture diffusion is ruled by complex laws of physics (see Fig. 4.12), but for long-term implants, some moisture will slowly slip through all plastic materials. The most resistant plastic materials, like Parylene, will nevertheless let some moisture in after a few years. When vapor concentration reaches saturation, then liquid water is formed by condensation, at the surface of the device, below polymeric coating, in bubbles and cracks, or inside of non-hermetic or near-hermetic housings.

It is not enough for an AIMD to be waterproof; it must be hermetic. Waterproof means that no liquid water will penetrate in the device. Hermetic means that no liquid and no gas (including water vapor, oxygen, hydrogen) will penetrate, either through a crack or by diffusing through bulk material. For example, a device with a rubber gasket will be waterproof, but not hermetic. I have seen waterproof Ti housings with a silicone rubber gasket becoming full of liquid water after 4 months implantation. Vapor had diffused through the gasket and condensed in the device.

One of the most common materials in implants, silicone rubber, is an excellent repellant of body fluid (in this sense, a silicone encapsulation is waterproof), but a very poor barrier to moisture (diffusion through silicone is easy for moisture). The dynamics of the flow, diffusion, exchanges, and migration of vapor through plastic materials surrounded by body fluids is complex, as additional movements are created by osmotic pressure, release of plasticizers, and continuing polymerization. Note that other gases present in the body, like O_2, H_2, H_2O_2, and CO_2, are also likely to diffuse through implanted polymers.

Diffusion of moisture through plastic materials is the reason why almost all AIMDs are encapsulated in metal casings with ceramic or glass FTs. Older devices, like some rechargeable neurostimulators, had their antenna and a part of the electronic embedded in plastic, but the battery was well protected by a double metal encapsulation. Some simpler AIMDs, like sensors or the inductive sacral root stimulation device manufactured by FineTech Medical (UK) [15], are encapsulated in plastic but include no battery and no highly integrated electronics.

One interesting case is SynchroMed (Medtronic), already described in Sect. 3.4.8, an implantable peristaltic pump for drug delivery. Rollers, activated by a step motor, push a liquid drug from a reservoir (titanium bellow) to a catheter, by squeezing a silicone tube. The device is encapsulated in a titanium case, with several hermetic compartments. In the cavity where the rollers squeeze the tube, some moisture has been found, coming from the diffusion of drug and excipient through the silicone tube material (called "sweating"). After a few years, the pump cavity is saturated with moisture; even some liquid may be present. For this reason, all the components of the pump head are designed to be resistant to corrosion and biocompatible (ceramic ball bearings, titanium parts).

Metals provide the best protection against moisture, followed by ceramics and glasses (see Fig. 4.12). It must be remembered that protection against penetration of moisture in the device is capital to avoid corrosion of the parts inside the encapsulation. If corrosion does anyhow occur, then a second role of the encapsulation is to avoid the corrosion residuals, like aqueous salts or electrolytes, to leak outside the device, affecting the patient.

Hermetic encapsulations have therefore a double mission:

- First, avoid moisture to penetrate the encapsulation and initiate corrosion.
- Second, in case of corrosion of some of the components inside the casing or in case of battery leakage, then the encapsulation must keep all nasty chemical inside.

Penetration of moisture depends on several factors:

- Type of material
- Thickness of the walls of the encapsulation
- Duration of implantation
- Gradient of pressure $P_{ext} - P_{int}$

The graph of Fig. 4.12 shows clearly why potting a device in epoxy or silicone will provide proper protection against moisture only for a short period of time. Thicker potting will extend the lifetime. Realizing thin BCI for long-term implan-

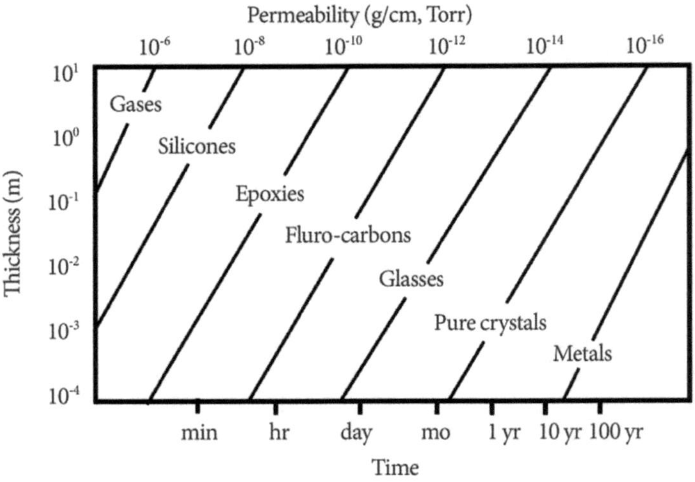

Fig. 4.12 Permeability of materials. (Source K. Ely [49], reprint approved)

tation requires thin walls of highly moisture resistant material. Only metals, pure crystals, and possibly glasses will fill these criteria. Adding the constrain of impact resistance, it leaves one single choice: metal. This explains why subcutaneous above-the-neck BCI implants are likely to be encapsulated in titanium housings.

4.9.1 Feedthrough (FT)

Feedthroughs are key components of hermetic encapsulations for AIMDs. The main function of a FT is to allow electrical connection between the inside and the outside of a housing. In addition, the FTs must guarantee electrical insulation between the conductive wire(s) and the housing, plus hermetic sealing.

The apparent simplicity of this component is misleading. In fact, FTs are one of the keys of high-quality encapsulations and, as such, deserve full focus. For decades, the design and the assembly processes have not progressed much. Recently, new applications, multiplication of the number of channels, trends to miniaturize, pressures on costs, requirement to integrate the various components on automated assembly lines, and needs to protect medical devices against electromagnetic disturbances have fundamentally changed the environment. Consequently, future generations of FTs will be different from their ancestors.

The historical suppliers of FTs did not actively drive the needed innovation and the evolution of technologies. Rethinking the concepts and starting with a “blank sheet of paper” have been perceived as an opportunity for new creative suppliers to enter this field. FTs are intimately linked to the overall hermetic encapsulation of AIMDs. In consequence, FTs should be considered as a key functional part of the encapsulation, and not as a mere component. Even if FTs have a simple function (basically a conductor insulated from the main body of the encapsulation), the complexity of manufacturing and assembling such parts is a serious challenge.

Hermetic FTs are also key components of the batteries encapsulated in AIMDs. Compared to AIMDs FTs, battery FTs have slightly different specifications, as they don’t need to be biocompatible, but must resist highly aggressive chemicals trapped inside the battery.

Major developments of hermetical FTs have their origins in the military, nuclear, and space industry. Total hermetic enclosures were a must for the space missions. Most of current medical FTs are in fact related to the space industry. Another important driver of hermetic encapsulation is the highly integrated electronic circuits. When the density of the ICs increased, the risks of building dendrites on the surface of the chip increased accordingly. So, the electronic industry developed better protection of the chips. This is the origin of ceramic IC packages, with gold-brazed pins and cover lids.

Reliability is the keyword for medical device components. FTs, as interfaces between the electronic sealed in the casing and the external environment, are especially exposed and become therefore one of the main actors of high reliability products.

On the other hand, we can see that several major issues occurred in the vicinity of FTs, mainly loss of electrical connection externally or internally (poor welds, broken wires) or short circuits due to header detachment or poor insulation of the external FT wires. This shows the importance of designing not only reliable FTs as stand-alone components but also of properly integrating the FTs in the device. In my opinion, successful future FT designs will include innovative features for connection, insulation, automatic assembly, and post-assembly testing.

In the 1970s, pacemakers were the first medical devices with a metallic casing (first stainless steel and then titanium) and therefore hermetic FTs. At that time, most pacemakers were "single chamber" (only one stimulation electrode), and stimulation was unipolar (current returning to the pacemaker through body tissues and to the can). In consequence, early pacemakers required only one single wire FT.

In the 1980s and 1990s, pacemakers become more sophisticated with the introduction of dual chamber stimulation and bipolar leads, requiring two or four connections. Today, a large majority of bradycardia pacemakers are dual chamber and bipolar, with four connections. Some resynchronization devices also stimulate the right side of the heart, requiring a total of six wires. For the sake of automation, Medtronic decided to use four single wire FTs placed side by side; the other manufacturers of pacemakers have two dual wire FTs or one quad wire FTs.

Apart from the tight requirements on hermeticity, biocompatibility, and reliability, FTs used in pacemaker applications are not very demanding in terms of electrical specifications. The voltage between wire and casing or between two wires is a few volts. So, insulation is not too difficult to achieve, and multiple wires FTs may be quite dense. The electrical current delivered to the bradycardia stimulation leads is in the range of a few milliamps, so the diameter of the conducting wire of the FTs is not critical.

In the 1990s, the first implantable defibrillators were developed. As for pacemakers, the evolution went from single chamber unipolar defibrillation chocks to dual chamber bipolar devices. A new type of FTs needed to be developed for defibrillators, as the voltage (up to several hundreds of volts) and the current (up to several amps) of the chocks require a much better insulation and a larger diameter of the conductors.

The third large category of commercial AIMDs using FTs is neurostimulators. The first devices appeared in the late 1980s and were very similar to pacemakers, with a titanium can and a few (up to four) external wires. The main difference was the nature of the electrical signals carried by the FTs and of course the location of the stimulating leads in the body. Until recently, FTs for neurostimulators were similar or identical to the ones used in pacemakers. New applications of neurostimulation, in the field of pain control, epilepsy, movement disorders, functional stimulation, and other exciting therapies, require more and more external connections or stimulation/sensing channels. Devices with 32 or 64 channels are already available, and some future applications might require several hundreds of connections. This will lead to a fundamental redefinition of the function and design of FTs. In terms of space, assembly time, and costs, it is not realistic to use conventional cardiac FTs for more than a dozen connections.

For cardiac applications, the size of the FTs has not been of high priority. Future multichannel neuro-applications will require high-density miniaturized FTs. Furthermore, the cost per connection must remain affordable.

In addition to the three large categories described above, many new medical devices and niche applications are in deep need of special FTs, adapted to their specificities. In the recent past, I have seen several exciting developments which could not be finalized because nobody was able to supply appropriate FTs. This is especially the case for highly miniaturized devices (implants in the eye or the inner ear, sensors in blood vessels, etc.). As we will see later, the small number of suppliers of FTs and the dominance of the cardiac applications makes the types and sizes of available FTs extremely limited.

4.9.1.1 Role of FTs

As part of the hermetic encapsulation for an AIMD, FTs have the following roles:

- Provide a high electrical insulation between the conductive wire(s) and the body of the device, with high insulation resistance (>xx MΩ) and a low leak current (<xx pA).
- Assure proper conduction of the specified electrical signals.
- Keep cross talk and electromagnetic coupling between wires (for multiple wire FTs) under a specified level.
- Assure a hermetic sealing (leak < xx 10^{-9} std cm^3/s).
- Be biocompatible and biostable on the external side of the FT.
- Resist cleaning and sterilization processes.
- The flange must be weldable to the casing.
- The wire or pin must be connectable to the inside electronics and to the external connection.
- The above should not be impaired by the forming or shaping of the wire during assembly.
- The above should remain true for at least 10 years implanted in the body.

4.9.1.2 Types of Electrical Signals Carried by FTs

FTs are used to carry a large variety of electrical signals. Here are some of the classifications:

- Direction:
  - Signals generated inside the AIMD and carried to a stimulating lead (majority of the applications, like pacemakers)
  - Signal collected outside the device and forwarded to the AIMD (BCI, sensors, measuring body potentials, inductive recharging, etc.)
  - FT used for both the above, in alternance

- Voltage:
  - Low voltage (like pacemakers)
  - High voltage (like defibrillators), requiring high insulation levels
- Current:
  - Low amperage (like pacemakers)
  - High amperage (like defibrillators), requiring larger diameter of wire and highly conductive materials
- Frequency:
  - Low frequency (like pacemakers or defibrillators)
  - High frequency (BCI and neurostimulators, RF communication)

This wide range of signals means that a given technology of FT must be declined in several models in order to cover the diversity of applications.

4.9.1.3 Two Main FT Technologies

In the field of FTs for AIMDs, there are basically two ways (see Fig. 4.13) to assure a hermetic join between the insulator and the metallic parts (flange and wire(s)):

- Brazing, meaning fusion of a metal layer at the interface between the insulator and the metallic part. We will call them ceramic FTs or gold-brazed FTs.
- Fusion of the insulator or of a part of it. We will call them glass FTs.

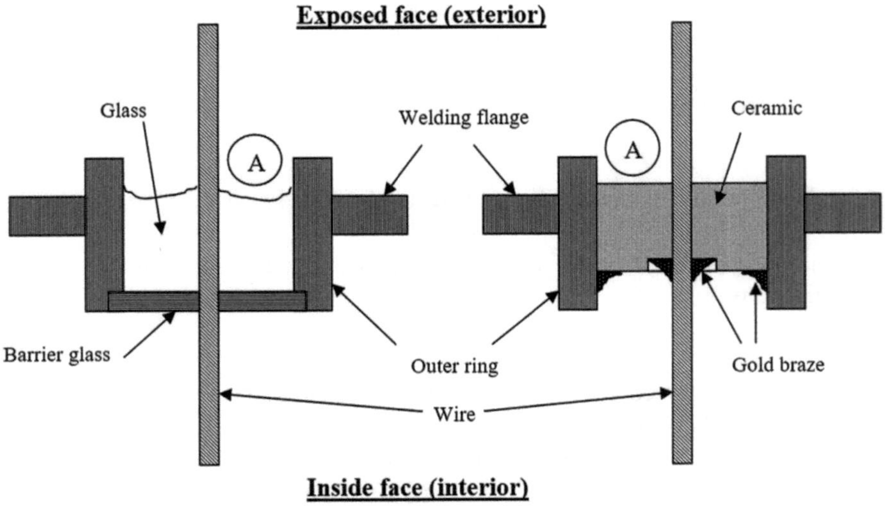

Fig. 4.13 Glass (left) and ceramic (right) FTs

These two technologies are described more in details below. I don't know any example of FT mixing both these technologies. Other technologies, like gluing or plastic insulators, will not be covered by this study, as not suitable to AIMDs. In medical applications, I have not seen coaxial FTs.

4.9.1.4 Ceramic FTs

A ceramic FT consists in the following parts:

- One or several electrical conductors, in the form of wires, ribbons, or pins
- An insulator in ceramic material, with holes for the insertion of the conductors
- A flange

Hermetical joining of the parts is assured by brazing, which consists in fusing a brazing metal (mainly gold) in the interfaces wire(s)-ceramic and ceramic-flange. The brazing material must have a fusion temperature lower than the other materials of the FT. When the brazing material reaches its melting point and becomes liquid, it "wets" both sides of the interface and adheres tightly to the surfaces. The melted material also penetrates in the gap by capillarity. Penetration depth in the gap depends on many parameters (temperature, surface treatment and potential, environment, geometry, etc.). When the brazing material cools down and resolidifies, parts are solidly joined and hermetically sealed.

The assembly of ceramic FTs can be done in three ways:

- Deposit of a layer of brazing material at the interfaces, prior assembly:
  - The ceramic must be metalized on the inside and outside diameters, for the gold braze to "wet" properly and adhere to the surface. One of the difficulties is to selectively metalize the area for gold brazing while avoiding having any metal on the top and down surfaces of the ceramic.
  - The brazing material is applied on the ceramic side of the interface or on the metal side of the interface or both. The deposit may be a solid metallization layer or a softer coating in the form of paste.
  - Assemblies are placed on a fixture which maintains the parts in position during brazing.
  - Brazing (fusion of the brazing material) in an oven under controlled atmosphere (protective gas or vacuum).
- Assemble the parts prior deposit of a join of brazing paste:
  - Insert the wire(s) in the holes of the ceramic and the ceramic in the flange.
  - Hold them in a fixture, as there is gap at the interfaces to allow penetration of the brazing material.
  - Dispense brazing paste on the top of the interfaces. Groves are usually machined in the ceramic to better keep the paste in place and facilitate penetration/wetting during fusion.
  - Brazing in an oven under controlled atmosphere.

- Insert rings of brazing material at the interfaces:
  - Instead of dispensing a soft paste, rings or inserts of brazing material are inserted between the parts or in groves above the interfaces:
    - Either rings of solid gold, machined, punched, or section of tube
    - Or "preforms" of gold powder sintered or joined in a solid or semisolid "glue" (this linking material will be melted and evaporated in the oven)
  - Brazing in an oven under controlled atmosphere.

The last method is best adapted to automated assembly and large volumes of production.

Other FTs are only brazed at the flange interface, but wires are not brazed but simply compressed in the ceramic by heating the subassembly and choosing different dilatation coefficients. This requires extreme precision of the dimensions of the hole and wire.

Machining of the flange and the wires is done by conventional means, with no specific difficulty. The critical steps of manufacturing ceramic FTs are:

- Machining the ceramic insulator: high precision holes and groves. This explains why several manufacturers of ceramic FTs are originally manufacturers of ceramic components.
- Dispense of brazing paste or insertion of brazing preforms.

Also critical are the steps of cleaning, surface treatment, etching, or dispensing primer on the various parts. The quality of the brazed join relies a lot on those preparation steps.

After assembly, inspection and testing represent a large part of the manufacturing costs of implantable grade FTs. Here are some of the inspection steps and tests, which most manufacturers have put in place, either because required by standards or as part of their own reliability and QA procedures. Note that all those tests are done for 100% of the FTs and that nobody takes the risk of testing only a subset of the production batches. In this sequence:

- Thermal shock: done in an oven, usually five cycles between 200 °C and −65 °C. The goal is to release stresses and open the potential cracks before leak test.
- Leak test or hermeticity test: helium leak test done on FTs placed in appropriate fixtures. This is the most challenging test, due to the difficulty to seal properly the flange with a gasket on the fixture.
- Current leak test or insulation resistance test: measuring current leaks between wires and wires/flange.
- Disruptive test or high-voltage test: application of a high electrical voltage between wires and wires/flange.
- Visual inspection: under binocular, searching for:
  - Cracks, inclusions, bubbles, holes, change in color of the insulator
  - Cracks, inclusions, bubbles, holes, change in color, discontinuity, short or bridge in the gold braze

- Cracks or other injuries of the wire(s)
- Cracks or other injuries of the flange
- Evidences of surface contamination, fingerprints, oil, dust

4.9.1.5 Glass FTs

A glass FT consists in the same parts (wires, insulator, and flange) than a ceramic FT, but joining and hermetical sealing are assured by the fusion of a part of the insulator itself (glass). There is no brazing or additional joining material at the interface.

The assembly is introduced in an oven, in order to melt the glass insulator. The fused glass will become liquid enough to wet against the metallic surface by capillarity. But such liquid glass will also tend to flow away under the pull of gravity. In consequence, a so-called barrier needs to be placed under the glass preform, with the objective to keep the fused material in place. This barrier should be a good insulator and remain solid when the glass is fused. Materials of choice for the barrier are:

- Sapphire or ruby (single crystals), like the "watch stones" used in high-end watches. Suppliers are from the watch industry. These are expensive components, as they are machined with diamond tools.
- Glass with a higher fusion temperature than the preform. It requires a perfect control of the oven temperature cycle, to get a good fusion of the glass insulator, without melting the glass barrier.
- Ceramic.

Some configurations have two barriers, one below to prevent the fused glass to flow by gravity and one above to add a certain pressure (weight gliding around the wire during the fusion process) in order to compress the fused glass against the metal parts and improve wetting.

The assembly steps of a glass FT are:

- Insert the flange in a fixture.
- Insert the wire(s).
- Insert the barrier in the flange.
- Insert the glass preform.
- If applicable, insert the top barrier and the weight.
- Place in an oven under cover gas or vacuum.

Assembling glass FTs is easier than assembling ceramic FTs, as it escapes expensive machining of ceramics and dispense of gold braze. Glass FTs also allow high-density configuration with many wires.

The inspection and tests steps are alike the ones used for ceramic FTs (described above). If the barrier is transparent (sapphire or glass), it presents the advantage to be able to inspect the insulator in transparence, with backlight. This facilitates the detection of cracks.

The weak point of glass FTs is the propagation of cracks in the glass insulator. After the melted glass has resolidified and adhered to the metallic parts, tiny cracks may be initiated by thermal stresses (during the cooling down of the oven, during thermal shock test, during laser welding of the flange) or by mechanical stresses on the wires, especially when bending and forming them.

At first, these tiny cracks in the glass will be local, creating no loss of hermeticity and therefore not detectable during helium leak test (of the FT alone or the entire device). Cracks might be so small that they will be invisible, even under binocular, even back lit. *It is the Achilles' heel of glass FTs.*

Cracks in glass tend to grow and propagate with time. It is a slow process, comparable to corrosion and diffusion. We don't understand all the physics around the phenomenon, but it is a reality. An originally minor crack may, after several years, propagate through the insulator, opening a path for moisture. As this propagation is slow and unpredictable, it represents a serious risk for AIMDs featuring glass FTs, in case some undetected initial crack is present during assembly.

The main root causes of cracks are:

- Thermal stress during laser welding of the flange.
- Shaping and folding the wires during the assembly process. The fused glass forms a wetting radius along the interface with the wires. The edge of this meniscus is very thin and fragile. When moving the wires laterally, the glass around the wires may break or peel, initiating a crack with a potential slow growth.

The fragility of glass also implies that conductors must be thin and flexible. If the wires are too rigid, modern assembly processes (like resistance welding, laser welding in fixtures) to connect the wires may generate stresses and shock waves in the glass. For example, short rigid pins and "top hat" configurations are not advisable in glass FTs.

In consequence, we see that selecting glass FTs will require special care in the design of the product, in the choice of the assembly processes, and in the testing procedures.

Because of their high resistance over a large range of pH, glass FTs are often the best choice for battery encapsulation.

4.9.1.6 Comparison, Advantages, Disadvantages, Specificities, and Trends (Table 4.1)

Most manufacturers dispense a drop of epoxy on the external face of the FT (both ceramic and glass), noted with the letter "A" on Figs. 4.13 and 4.14. This is usually done when the FT is fully assembled, the wires formed and connected. The drop of epoxy will provide stress release and will delay exposure of the FT to body moisture.

Because of its advantages of robustness and due to a better adaptation to automated assembly processes, ceramic FTs represent today the largest market in AIMDs (almost all manufacturers of pacemakers and defibrillators use ceramic FTs). Some special products, where miniaturization is a must (cochlear implants, implanted sensors, etc.), find advantages in the glass FTs segment.

Table 4.1 Ceramic and glass FTs in AIMDs applications

	Ceramic FTs	Glass FTs
Advantages	Robust Fits most applications Adapted to automation Long term reliable	Lower cost High density possible Resistant to corrosion Easier to miniaturize
Disadvantages	Expensive Difficult to miniaturize Max 10–12 wires (some exceptions)	Fragile Slow propagation of cracks Flexible wires only Poor for high current
Specificities	High voltage (ICDs) High current (ICD)	High density (neuro, BCI) Miniature (sensors, etc.)
Trends	High volumes Automated assembly	Special applications Miniaturization

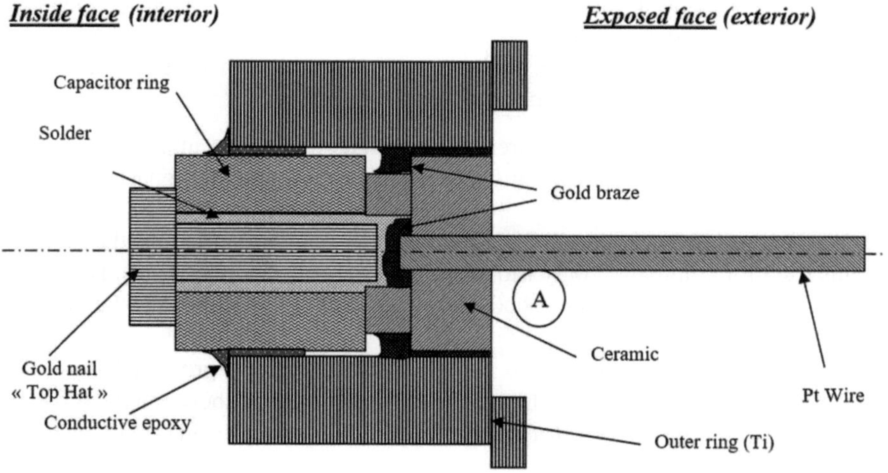

Fig. 4.14 Filtered FT (single wire)

4.9.1.7 Filtered FTs

In most cases, metallic encapsulations provide an adequate shield to incoming electromagnetic waves, especially in the range of radio frequencies (RF). Unfortunately, the sensitive electronics of AIMDs is in contact to the external electromagnetic environment through the appropriately called FTs, connected to stimulation or sensing leads. These leads behave like antennas and carry a certain quantity of electrical energy, collected by the "antennas," to the inside of the implanted device.

Until the 1990s, the impact of electromagnetic disturbances (EMD) on pacemakers were negligible because:

- The electronics of the implants were simple and robust.
- Pacing was the main function; sensing was only rare.

- The electromagnetic smog was small compared with today (no mobile phone, no wireless consumer products).
- MRI was not yet widely spread in hospitals.

Today, the situation of AIMDs regarding EMD is different:

- Many devices measure tiny signals (a few mV or even μV) in various locations in the body. These small signals need to be amplified to extract useful information. Leads are susceptible to absorb surrounding electromagnetic waves and noise, which also enter the amplifier.
- Over a large spectrum, the electromagnetic smog is increasing exponentially in time, most of it due to cell phones, wireless links, and anti-theft systems in shops. A large increase has been seen in a range of wave lengths that correspond to the length of implanted leads, which, therefore, may resonate and absorb a maximum of energy.
- MRI has become a widespread diagnostic tool, with high-intensity magnetic fields. Several adverse events with patients wearing an AIMD have been documented. Due to the nature of their diseases, patients with neurostimulators or BCIs are especially prone to be investigated with MRI.

In consequence, all manufacturers try to reduce the above impacts by adding filters at the lead connections to remove incident electromagnetic signals outside the measuring spectrum.

The first actions were to add filters on the PCB or hybrid circuit of the implant. Leads used for stimulation only (no sensing) were protected by diodes. But this is somehow "too late" as the unwished energy has already penetrated in the titanium shield, through the FTs.

Ideally, the filter should canalize the unwanted noise on the grounded shield. So, the best physical location of the filter is at the level of the FT, between the wire and the flange. In the mid-1990s, Medtronic started to weld (by hand) small capacitor chips on the side of the FTs, between the wire and the flange.

The ultimate configuration is to integrate the filter in the FT. Medtronic patented and developed such an integrated filtered FT more than two decades ago [16, 17]. The principle is described in Fig. 4.14.

This filter configuration is a simple capacitor (order of magnitude 10 pF to 1 nF) connecting the wire to the grounded flange. It will therefore filter away the high frequencies. Unfortunately, such a first-order passive filter has a poor rejection rate and no sharp cutoff frequency. This filtered FT offers some basic protection for pacemakers but is not enough for highly sensitive sensing leads (like in some advanced neuro-applications and BCI).

More sophisticated filters (LC filters, active filters) cannot be integrated in the core of the FT. There are numerous patents describing filtered FTs, some including inductors or active components. Some manufacturers also study the ways to integrate the filter, or part of it, in the lead or in the connector block. The main difficulty, if elements of the filter are outside the titanium case, is to preserve biocompatibility.

In principle, filters are needed only for sensing leads or combined sensing/stimulating leads. Leads stimulating only (no sensing function) have their output drivers well protected by diodes. Most cardiac pacemaker leads are bidirectional and therefore require some filters to protect the electronic. Neurostimulation is mostly unidirectional (stimulation only) and is less critical in terms of resistance to electromagnetic disturbances. BCI electrode arrays collect tiny signals and are therefore connected to high gain amplifiers, rendering the device even more sensitive to EMD. Electrical protection of BCIs will be covered in more details under Sect. 4.11.2.

Considerable progress has been done to make AIMDs MRI compatible. Improved FTs which integrate filters and protections are part of the strategy for MRI compatibility. See Sect. 4.11.3 for more details.

4.9.1.8 Single Wire or Multiple Wires

We have discussed above so-called single wire FTs, with only one conductor surrounded by insulating material. The evolution of the pacemaker industry rapidly induced the need for two or four conductors per device. Manufacturers took two fundamentally different design approaches to meet the needs of four connections per pacemaker:

- *Side-by-side*: design including four single wire FTs, place side by side, in a raw (see Fig. 4.15). They are located at a lower area of one of the shields, perpendicular to the surface. The FTs are laser welded on the top shield, prior the assembly of the pacemaker. This offers the advantage of making possible a leak test of the shield with four welded FTs, before starting the assembly of the device. Having four separate FTs has the disadvantage of occupying more space on the surface and more volume inside the pacemaker to connect to the electronics. This configuration is one of the keys of the success of Medtronic in automating the assembly lines. The manufacture of the component "single wire FT" has also been highly automated. The cost of four such parts is much lower than a quad FT. This "side-by-side" strategy was later extended to devices with large numbers of channels (up to 32). The footprint and volume occupied by FTs become then prohibitive.
- *Multiwire FTs*: manufacturers of pacemakers also use two dual wire FTs or a quad FT, placed on the flat top of the can, squeezed between the two clamshell-shaped shields. When all the components are assembled and connected in the pacemaker, the two shields and the FT(s) are laser seam welded in one single operation, under cover gas or in a glovebox. This configuration permits smaller and compact designs, but it is more complicated in terms of assembly. It is also more difficult to automate and more expensive.

Most of the suppliers of FTs have also developed "in-line" FTs (see Fig. 4.16), with a variety of shapes and number of wires (4–15) in one or two rows. The ferrule or flange must be machined at the right shape, generating high costs.

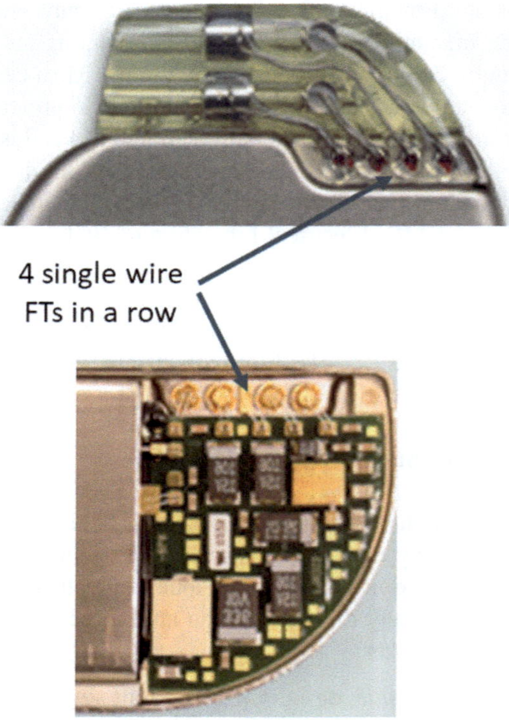

Fig. 4.15 “Side-by-side” configuration, four single wire FTs

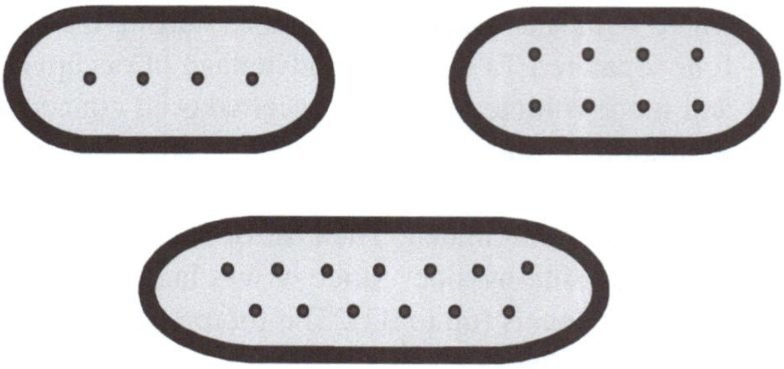

Fig. 4.16 “In-line” FT configurations (4, 8, 13 wires)

For very high numbers of connections (20+), like neuro- or functional stimulators, we can identify four trends:

- Miniaturize single wire FTs and weld them side-by-side on the shield or in arrays.
- Use several high-density multiple wires FTs (mainly glass FTs).
- Design special FTs, based on flat connectors placed in a row, like the pins or pads of ceramic encapsulated ICs.

4.9.1.9 Wire Conductors or Shaped Pins ("Top Hat")

Traditionally, the conductors of FTs are made of long flexible wires, which are trimmed at length and formed to be welded on the electronics (PCB or hybrid circuit) inside the device and on connector blocks outside the encapsulation. It is well adapted to manual assembly and permits a lot of complex designs.

Until recently, the wires were resistance welded on gold blocks (placed on the hybrid) and directly on the connector blocks on the header side. It consumed a lot of labor and was sometimes difficult to inspect or pull test.

In order to reduce assembly costs, manufacturers (Medtronic first) introduced step-by-step some automation or more flexible processes:

- "Top hat": replacement of the wire conductor on the interior side by a round shape block of gold. On the exterior side, the conductor remains a wire (see Fig. 4.14). As the four FTs are welded perpendicularly to the shield, the four gold blocks are placed at a small distance of the corresponding gold blocks on the hybrid. Connection between the gold blocks is then done with an automatic wire bonder, as for connecting an IC to its board. Double bond per connection is often used to increase reliability. Automatic pull test is done 100%, straight after bonding.
- "Grid": the automatic wire bonding for linking the FTs to the electronic is replaced by laser welding a grid. The grid consists of four ribbons linked with small bridges. After welding the four ribbons at both ends, then the laser cuts the bridges.
- Extension of the grid concept on the external side of the pacemaker, to connect the four connector blocks to the FTs. For this, the FTs should have a "top hat" on both sides. This would be a FT without wire, just two gold blocks linked by a pin.

Another possible evolution of the design of FTs, with the objective to be better adapted to automation, is to replace the flexible wires by rigid pins or flat ribbons. The challenges there are numerous, as rigid pins, during the welding process, may damage the insulator of the FTs. Design for automation is also more challenging for multiple wires FTs or for small miniaturized FTs.

4.9.1.10 Various Types of Flanges

There are two main configurations of flanges:

- *Single lip flange*: (Fig. 4.17) fits in a hole in the casing (type of assembly used by Medtronic) or a hole in a base plate. The flange overlaps the casing for subsequent laser welding. Thickness and diameter of the overlapping part are important parameters for the quality of the laser weld and are dependent on the laser welder characteristics (focal length, spot size, shooting angle, etc.). Usually, single lip flange FTs are welded on the can (or base plate) early in the assembly process. Seam weld of the can comes at later stage.
- *Double lip flange*: (Fig. 4.18) the two halves of the clamshell-type casing are inserted between the two lips. It is the design of choice for pacemakers with dual

or quad ceramic FTs. The FT itself has a "self-fixturing" function as it facilitates the positioning and alignment of the shells. Alike single lip flanges, only the top overlap is laser welded on the casing. Usually, welding the FT(s) is done in the same laser welder and at the same time as welding the two halves of the can. Some manufacturers weld the FT(s) first and then seam weld the can. Some weld the FT(s) after the can. The cost of double lip flange is higher than the one of single lip flanges, due to the difficulty to machine the high precision groove between the "lips."

4.9.1.11 Materials

Here below, a list of the various materials most commonly used in FTs. Some special applications may require more exotic materials. Any material outside the hermetic sealing must be biocompatible, even if they are covered by the header. The header is not hermetic. Moisture will creep through the plastic material, especially through silicone rubber and glue. So, materials will be exposed to some moisture after a while, and any non-biocompatible material may diffuse back to the body fluids.

Fig. 4.17 Single lip flange

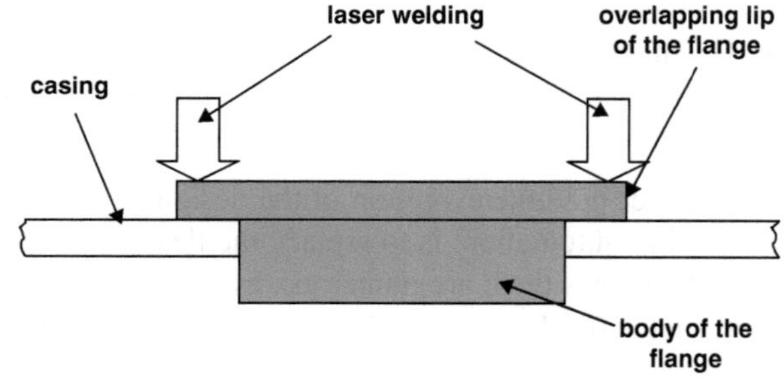

Fig. 4.18 Double lip flange

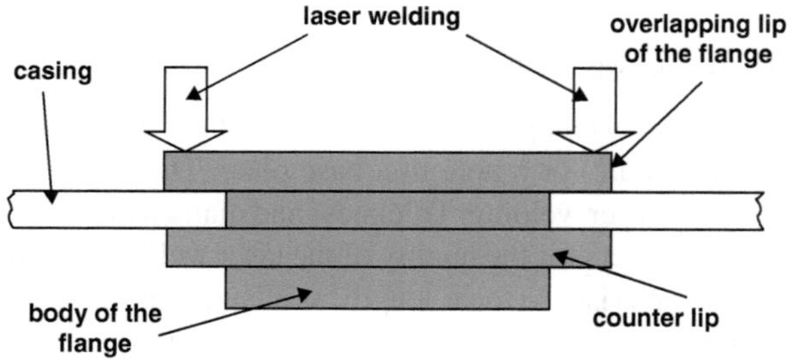

4.9.1.12 Insulators

- "Glass FTs":
  - Glass (fusion temperature in the range of 800 °C)
  - Barrier of sapphire, ruby, ceramic
- "Ceramic FTs":
  - Alumina Al_2O_3, 99% pure
  - Zirconia ZrO_2
  - Synthetic ruby (monocrystal of alumina)
  - Synthetic sapphire

Glass preforms are molded or cut out of glass tubes. Sapphire and ruby are extremely hard materials, which can only be machined by diamond grinding. Ceramic insulators are first sintered, but the shrink is too gross to give exact dimensions. So, the external diameter must be grinded to exact dimensions, and some other machining is needed for potential groves and edges and to bore the holes. Gold braze is a rather "forgiving" process, meaning that variations in the gap size can be partly compensated by gold penetrating in the gap. In any case, the insulator is an expensive part.

4.9.1.13 Conductors

- Pure platinum (Pt) 99.95%
- Platinum/iridium (90%/10% or 80%/20%)
- Niobium
- Tantalum
- Palladium
- Titanium
- Tungsten
- Gold (pins and "top hats")
- Some special alloys

The choice of material for the wires depends on the requirements of flexibility, formability, welding process and parameters, electrical resistivity, cost, and so on. It is also possible to use stainless steel (316L) as long the current to be carried is modest. Note that the conductive material on the external side of the FT must be biocompatible. The rigidity of stainless steel might be an issue, creating stress of the welds, by "spring effect."

One possibility is to have a wire with a highly conductive core (silver or copper) and an external mantel of biocompatible material. But then, the mantel may be damaged when welding on connector blocks, exposing the core material. This would only work with crimping instead of welding. Gold wires may be a possibility.

4.9.1.14 Flanges

- Pure unalloyed titanium (Ti), grade 1 or 2
- Ti-Al-V alloys, grade 4 or 5
- Niobium

Pure Ti is rather difficult to machine, so some manufacturers use alloyed Ti (grade 4 or 5), which is harder. But this has an impact on the laser welding parameters.

4.9.1.15 Other Materials (Filters, Coating, Brazing, Protection, etc.)

- Brazing material:
  - Gold 99.99%
  - Some special alloys (Ni/Au, In/Cu/Ag, etc.)
- In filtered FTs:
  - Special ceramics or tantalum for capacitor rings
  - Epoxy and conductive epoxy
  - Solder
- Coating of the wires:
  - Gold flash or other metallization to facilitate welding
- Protection of the ceramic:
  - Epoxy
  - Parylene

As this type of protection does not stand the high temperature of laser welding, the addition of a protective layer is usually done after welding.

4.9.1.16 Assembly on the Device Casing

The external face of the FTs is not in direct contact with the body fluids. The connecting head of the device covers and protects the FTs. But the level of protection varies much depending on the header technology (see Fig. 4.19). Here are some examples of assembly of the header on the housing:

I. *FTs perpendicular to the housing, glue-on-header*:
   - Cavity "A" on Fig. 4.13 filled with epoxy prior attaching the header
   - Injected header, glued on the housing with silicone adhesive
   - The wires are bent in groves on the surface of the header and welded to the connector blocks. After welding and pull test, all the openings and cavities

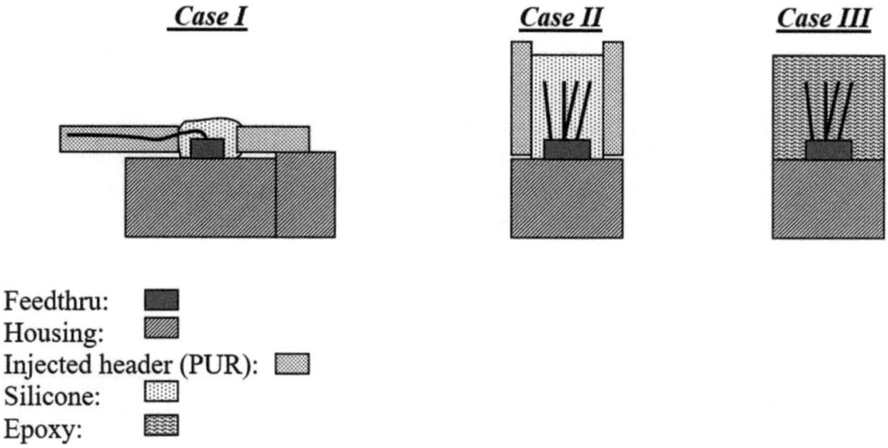

Fig. 4.19 Three configurations of FTs protection

in the header are filled with silicone (injected with a syringe). The groves (where the wires are) are also filled with silicone.

- The wires and the FTs are covered with a thin layer of silicone and therefore not well protected against moisture.

II. *Multiwire FT(s) perpendicular to the edge of the housing, glue-on header*:

- In majority ceramic FTs, with two or four wires.
- Cavity "A" on Fig. 4.13 filled with epoxy prior attaching the header.
- Injected header, glued on the pacemaker with silicone adhesive.
- The header is inserted above the wires and FTs, protecting the FTs on all sides. The wires are formed and routed on a side of the connector blocks.
- After welding the wires on the blocks and pull test, the cavity around the FTs is filled with silicone (injected with a syringe). The opening on the side of the header, where the wires have been welded to the blocks, are also filled with silicone and/or closed by a PUR lock glued with silicone adhesive.
- Compared to configuration "I," the FTs are much better protected against moisture.

III. *Multiwire FT(s) perpendicular to the edge of the housing, cast epoxy header*:

- Ancient configuration sill used by some manufacturers for specialty products. It is also the design of choice for highly complex connections or products manufactured in small quantities, like neuro-devices and BCIs.
- The wires are welded first to the connector blocks.
- The pacemaker is then placed in a one-time silicone mold and epoxy is poured over the connector blocks, wires, and FTs.
- The FTs are totally encapsulated in epoxy, providing a barrier to moisture several orders of magnitude better than silicone.

- This over-molding or cast method is also appropriate when the electrode cables are directly attached to the FTs, without connector.

Alternative *III* is clearly the best regarding moisture protection, corrosion, and therefore reliability. But it is more difficult to automate and requires much labor.

Glue-on headers (*I* and *II*) are injected in polyurethane (PUR), with cavities in which connector blocks (of Ti or stainless steel 316L) are inserted. In both *I* and *II*, there are openings on the side of the header to access the area where the wires are welded on the blocks.

Even being totally encapsulated in epoxy, FTs used in *III* need to be biocompatible, as epoxy is not a perfect barrier to moisture, which will diffuse slowly through the header with time.

In the three cases, the connecting pin of the stimulation electrodes is maintained by a so-called set screw in the connector block. The set screws are fastened in position by the surgeon during the implant procedure. The screwdriver reaches the screw through a soft silicone septum, which will seal back when the screwdriver is pulled out. The septum is a weak point regarding moisture protection, meaning the blocks are likely to be rapidly exposed to moisture, which in turn will creep along the wire, down to FT area.

Wires are usually shaped to be well positioned with regard to the connector blocks and internal connection pads. For this, they must be flexible and deformable. After welding, both internally and externally, wires are pulled to test the quality of the weld. This is the only mechanical stress, beside some induced by the welder and the fixtures. In the finished product, there is almost no stress on the wires. Therefore, there are no specific requirements regarding fatigue of the wires.

4.9.1.17 Connection of the Wires Inside the Casing

How the wires are connected to the electronics inside the hermetic housing has already been briefly discussed above.

Until recently, most designs were based on rather long flexible and formable wires (Pt) which were shaped and routed in place by hand with tweezers. Then, the wires were resistance welded (mainly parallel gap resistance welding), usually on gold blocks or pads located on the hybrid circuit. Quality of the weld was checked by visual inspection and manual pull test. Some manufacturers still use this approach, which is labor intensive. This method remains acceptable for small volume production, like neuro-devices.

Already in the 1990s, Medtronic introduced a more automated connection method for dual chamber pacemakers, with four single wire FTs with gold "top hats" on the inside of one of the shells. Then, the hybrid or PCB was placed beside the FTs, and an automatic wire bonder (thin gold wires) was making the connection between the gold "top hats" and the gold blocks on the circuit. Pull test was done automatically with the same machine. It was even possible to make two wire bridges per connection, to increase reliability. This concept remains the best solution in terms of cost, assembly time, and labor, but the use of four FTs side by side consumes more space. This is also the best example of a design where FTs were adapted to the assembly processes. The concept has been later extended to neuro-stimulators with up to 32 single wire FTs.

More recently, several newer approaches have been tested or even implemented, like replacing the gold bonds by a grid of ribbons. Both ends of the ribbons are laser welded, on one side to the top hat or the wire of the FT, on the other side to the electronics. Then, the bridges, holding the ribbons together, are cut by the laser.

4.9.1.18 Suppliers of FTs

In the early days of the pacemaker industry, a large majority of FTs was supplied by Greatbatch Inc., known today as Integer [18]. This company was founded by Wilson Greatbatch [19], one of the pioneers of the implantable devices industry. For several decades, Greatbatch Inc. supplied FTs for almost all manufacturers, with exception of Medtronic, manufacturing their own FTs (mostly under license of Greatbatch). Later some competitors appeared in the field of ceramic FTs, like Morgan Technical Ceramics [20], SCT (Société de Céramiques Techniques) [21], or Hermetic Solutions Group [22].

Half a dozen companies supply glass FTs to the medical industry.

Section 7.1 will include the description of breakthrough new FT technologies which will be the key for miniaturizing future above-the-neck BCI implants.

4.9.2 Hermetic Sealing

Hermeticity is never perfect. As stated by Prof. Anne Vanhoestenberghe (University College London) [23–25], “Ultimate hermeticity is practically unachievable. What matters is that your package is sufficiently hermetic for your application.” In consequence, the fundamental question to be answered by designers of AIMD is:

“What is sufficiently hermetic for my application?”

The level of hermicity of the housing is only one parameter of the overall moisture control (see Sect. 4.9.4). We also need to understand what is happening inside a sealed housing. As soon as a hermetic housing is sealed around an electronic circuit and a battery, the inside of the box becomes a confined environment which is far from static and stable. Dynamic chemical, physical, and even biological phenomena will change the characteristics of this confined environment. This evolution may last for years. The first step to keep these changes under control is to have a proper hermetic sealing of the housing, preventing exchanges of gas and liquid across the envelop, in both directions.

As described above, hermetic sealing is a capital element of moisture control, corrosion, and leakage prevention related to the content or the inside of an AIMD

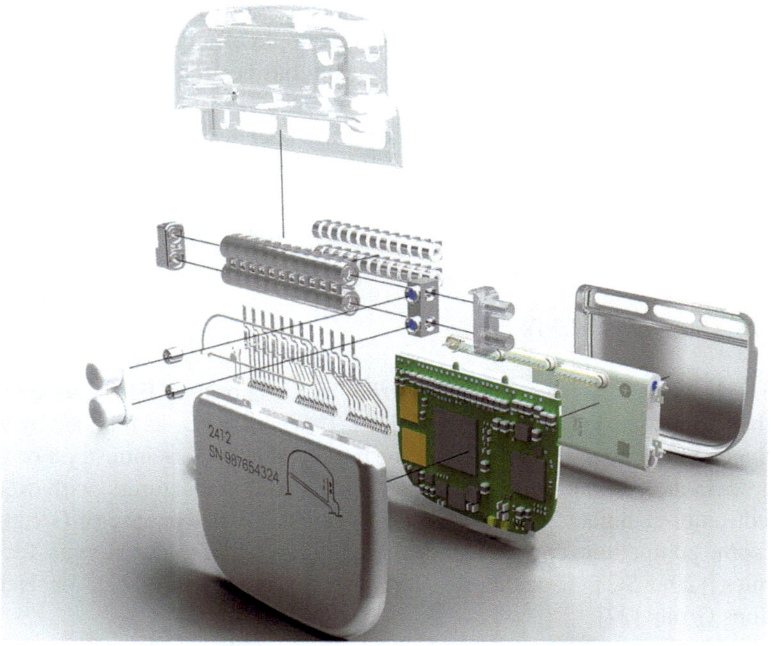

Fig. 4.20 Algovita, example of a modern design of neurostimulator, with titanium housing and 2 × 12 in-line connectors. (Courtesy of Nuvectra Inc.)

housing. Hermetic sealing is the process which allows the closure of housings around the electronics/battery subassemblies.

Most of the neurostimulators on the market today are inspired from the pacemaker technologies, with a two-shell titanium housing, FTs, and modular electronics, as seen on Fig. 4.20.

Housings consist in the following components:

- At least two shells made of intrinsically hermetic materials (metal, ceramic, sapphire, glass).
- In most cases, FT subassembly(ies) with insulated wires or pins.
- Alternatively, FTs or vias might be integrated in one of the shells.
- The shells may also include other elements, like windows for optical or RF communication, passages for fluids or deformable structures.

Two main assembly configurations are used in the industry:

- FTs pre-assembled in one of the shells (Fig. 4.21a). See also Fig. 4.15.
- FTs are squeezed between the two shells (Fig. 4.21b).

There are two fundamentally different sealing processes for hermetic housings:

- *Welding* the seam between the two shells by melting the material of both shells, at their interface, without additional joining material. Local melting is created by a focused laser ray, in a succession of spot welds with overlap. The local

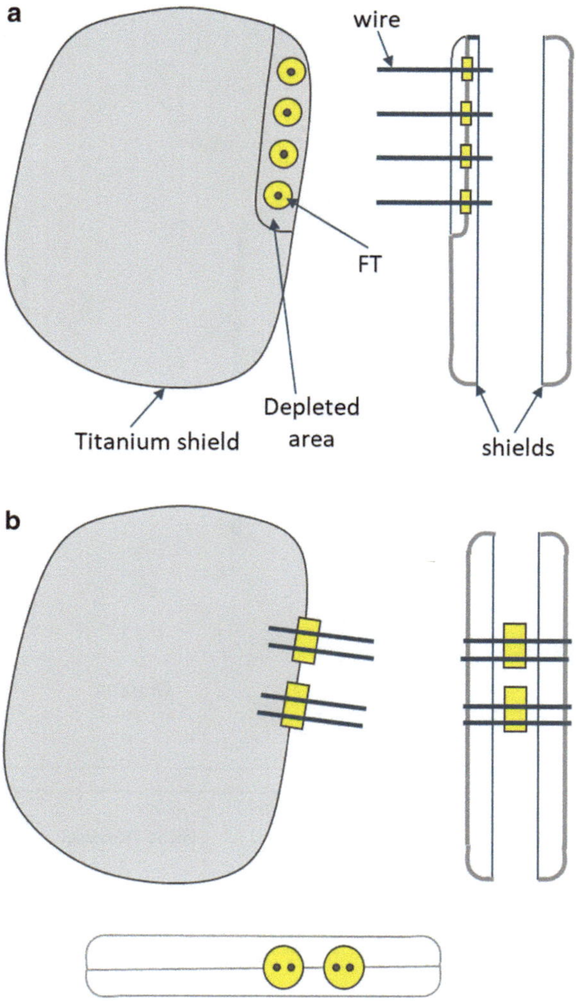

Fig. 4.21 (**a**) FTs pre-assembled in a shield. (**b**) FTs inserted between the shields

temperature at the focus point of the laser ray is appropriate to fuse the material, but the global heat remains low and prevents any damage to the electronics inside the can. Some examples are:

(a) Ti-Ti weld with a pulsed yttrium aluminum garnet (YAG) laser ray focused at the edge of the gap between the two shells or along the edge of the Ti ferrule surrounding FT subassemblies. In some configurations, an overlap is provided to facilitate welding (Fig. 4.22). >99% of AIMDs are sealed this way.
(b) Glass-glass weld with a femto laser (Fig. 4.23). The laser energy is shot through the glass cover and focused at the interface, locally melting the

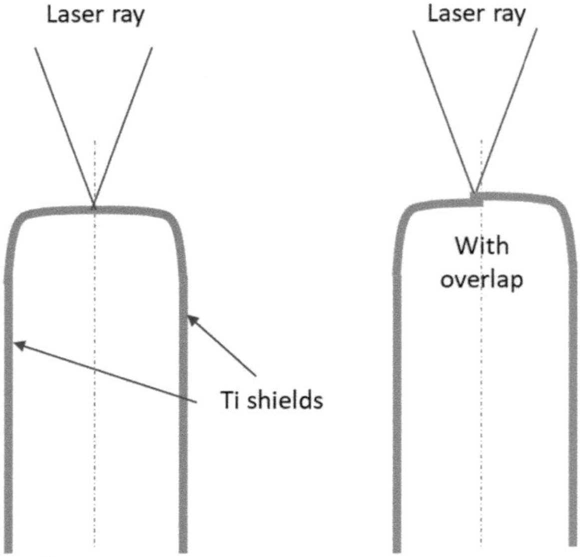

Fig. 4.22 Edge-to-edge and overlapping shields laser seam welding

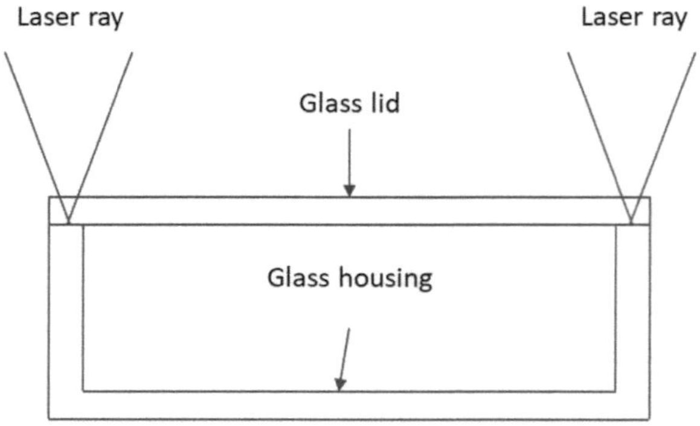

Fig. 4.23 Glass-on-glass laser seam welding

glass. Alternatively, the same process can be used for joining a glass shell or lid to silicon or sapphire substrates. Primoceler (Finland) [26], recently merged in the glass company Schott (Germany) [27], is leading the field of glass encapsulation for medical devices.

- *Brazing* (additional brazing material added between the two shells) in an oven. A gasket, preform, or paste of brazing material is squeezed between the two shells, and the assembly is introduced in an oven at a temperature high enough to melt the brazing material. Three categories must be distinguished:

  (a) Brazing of subassemblies without electronics inside (e.g., a ceramic FT in a Ti shell): gold brazing (Fig. 4.24). The high fusion temperature of Au (about 1060 °C) excludes insertion in the oven of any electronic parts. Gold brazing is a mature and stable process used since decades in the AIMD industry.

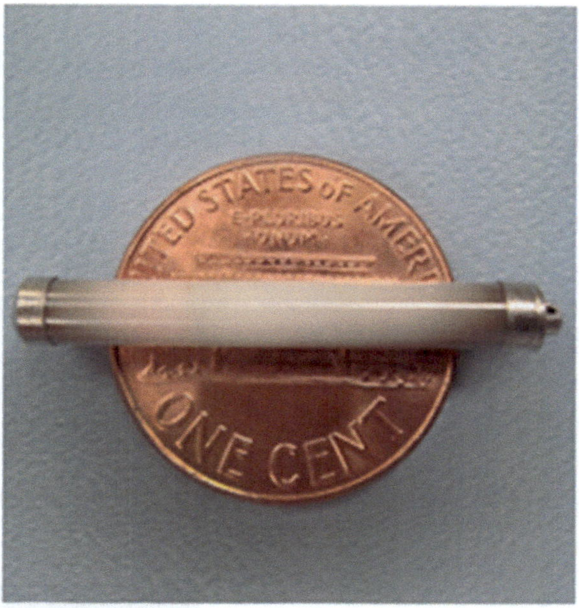

Fig. 4.24 Cylindrical brazed ceramic-Ti hermetic encapsulation, battery-less, Bion. (Courtesy of Alfred Mann Foundation)

The gold joint is biocompatible, biostable, and resistant to corrosion. Gold brazing is a forgiving process, as melted gold may fill the gaps even if there are some misalignments.

This process is also used to add a Ti ring or flange to a ceramic shell, later laser welded to the other shell or lid as described above. This process has been used since two decades, for example, by the Alfred Mann Foundation (AMF) [28] for a FES device called Bion (Figs. 4.24 and 4.25) [29]. The technology has been further developed by Hermetic Solutions Group [30], and some human products are on their way to approval, like a 32-channel battery-less device (Figs. 4.26, 4.27, 4.28, and 4.29) from Ripple Inc. [31].

(b) Brazing of housing incorporating electronic boards: low-temperature brazing (around 250 °C). As the electronics is already inserted in the device prior to brazing, high-temperature gold brazing will damage PCB components. Therefore, the temperature of the brazing oven is kept to low temperature. Unfortunately, these brazing alloys contain non-biocompatible materials (Sn, Ag, In, Pb, etc.). To my knowledge, there are no low-temperature brazing materials proven to be biocompatible on the long term. Using such brazing materials breaks the law of "only biocompatible material outside the hermetic housing," as part of the brazing join is facing outside. Manufacturers using this technology claim that they provide a good protection over the join (silicone rubber, Parylene, epoxy, or a combination of them). Long-term stability of these coatings remains to be proven. It may delaminate, erode, or be damaged during the implantation

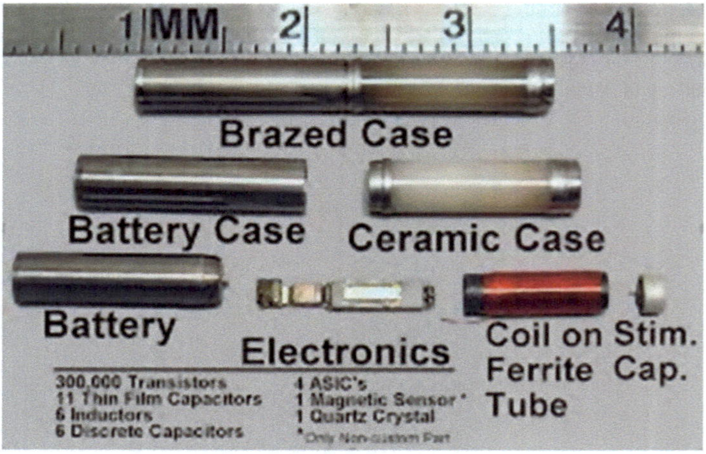

Fig. 4.25 Cylindrical brazed ceramic-Ti hermetic encapsulation, including battery, Bion. (Courtesy of Alfred Mann Foundation)

Fig. 4.26 Example of flange brazed on ceramic. (Courtesy of Ripple Inc.)

procedure. More discussion will be done in Sects. 4.9.6 and 4.9.7. Note that the process of brazing in an oven is limited to battery-less devices, as implantable batteries cannot be exposed to temperatures above 55–60 °C. Other electronic components like supercaps or electrolytic capacitors are also excluded of oven brazing assemblies. Pioneer work in this technology has been done at IMTEK [32], University of Freiburg, Germany, in the labs of Prof. T. Stieglitz [33]. M. Schüttler [34, 35] from CorTec GmbH [36] followed up. CorTec developed Brain Interchange (Fig. 4.30) a multichannel read/write device encapsulated in low-temperature-brazed ceramic.

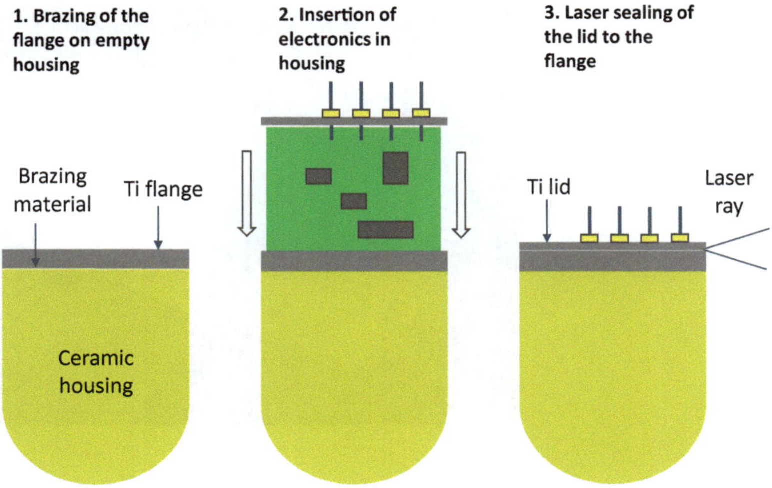

Fig. 4.27 Brazed TI flange on ceramic, laser sealing of the lid

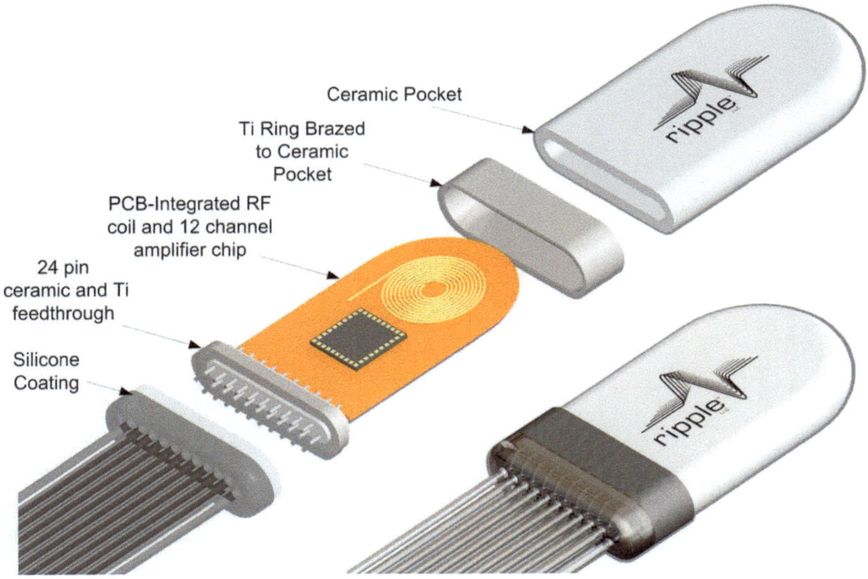

Fig. 4.28 Example of brazed ceramic-Ti hermetic encapsulation. (Courtesy of Ripple Inc.)

(c) Biocompatible brazing material like Au or Pt melted by local heating (focused laser energy) instead of being inserted in an oven (see Fig. 4.31). It merges the advantages of both (a) and (b) above: biocompatibility and possibility to seal a ceramic housing containing an electronic and even a battery. One or two companies have been successful in melting Au join by shooting laser pulses through a glass lid.

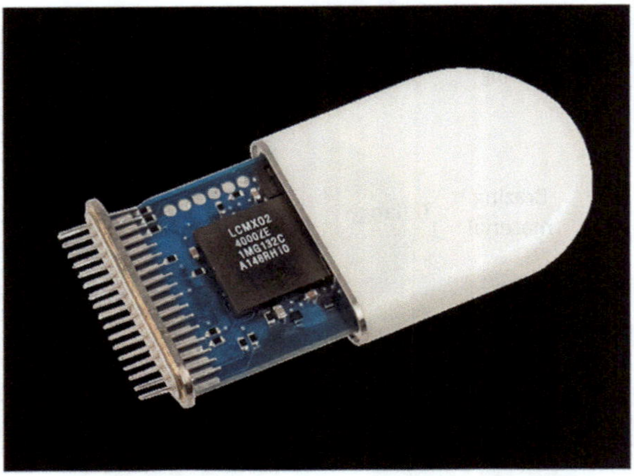

Fig. 4.29 Example of brazed ceramic-Ti hermetic encapsulation. (Courtesy of Ripple Inc.)

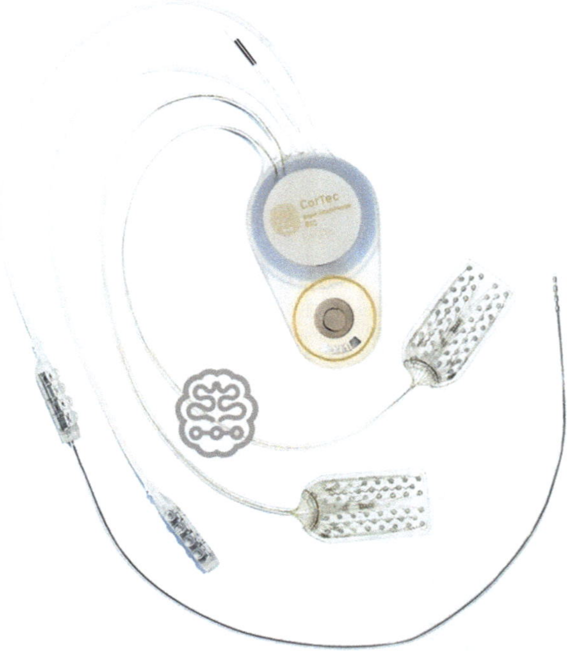

Fig. 4.30 Brain Interchange in a configuration for close-loop DBS. (Courtesy of CorTec GmbH)

Regarding electromagnetic transparency, several alternatives to titanium have been tested in various configurations. The first goal of these encapsulation methods is to permit RF communication through the envelop, keeping the antenna inside. The second objective is to preserve hermeticity, for long-term reliability. There are five main categories of electromagnetically transparent hermetic encapsulations:

- *Addition of a window in the titanium housing*: a ceramic, glass, or sapphire window is brazed in one of the titanium shells. The RF antenna is located behind the

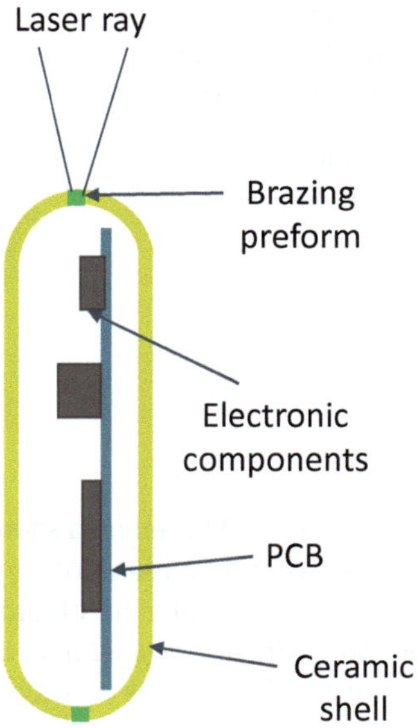

Fig. 4.31 Laser-melted biocompatible brazing joint for hermetic encapsulation

window, inside the housing. As the window is brazed in the titanium shield before final assembly and laser seam welding, the brazing process can be done at high temperature. Gold is the material of choice to properly seal the window in its titanium flange. An example of such a windowed device is SBNC [37] (Fig. 4.32), a wireless BCI interface for cortical recording, tested on animals at Brown University, Providence, Rhode Island, in the labs of Prof. A. Nurmikko [38]. On Fig. 4.32, the RF antenna is the blue component in the center of the window. The surrounding coil is used for inductive energy transfer. The presence of metal (copper coil and titanium housing) around the RF antenna influences deeply the electromagnetic performances of the radio system. This specific design cannot be considered as totally electromagnetically transparent. The window facilitates RF communication but does not create an optimal surrounding for the implanted antenna.

- *Hybrid ceramic-titanium*: a part or most of the housing surface consists in a ceramic surface which let RF in and out. The edge of the ceramic shell is brazed to a titanium ring or flange. Brazing is done prior final assembly and can therefore be done at high temperature (usually gold brazing). Then, the electronics is introduced in the ceramic-titanium flanged shell, and a titanium lid is laser seam welded with the same technologies developed for titanium encapsulations. This hybrid configuration combines the advantages of the transparency of the ceramic and the robustness of the Ti-Ti laser seam weld. See Figs. 4.24 and 4.28. As for the window concept, presence of titanium introduces perturbations in the electromagnetic environment which are susceptible of making RF communication more difficult. Influences of metal parts on the RF depend on the location and the directivity of the antenna, the frequency, the various reflection interfaces, and the

Fig. 4.32 SBNC, 100 channels wireless BCI, with sapphire window. (Courtesy of Prof. A. Nurmikko, Brown University)

materials. The antenna should be as "free" as possible from metals in the vicinity. The Bion configuration (Fig. 4.24) is favorable, as only the extremities of the ceramic cylinder are obturated with metal.

- *Sapphire housing with laser-melted brazing joint*: the electronics is enclosed in two hollow shells made of monocrystalline sapphire (Al_2O_3). A metallic preform (Au or Pt) is placed between the shells, as a gasket. This metal preform is locally fused by a laser ray shot through the transparent material and focused at the interface. It provides hermetic sealing with only local temperature increase, preserving the integrity of the electronics inside the housing. Such an assembly is both hermetic and transparent to RF waves. Nevertheless, it presents a disadvantage. The metallic-fused gasket is a highly conductive ring in short circuit, which will induce counter-electromagnetic field, limited radio communication, and potentially generating heat. Sapphire is very inert and biostable when exposed to body fluids. As sapphire is extremely hard, hollow shells must be done by diamond grinding, an expensive and not scalable process. A Swiss research institute, Centre Suisse d'Electronique et de Microtechnique (CSEM) [39], has developed this technology for miniature implanted laser sources (Fig. 4.33).
- *Glass-glass sealing*: (see Fig. 4.34a) as above, the electronics is enclosed in two hollow shells made of glass. No additional material is joining the two shells. An appropriate laser ray is shot through the glass and focused at the interface, locally melting the material and joining both parts. The seam weld consists in a succession of melted spots. This assembly has the advantage of avoiding the induction ring of the sapphire encapsulation described above, but glass is less biostable and more fragile than sapphire. Through glass vias (TGV) are made of tungsten wires pushed through the glass bottom following a proprietary process. This type of FT differs from the glass FT described in Sect. 4.9.1 in the sense that the tungsten wires are inserted in the bottom glass plate at a temperature approaching the melting point of glass, when glass is in a plastic paste-like phase.
- *Low-temperature brazing*: two shells of ceramic, or one ceramic shell and a metallic lid, are brazed together with a low fusion temperature alloy. As the

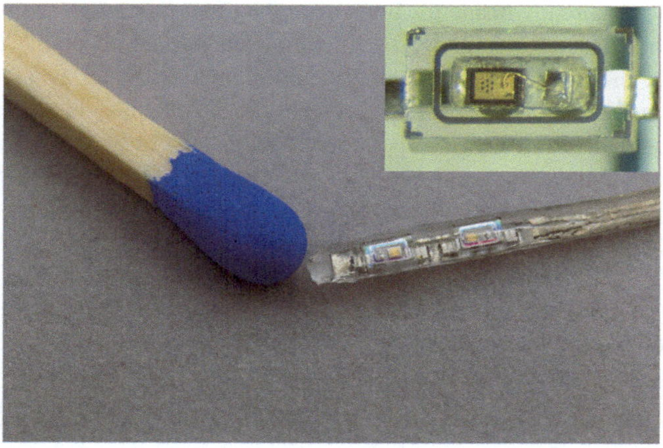

Fig. 4.33 Sapphire hermetic encapsulation of a laser source (<1 mm long). (Courtesy of CSEM)

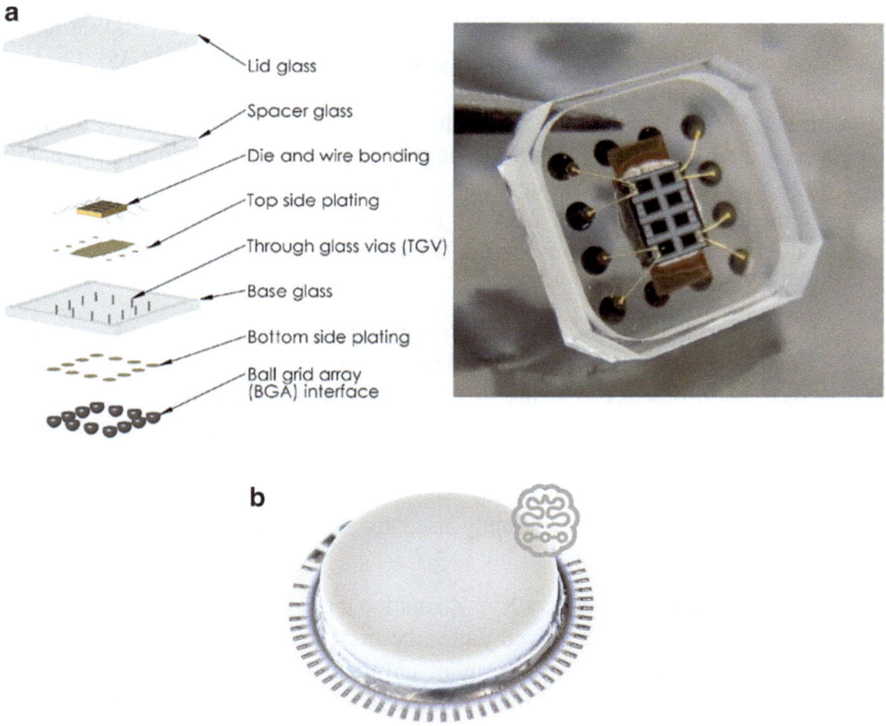

Fig. 4.34 (**a**) Glass encapsulation. (Courtesy of SCHOTT Primoceler Oy). (**b**) Ceramic encapsulation. (Courtesy of CorTec GmbH)

electronics is already inside the housing at the time of brazing, sealing temperature cannot exceed 200–250 °C (depending on the nature of the components inside the housing). Such housings cannot include batteries or electrolytic capacitors. The main drawback of this technology is the non-biocompatibility of the brazing material. Low-temperature brazing materials all contain non-biocompatible metals like lead (Pb), tin (Sn), silver (Ag), or indium (In). This assembly is hermetic and radiotransparent while presenting the same drawback (induction ring in short circuit) as the sapphire encapsulation. The main concern regarding this encapsulation is finding an appropriate protection of the sealing metal, avoiding long-term exposure to body fluids. With regard to RF communication and inductive charging, the brazing ring represents a source of losses, as it has been seen before for the sapphire encapsulation (Fig. 4.34b).

As we will see in Sect. 4.9.4, hermetic sealing must be done in a way which minimizes the quantity of moisture and oxygen trapped in the can. Prior to sealing the housing, a drying process called "bake oven" must be applied (described later under Sect. 4.9.4). When the content of the housing is judged as being dry enough for being sealed, then manufacturers have basically two choices:

- *Sealing in a glovebox.* A glovebox is a confined workbench accessible to the operator by putting hands/arm in airtight rubber gloves (see Fig. 4.35). For hermetic laser sealing encapsulation, the glovebox is an ultradry chamber, filled with helium (He) or a mixture of helium and argon (20%He–80%Ar) which is cheaper than pure He. This atmosphere is kept very dry down to dew points <−42 °C (100 ppm of water vapor). MIL-STD-883, Method 1013, even recommends a dew point of −65 °C. The glovebox has two antechambers with two doors each (one opening on the outside and one opening in the glovebox) (Fig. 4.35):
  - Incoming antechamber: also used for drying process. Assemblies to be sealed are placed in the chamber/oven and remain there for 12–24 h of drying, called "bake oven process," usually at 55 °C (maximum temperature tolerated by batteries) under a deep vacuum. When assemblies are dry, they are moved in the glovebox through the inner door and are laser welded in the glovebox, trapping dry He-Ar in the device.
  - Outcoming antechamber: simply used to extract the sealed units from the glovebox.

  In the center of the glovebox, a X-Y-Z-rotation table with appropriate fixture will allow the assembly to be moved under the laser ray for proper welding. It is a complex setting, with optical alignment and focusing capabilities. Ti soot resulting from the welding process must be filtered away and handled in airtight container (Ti soot is highly explosive when in contact with oxygen). The pressure inside the glovebox is maintained slightly (10–30 mbar) above the atmospheric pressure for two reasons:

  - The overpressure in the ultradry atmosphere prevents ingress of humid air from the room.

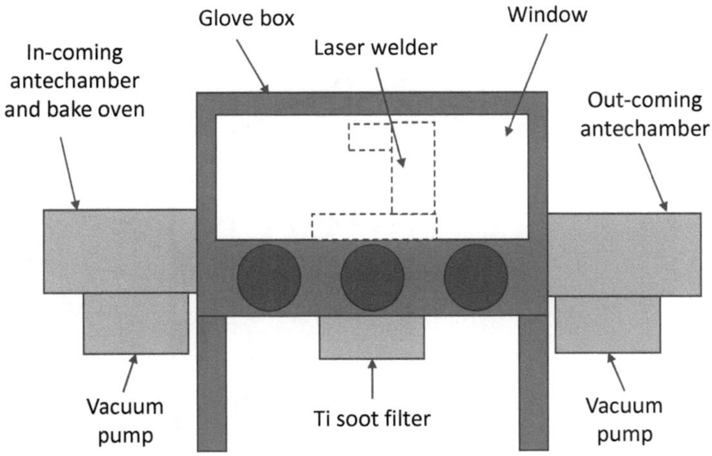

Fig. 4.35 Laser seam welding glovebox

- The gas trapped in the device after hermetic sealing will be at a small positive pressure with regard to atmospheric pressure, providing two advantages:
  - Facilitates gross leak test
  - Postpone penetration of body fluids in case of late opening of a minor leak

Laser seam welding in a glovebox is the most difficult and demanding process of the AIMD industry. Being able to keep tight controls on the bake process, on the level of moisture and oxygen in the chamber, and on the accuracy of the spot welding is the best promoter of quality and long-term reliability. Such a glovebox with laser, displacement table, cameras, filters, antechambers, cameras, and sensors is a large investment (1–3 M$) and requires expert knowledge to operate. Large manufacturers of AIMDs own such equipment and master their operation. Only a handful of original equipment manufacturer (OEM) companies provide this service to third parties and start-ups. It is very rare to see a start-up company making seam welding in their own glovebox.

- *Sealing under cover gas with a purge hole*. Compared to the glovebox approach, this process is cheaper but provides results which are less stable and less predictable. The to-be-sealed unit is placed in a fixture of the X-Y-Z-rotation system of the laser welder (Fig. 4.36). One of the Ti shell has a small hole, called "purge hole." The unit is then laser welded in room atmosphere, but a "cover gas," usually Ar, is blown on the laser spot welding area to minimize oxidation and nitrification of the fused Ti. After welding, the unit is placed in a small chamber (Fig. 4.37) for 12–24 h at 55 °C for vacuum drying (Step 1, Fig. 4.37). Moisture exits the device through the remaining tiny purge hole. At the end of the drying process, the chamber is filled with ultradry He (Step 2, Fig. 4.37). Dry He enters in the device through the purge hole. Then, the purge hole is hermetically sealed by shooting with a laser through a window of the vacuum chamber (Step 3, Fig. 4.37).

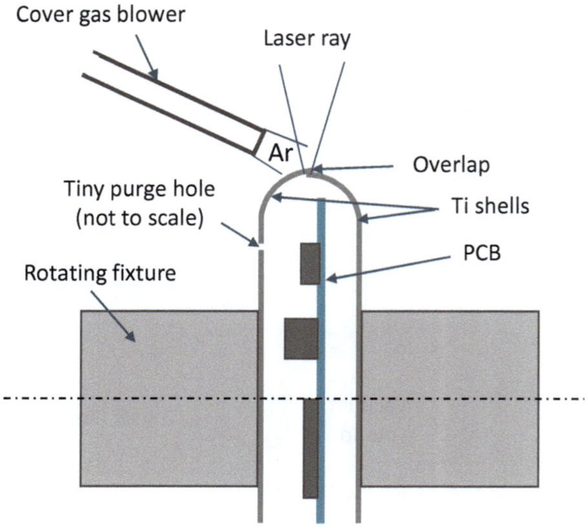

Fig. 4.36 Laser seam welding in room atmosphere with cover gas

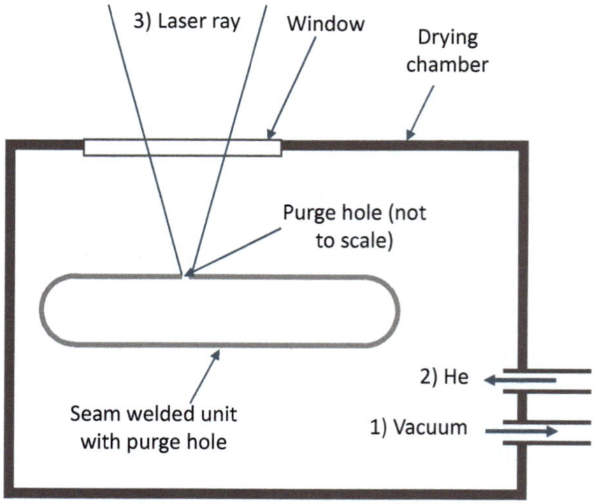

Fig. 4.37 Drying, admission of dry helium, and laser sealing of the purge hole

4.9.3 Leak Testing

Section 4.9.2 explained how devices were hermetically sealed. How do we assess and measure hermeticity after sealing?

Hermeticity, or rather lack of hermeticity, is measured as "leak rate." A leak can be defined as a default in the housing letting gas sip in and/or out. In literature and applicable standards, there are some confusions regarding units and definitions of leaks. A coherent approach would be to quantify the flow of gas through a hole in [mole × atm/s]. In fact, the industry commonly refers to leak rate referenced to leaking He at 37 °C in (cm^3/s) for a pressure gradient of 1 atmosphere (e.g., inside pressure of 1 atm, unit in a vacuum chamber). In industry, for being able to efficiently test units on the production line, we distinguish between two categories of leaks:

Gross leaks: We usually speak about leak rate above 1×10^{-5} cm^3/s as being gross leaks. This corresponds to 1 cm^3 every 28 h in standard conditions. A tiny hole of diameter 0.1 μm (one thousandth of a hair!), impossible to detect visually, even under binoculars, will leak at 4.6×10^{-6} cm^3/s or hundred times faster than acceptable MIL specs. The diameter of a molecule of water is 4×10^{-10} m. 250 molecules of water, side by side, fit across the diameter of this tiny 0.1 μm leak. These figures show how difficult it is to reach perfect hermeticity.

Helium trapped in the enclosure and measurement made by a mass spectrometer is appropriate to detect fine leaks (2×10^{-10} to 1×10^{-5} cm^3/s). For smaller devices, gross leaks will not be detected, as most or all He would have leaked out before setting the device in the fine leak detector. Visual inspection is not enough to detect gross leaks. MIL-STD-883, Method 1004, page 83, and MIL-STD-750, page 91, define the procedures to detect gross leaks. There are basically two main methods to identify gross leaks:

- *Bubble test* (nondestructive): aerospace leak test requirements (bubble test without bombing if internal pressure is >84.6 mbar above atmospheric pressure) and MIL-STD 1576 [47] are less stringent than MIL-750. If no positive pressure is entrapped in the device, a bombing (applying a substantial pressure of gas outside the device, forcing some gas through the leak) should be done. Regular bubble test (test condition D, MIL-750) is not recommended for internal volume less than 1 cm^3. Instead, we can use the fluorocarbon gross leak method, test condition C, MIL-750, page 93):
  - Reduce the pressure (part vacuum at <670 Pa) for >30 min.
  - Cover the device with fluorocarbon fluid before breaking vacuum.
  - Bomb at 5 bar during 2 h (or other bomb pressure/time combinations)
  - Remove pressure.
  - Wash and dry.
  - Immerse in type II fluorocarbon indicator.
  - Observe bubbles, if any.
  - Reject if two bubbles or more at the same spot.
- *Dye penetration test*, also called *Zyglo test* (destructive): used mainly for the analysis of failures or field returns. The device is immerged 3 h in liquid dye at 7 bar. Then, the device is broken open, and trace of dye inside is search under UV light and binocular. More details in MIL-750.

Fine leaks: Helium is used to detect fine leaks, because it is a very small molecule (the smallest after H_2), that will sip easily through the smallest leaks. It is also inert and rare in the air (4–5 ppm) avoiding false readings. Leak rates are measured by the volumetric flow by unit of time, in cm^3/s. Because of variation of the viscosity and atomic size, various gases flow at a different speed through the same leak.

For example, He leaks 2.8 times faster than air. For the sake of standardization, leak rates are measured in atm × cm^3/s of air at 25 °C, meaning the pressure differential is 1 atmosphere (1 bar). This unit is called "std cm^3/s." Standard leak test is defined as the quantity of dry air at 25 °C, in cm^3, flowing through a leak per second, when the high-pressure side is at 1 atm (760 mmHg) and the low-pressure side is <1 mmHg. The smallest leaks measurable with a regular industrial mass spectrometer are in the range of 2×10^{-10} std cm^3/s. To reach such sensitivity, the size of the vacuum chamber should be as small as possible. MIL-STD-883 sets the maximum leak, for devices with internal cavity of less than 50 mm^3 to 5×10^{-8} std cm^3/s.

The flow through a leak will be roughly proportional to the differential pressure. Nevertheless, it does not mean there is no flow when the pressure is identical inside and outside. Leaks must be considered as an exchange area, where the various gases on each side diffuse to the other side. The speed of diffusion depends on temperature and atomic mass. Even with zero pressure difference, He will leak out and water or vapor will leak in. It has been established that it takes only 12 days to exchange the gas of a 100 mm^3 cavity (pacemaker) with a leak of 1×10^{-7} std cm^3/s, with no pressure differential.

There are four basic methods to test fine leaks with helium:

- Vacuum pump the interior of the device toward the spectrometer and spray He on the outside. This is applicable to test FTs pre-assembled on a shield.
- Pressurize the interior of the device with He, and vacuum pump the chamber outside the device toward the spectrometer. Not applicable to sealed device.
- Seal the device in a glovebox containing He or backfill through a purge hole with He, as described in Sect. 4.9.2. Then, the sealed device is placed in a chamber, which is vacuum pumped toward the spectrometer. This method is used by the most manufacturers to test pacemakers and other IPGs. Note that the helium leak test must be done rapidly after sealing. Waiting too long presents the risks that most of the helium has leaked out and the He detector will not be able to identify the failure. Unless fine leak test is done immediately (within a few minutes) after sealing, a gross leak test must always be combined with a fine leak test and a visual inspection of the seal under binocular.
- The sealed device does not contain He. The device must then be "bombed," meaning exposed to He under high pressure for a rather long time, so some He will be pushed in, through potential leaks. Immediately after "bombing," the device is tested as above. If there are leaks, some He has penetrated in the device and will be detected (as described above) when it leaks out again. MIL-STD 750 (page 98) recommends bombing 2 h at 5 bar and testing for He leaks within less than 1 h after bombing. If the device does not stand such high pressures, less violent bombings are possible but over much longer periods: 23.5 h at 2 bar, 8 h at 3 bar, and 4 h at 4 bar.

The most famous adverse event regarding loss of hermetic sealing in AIMDs happened with cochlear implants, in 2004 [48], when several devices showed entrapped moisture in the range of 200,000 ppm (relative to 5000 ppm recommended as a limit by the standards). Note that the FDA reported values in the range of 200,000 ppm, but saturation moisture at body temperature cannot exceed 58,000 ppm. It is one more sign that the physics of moisture control is not always fully understood. One explanation of the discrepancy may come from the fact that RGA is done at 100 °C (see Sect. 4.9.5).

4.9.4 Moisture Control

The two previous sections describe how to build a hermetic encapsulation and how to test sealed devices. It is an important part in our efforts to control moisture levels in our devices. Inside the hermetic housing, electronic components occupy most of the space available. There is also an "empty" space between the parts, filled by He-Ar trapped inside the housing during the laser seam welding operation. The pressure inside the sealing glovebox is maintained at a few mbar above the atmospheric pressure, so that, in case of a minor leak when the device is already implanted, the leak flow will be from the inside to the outside, avoiding sucking body fluids in, at least for a certain time.

Ideally, the gas trapped inside the device should be as dry as the gas in the glovebox. But reality is different. The moisture content in the sealed empty space will keep growing for different reasons (Fig. 4.38):

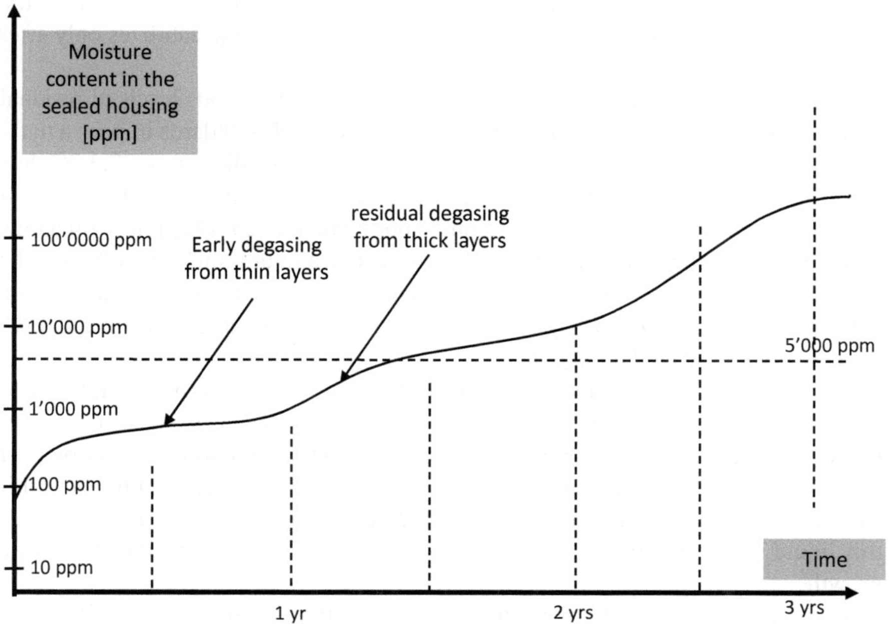

Fig. 4.38 Moisture increase in a sealed housing over time (due to various causes)

(a) Laser welding will heat and melt Ti locally and free some water molecules absorbed in or on the surface of the metal.
(b) Moisture may sip in through tiny undetectable leaks in the sealed housing (perfect hermeticity does not exist).
(c) Moisture absorbed in the housing, including feedthroughs and battery housing, may also degas slightly.
(d) Even if properly dried during the bake oven process, the electronic components (especially if packaged in epoxy or in silicone rubber) and the PCB substrate will continue to release some moisture (in most cases, the major post sealing source of moisture)
(e) H_2 may be released from metal-plated surfaces, like the gold or copper traces of the PCB, combining with oxygen trapped in the can and forming molecules of water.
(f) Components like electrolytic capacitors or supercaps are not hermetic and may release moisture and other gases. Including such components in hermetic housings is not recommended.
(g) Substantial sources of moisture are found in plastic retainers, nests, wire isolations, spacers, glove-top potting of bare dyes, backfill material under flip chips, and various glues (epoxy, silicone) used for the assembly.

The above potential causes of increased moisture content in a sealed implant have different time scale and importance, making it a complex dynamic evolution. It is capital, in the design phase, to minimize the potential sources of moisture. I have seen poor designs, with large relative volume of plastic materials inside the housing, leading to substantial post-sealing moisture diffusion. As most of the released moisture is linked to diffusion, the process is slow and stabilizes only after months or years of implantation.

In consequence, hermetically sealed exposures present various levels of residual moisture. For hermetically sealed electronic circuitry, MIL standards impose a maximum of 5000 ppm of water vapor, per volume, corresponding to a −2.3 °C dew point. 10,000 ppm corresponds to a +7.2 °C dew point and 20,000 ppm to 17.7 °C. At room temperature (20 °C), the maximum vapor content is 23,000 ppm, with a dew point of 20 °C, an absolute humidity of 17.5 g/m^3, and a saturation level of 20 °C. The correlation between pressure, temperature, and vapor content can be visualized on the dew point nomograph of Fig. 4.39. On this chart, we see that at 1.0 atm and 0 °C, the dew point is around 6000 ppm.

Above dew point, the moisture trapped in the enclosure is vapor. Compared to liquid water, vapor is less prone to create problems on the electronics. If the temperature passes under dew point, vapor condenses to form liquid water. Droplets of water will deposit anywhere inside the housing, on the electronics, or on other critical elements. Combined with potential ionic contamination or free metallic ions, this liquid water may induce short circuits, trigger corrosion, or facilitate dendrite growth.

Regarding impact of trapped moisture, the critical moment in the life of an active implant is between the hermetic sealing process and the implantation in the

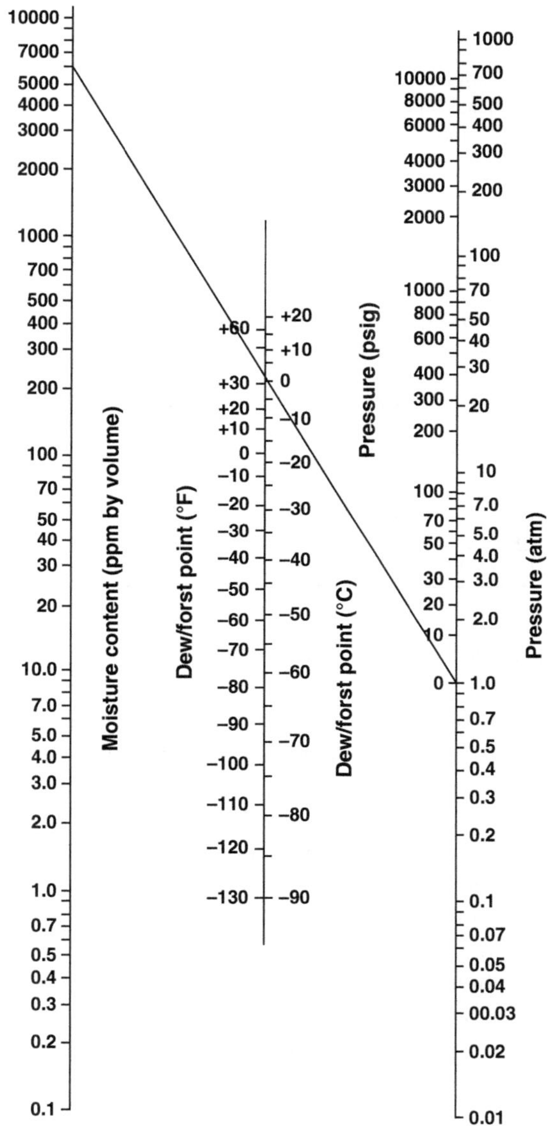

Fig. 4.39 Dew point graph, example at 1.0 atm, 0 °C, and 6000 ppm. (Source: Kevin Ely [49])

patient. During transportation from the factory to the hospital, the implant is likely to be exposed to a wide range of temperatures, including below freezing point, for example, waiting to be loaded in an airplane in wintertime. This is the reason why maximum moisture content was set to 5000 ppm. Below this limit, if temperature drops below 0 °C, vapor crystallizes in ice flakes, without passing by the liquid phase. Ice does not recombine with ionic contamination and does not present the same risks as liquid water. When the temperature passes again above 0 °C, the ice will sublimate in vapor, again without becoming water.

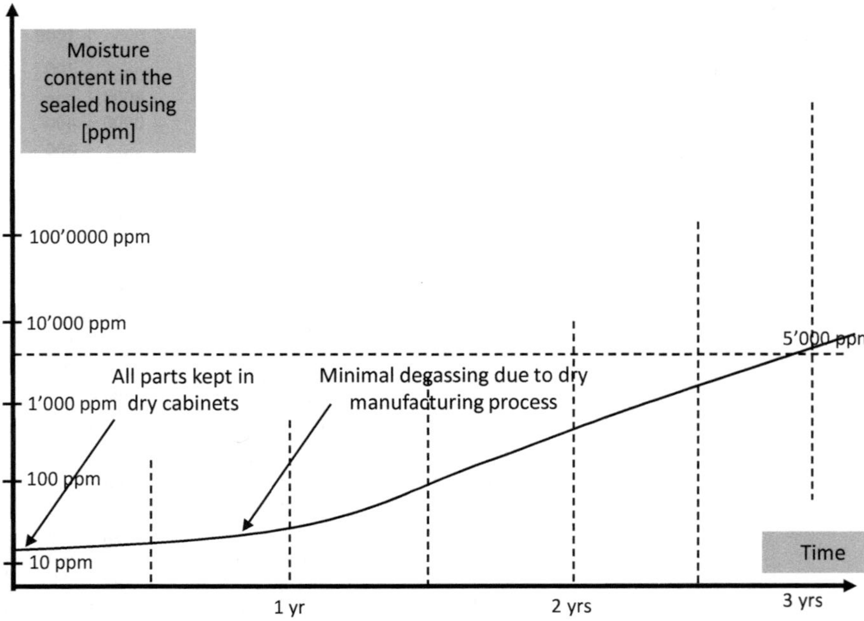

Fig. 4.40 Reduced moisture level resulting of proper pre-sealing drying

Therefore, it is critical to have a low moisture content for the few months before implantation, especially during transportation. After implantation, the stable body temperature reduces substantially the risks of condensation, even if moisture content rises slowly over the years. As described in Fig. 4.40, the critical point is to avoid condensation before implantation and during storage and transportation. If the implant is rapidly implanted, without being exposed to very low temperatures before reaching the hospital, then moisture content is not an issue. Implanting a device more than 1 year after sealing, and transporting it in cold situations, may generate risks of condensing water in the enclosure. Besides preservation of the sterility, limiting the shelf life of an AIMD to 1 year also prevents the exposition of units to cold weather when the inner moisture content is already too high.

4.9.4.1 Bake Oven Process

The purpose of vacuum bake is to release the moisture absorbed by various materials inside the device. Moisture is absorbed by most plastics, coatings, and glues. The speed of releasing the absorbed water will depend on the temperature, the vacuum, the type, and the thickness of the plastic parts.

For electronics without Li-ion batteries, baking 24 h at 125 °C is recommended. In the pacemaker industry, the bake process is usually done for 8–12 h at 50–55 °C (max temperature for the battery), vacuumed with a turbo-molecular pump. The rule of thumb says that an increase by 10 °C will half the bake duration.

The bake oven process is one of the critical steps in building reliable hermetic implantable devices. We lack scientific evidences showing that the conventional way of baking is optimal. Some experts are expressing doubts about the efficacy of applying a deep vacuum, as leakages at the level of the door will let humid air in. An improvement I suggest would be to place the bake oven in a second chamber, filled with dry nitrogen.

The time constant of moisture desorption is counted in weeks or months, depending on the material and its thickness. But back oven processes on the manufacturing floor cannot exceed 24 h; otherwise the assembly time becomes unsustainable. In consequence, a good assembly process must follow these fundamental rules:

- The quantity of plastic materials of all types, including packaging of chips, PCB substrates, retainers, spaced, insulators, and glue, must be kept to a minimum.
- The thickness of these plastic materials must be minimized.
- Prior hermetic sealing, all parts must be stored in dry cabinets.
- Transportation and handling of parts and subassemblies must be done in protective bags, avoiding exposition, even for a short time, to ambient moisture.
- Clean rooms used for the assembly must be as dry as sustainable related to ESD. Letting clean rooms be dryer than 30%HR induces risks of damages to electronic circuits due static electricity (see Sect. 4.11.1).
- The duration of the baking process should be, if possible, under the limitation of what is acceptable for the flow of products on the manufacturing floor.
- Baking temperature should be as high as possible, with regard to the maximum temperature tolerated by critical parts. Batteries, electrolytic capacitors, and supercaps are the most temperature-sensitive components, usually limited to 55–60 °C. Battery-less devices, like cochlear implants, can be backed at much higher temperature, for example, 100–150 °C. Compared to devices with battery, battery-less device will be much dryer at the time of sealing and release less moisture later on, leading to a better reliability on the long term. This is a specific superiority of battery-less device which has a heavy weight in the design of BCIs.
- The depth of the vacuum is an important factor, even if we lack evidence about the relation between vacuum and dryness. The quality of the gaskets around the door of the oven is certainly more important than the power of the vacuum pump.

Baking has been introduced to remove moisture previously absorbed by the electronic circuit. Wouldn't it be more appropriate to avoid or minimize exposure to humid air before sealing the housing? Some manufacturers have recently understood this trivial statement:

"If moisture hasn't been absorbed, it doesn't need to be removed."

I do not suggest eliminating baking right before sealing, but to preserve dryness of the electronic circuits from burn-in to encapsulation. By proper measures to keep dryness to an optimal level before sealing, post-sealing moisture will be greatly improved (Fig. 4.40).

This strict preservation process takes advantage of the great drying capacity of the burn-in process. After population and testing of the electronic boards,

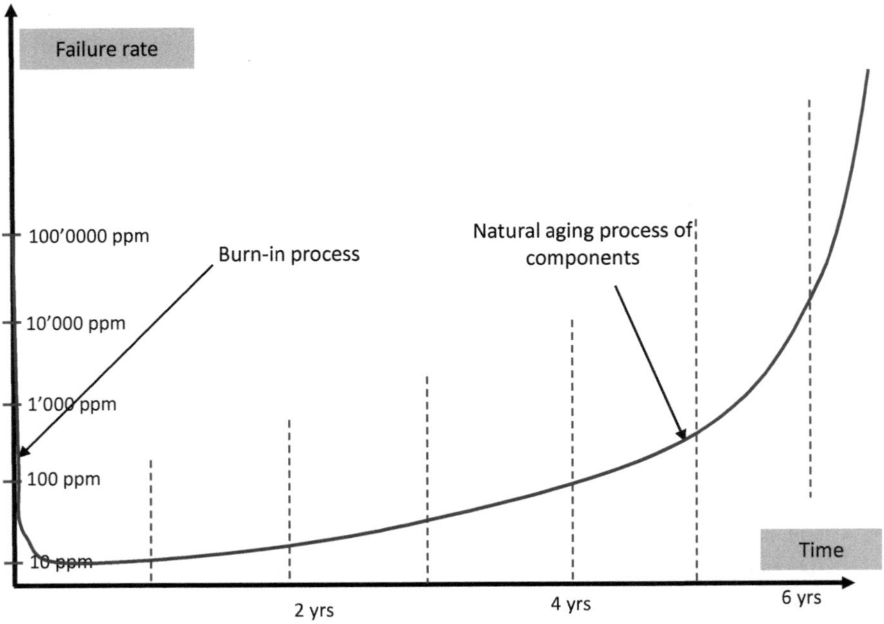

Fig. 4.41 Failure rate of electronic boards after burn-in

manufacturers impose severe functional (standard voltage applied) testing cycles, under elevated temperature (in the range of 150 °C) during several days. The purpose of this burn-in process (see Fig. 4.41) is to eliminate the "childhood" failures of the electronic assemblies.

After several days at high temperature, boards are extremely dry, and most of the moisture has diffused out of the plastic parts. In the past, at the end of the burn-in process, boards were stored in room atmosphere, starting immediately to again absorb moisture. The method described above prevents unnecessary exposure to moisture from the end of burn-in to introduction of the devices in the bake oven, by storing parts and subassemblies in dry cabinets (dry N_2).

Another interesting approach for a better control on moisture in sealed encapsulations is the use of desiccant or getter. These special materials inserted inside the hermetic housing have the property to absorb residual moisture. There are two categories of desiccants:

- Reversible moisture absorption: the chemical will absorb moisture until its saturation level. From that point, the desiccant may again release moisture in the concealed environment.
- Irreversible moisture absorption: the chemical reaction with moisture is permanent: $CaO + H_2O \rightarrow Ca(OH)_2$. The reverse reaction only takes place around 650 °C. When the desiccant is saturated, it will not be any longer able to absorb additional moisture, but it will not release any.

The use of desiccant is an efficient way to minimize potential damages resulting from excessive vapor content. It may substantially expand the lifetime of a device. Nevertheless, introducing desiccant in implantable devices generates several new challenges:

- There must be available space in the housing to add an appropriate volume of desiccant.
- Selecting the right volume of desiccant to assure long-term moisture absorption is tightly linked to the mass and type of plastic material trapped in the housing. We do not have today any established method to properly estimate the volume of desiccant necessary to assure long-term reliability.
- RGA of devices including a desiccant becomes difficult to assess.
- The validation of the effectiveness of the RGA is heavy and time-consuming.
- The desiccant must be dried in an oven just before insertion in the device. After being dried in an oven, the desiccant should not be exposed for more than a few minutes to clean room atmosphere; otherwise it will start absorbing moisture. In consequence, the process flow and its validation require much attention.

Companies like Alpha Advanced Materials [50] have developed various getter materials optimized to absorb moisture (H_2O) but also O_2, H_2, and CO_2. Electroplating (e.g., the traces of the PCB) usually leads to the release of H_2 on the long term. In a confined environment, H_2 may recombine with metal oxides and generate water molecules.

Several implantable products including a desiccant have been approved for commercialization. In most cases, the desiccant consists in a plate, strip, or film is inserted in the device right before hermetic sealing. A new interesting trend is to dispense or inkjet print the desiccant on the inside of the housing. Coating the electronics with a layer of desiccant might also be a promising topic to be explored.

4.9.5 Residual Gas Analysis (RGA)

RGA is performed by doing a deep vacuum around the device, preheated at 100 °C. Then, a puncturing mechanism opens a hole in the device. The gases trapped inside the housing go to a spectrometer which can quantify the percentage of and identify the various gases. At 37 °C, the saturated vapor concentration is 44 g/m^3 and 598 g/m^3 at 100 °C. So, at body temperature, the maximum vapor concentration is 58,000 ppm.

RGA also identifies other gases which may impact corrosion, like O_2. It must be remembered that corrosion of various pure metal or alloys can happen without condensed water. Vapor and/or oxygen can trigger corrosion processes, but to a lesser severity level than liquid water.

The RGA test is destructive and expensive and requires much expertise and skills. Only a few labs in the world master this test, like Oneida Research Services

(ORS) [51], present on both side of the Atlantic. Depending on the size of the cavity, various methods and equipment are used [52]. RGA measurements are done during the development of a new device or of an innovative hermetic housing. This is the only way to assess if the design is in line with the expectations of residual moisture at the time of sealing. RGA must also be done during the validation phase of the final device. It is also recommended to do several RGA tests over time, for example, one sample right after sealing and other samples 3 months, 6 months, and 1 year after sealing. The evolution of concentration of moisture and oxygen over time is a good predictor of long-term reliability. Most manufacturers have included an annual RGA tests for each of the products, as a post market surveillance routine.

When a device is submitted to an accelerated aging test, RGA tests at successive time periods have a great value.

The precision of RGA tests is coarse. Anyway, RGA is a good indication of the dynamic evolution of moisture and other gases in the confined hermetic microcosm. I have seen examples were a first RGA, after 12 h bake oven, showed a moisture content around 8000 ppm. We then decided to double the baking time to 24 h. RGA showed then a moisture content of 10,000 ppm! Doubling again the baking time to 48 h led us to be back to 8000 ppm. Such inconsistent results illustrate how difficult this measurement is. Some of these uncertainties could have been lifted by submitting more samples to RGA. But, remember that it is a destructive test. Ideally, we should pass many devices through RGA for having a statistically coherent result. Unfortunately, companies rarely can afford such costs.

4.9.6 Near-Hermetic Encapsulation

In the previous chapters, we have heavily insisted on the necessity to encapsulate implantable electronics in hermetic housings. We even stated that hermeticity was a "must" and that "only biocompatible materials should be found outside the hermetic housing." This strict approach is based on the history of AIMDs.

The early pacemakers, pioneers in the field of AIMDs, were simply potted in silicone rubber or epoxy. This was the best technology available at that time. Very soon, many failures and adverse events did show that potting had major limitations regarding reliability, long-term performances, and patients' safety. Due to the unavoidable diffusion of moisture through silicones and epoxies, high-density electronics could not be used. At the end of the 1970s, when the first fully hermetically sealed titanium encapsulations became available, it was a revolution. Suddenly, it was possible to seal electronics in a way it could possibly work for decades (unless the life of the battery will require a quicker exchange) and assure the best possible patient safety. Titanium housing was a major change for this industry.

Today, with the cumulated experience of several decades of annual implantation of millions of devices, with amazingly rare adverse events related to hermeticity, we may wonder if the fully hermetic sealing strategy was not overkill. Intrinsically, a pacemaker is encapsulated in a way that would prevent moisture ingress for 20, 30, or may be 100 years. But the battery gets depleted after 10 years, and the device gets replaced by a new one. Is hermetic titanium encapsulation too good for 10 years device life? We may at least say that pacemakers have a good safety margin regarding their resistance to moisture.

Neuro-devices and BCIs have specificities that pacemakers did not present. The two main ones are:

- BCIs have many channels, requiring many FTs. With the current available FT technologies, it becomes difficult to combine multichannel titanium encapsulation and miniaturization.
- BCIs communicate with RF at high frequency. Titanium housing is a shield or a barrier for electromagnetic waves.

In consequence, for BCIs and other complex neuro-devices, times have come to reassess the use of a hermetically sealed titanium housing.

There are three ways to protect an electronic board implanted in the body:

- *Fully hermetic encapsulation*: Insert the electronics in a hermetically sealed box with only biocompatible materials outside of the hermetic enclosure, as described in detail above (see Fig. 4.9)
- *Hermetic/near-hermetic hybrid encapsulation*: The electronics is hermetically sealed, but non-biocompatible materials outside of the hermetic housing are only protected by a near-hermetic coating or potting (see Figs. 4.30 and 4.42)
- *Near-hermetic* encapsulation: Coat and/or pot the electronic assembly (see Fig. 4.43).

The concept of sealing a box around the electronics has been thoroughly studied in previous chapters. In view of our needs of miniaturizing BCI devices and making them as thin as possible, "sealing in a box" has three major drawbacks:

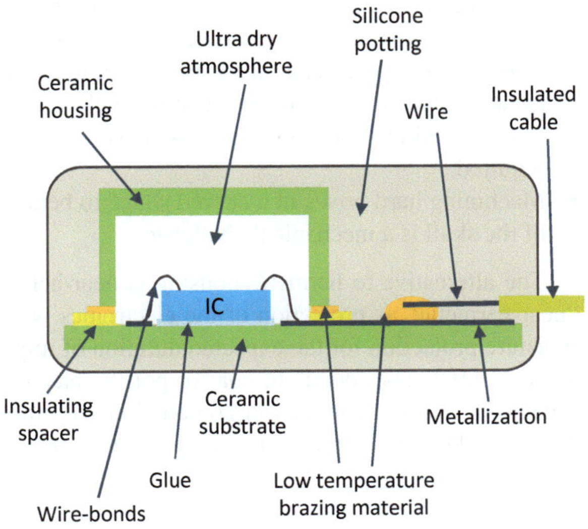

Fig. 4.42 Hermetic/near-hermetic hybrid encapsulation

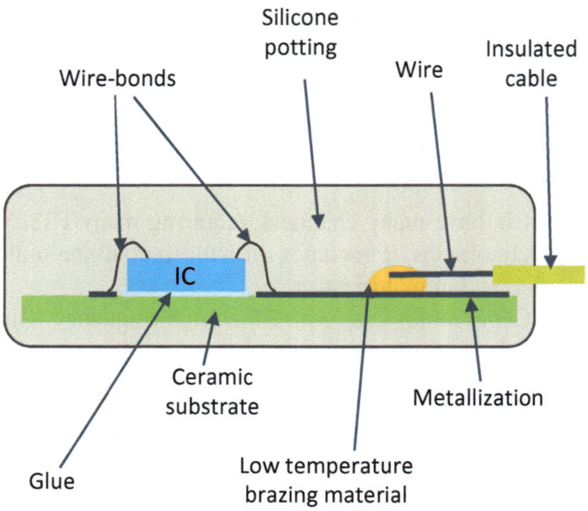

Fig. 4.43 Near-hermetic encapsulation

- The ratio "useful volume/overall volume" is low, in the range of 10–30%. Wasted volumes are due to:
  - Empty space between the components and the inner wall of the housing.
  - Thickness of the housing walls. Titanium housing has rather thin walls (0.2–0.4 mm), but fragile materials, like glass or ceramics, require thicker walls, especially if placed on the skull. In this case, depending on the shape and size of the housing, one must count with wall thickness of 0.5–1.0 mm for glass and 1–2 mm for ceramics.
  - Hermetic FTs are bulky in their conventional design. For large number of channels, the volume occupied by FTs becomes significant.
- The overall thickness of the implant is the maximum height of the electronic circuit, plus the thicknesses of the bottom and top shells. If the populated PCB is 2 mm thick, it will lead, in the best case, to a cranial implant of 3 mm thick if encapsulated in titanium, but 6 mm thick if encapsulated in ceramic. In this sense, titanium housings may still have a future in BCI, if thin profiles are required.
- Machining hard boxes in a curved shape to better adapt to the natural curvature of the skull is a mechanical challenge.

The alternative to hermetic housing is near-hermetic encapsulation. We define "near-hermetic" as protection of the electronics which provides "the appropriate" moisture protection for the expected duration of implantation, as already discussed in Sect. 4.9.2. The idea is to coat or pot the electronics and its connections with organic or plastic materials, in conjunction with inorganic thin layers. Multilayer coatings, alternating organic and inorganic materials, show promising moisture protection grades.

Early pacemakers were over-molded in silicone rubber or epoxy (see Sect. 1.4.1.1). This was already a near-hermetic encapsulation, providing "appropriate" moisture protection for the simple robust and forgiving electronics of that time. These devices had substantial distances between components and no high-density IC. Therefore, diffusion of moisture led to acceptable current leaks. Substantial improvements have been achieved by having a first protective coating of the electronics with Parylene. This thin layer of Parylene provides a good, if not perfect, barrier to moisture. A thick potting of epoxy or silicone around the Parylene-protected PCB adds a proper mechanical protection and an additional barrier to diffusion.

Several academic groups and companies are currently working on innovative processes to improve moisture resistance of conformal coating layers. The main principle for a better barrier to moisture diffusion is the *multilayers* approach. It consists in the deposition, in alternance, of organic and inorganic thin layers. Prototyping and long-term aging are currently evaluated in two main directions:

- Atomic layer deposition (ALD): extremely thin conformal layers are coated on the structures (IC, populated PCB, electrodes, cables, etc.) to provide protection from moisture and oxygen. Various materials, thickness, alternance, and number of layers are being tested in several labs with promising results. The group of Maaike Op de Beeck [57] has tested a triple layer of 8 nm HfO_2/20 nm Al_2O_3/8 nm HfO_2 in sandwich between two layers of polyimide (11 and 5.5 µm), which shows an amazing resistance to moisture penetration, about 8000 times better than a reference Parylene layer of equivalent thickness (see Fig. 4.44). These moisture-tight ALD triple-layer coatings (see Fig. 4.44) have been tested on

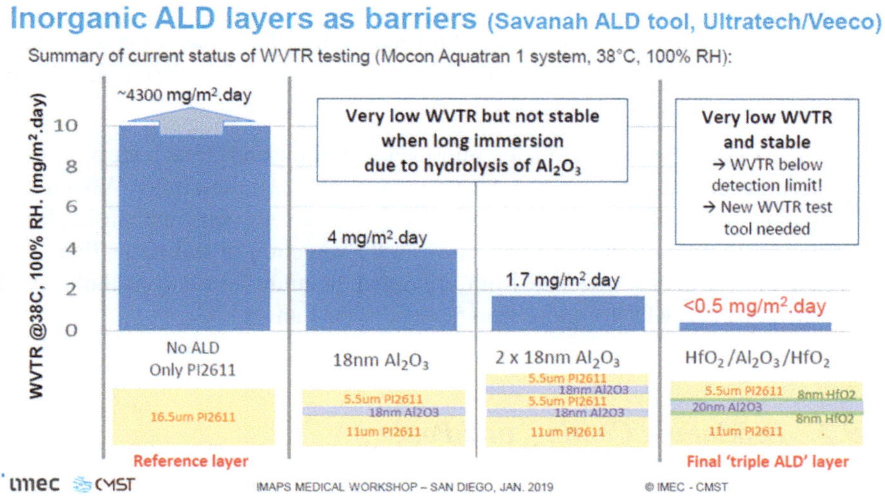

Fig. 4.44 Triple ALD layer for optimal moisture protection. (Courtesy of Maaike Op de Beeck, Centre for Microsystems Technology (CMST)_IMEC and University of Ghent, Belgium, extract of [57])

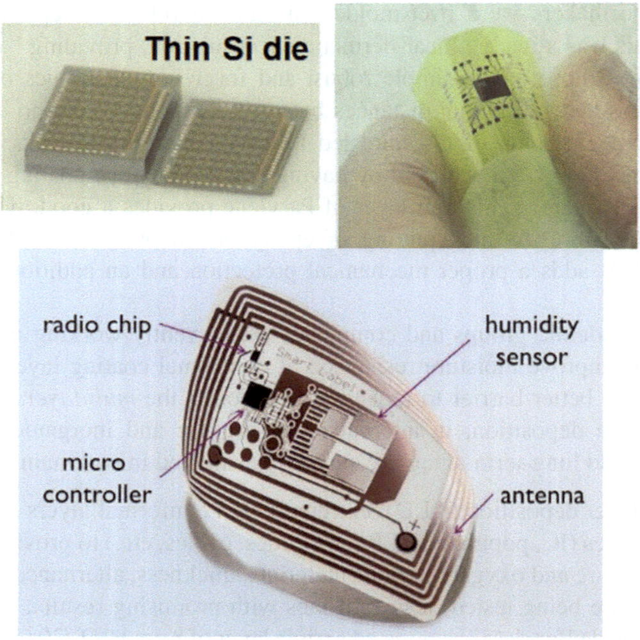

Fig. 4.45 Minced chips on polyimide substrates. (Courtesy of Maaike Op de Beeck, Centre for Microsystems Technology (CMST_IMEC and University of Ghent, Belgium)

electronic polyimide-based structures including minced chips. CMOS ICs have been minced down to 30 μm thickness, which provide flexibility to the chip itself (see Fig. 4.45).

- Multilayer coating of Parylene C and SiO_2 as developed by Coat-X in Switzerland [58– 61]. The process can be applied to rigid populated PCBs (see Fig. 4.46) or to flexible substrates (see Fig. 4.47a, b).

Both processes claim a significant improvement (thousands time better) of resistance to moisture compared to regular Parylene single-layer coatings. When these new coating encapsulations will be fully validated for long-term human implants, the frontiers of near-hermeticity will move in the direction of full hermeticity. We can envision, in about a decade from now, that titanium-less encapsulation will properly protect AIMDs for more than 5 years in the human body.

4.9.7 *Insulation, Coating, and Potting*

Previous sections of this chapter have covered the hermetic or near-hermetic encapsulation technologies used to assure the protection of implanted electronics. The active parts of the implant, the electronics, and the battery are only a subset of the entire implant, as seen in Fig. 4.48. Implanted systems may include the following elements:

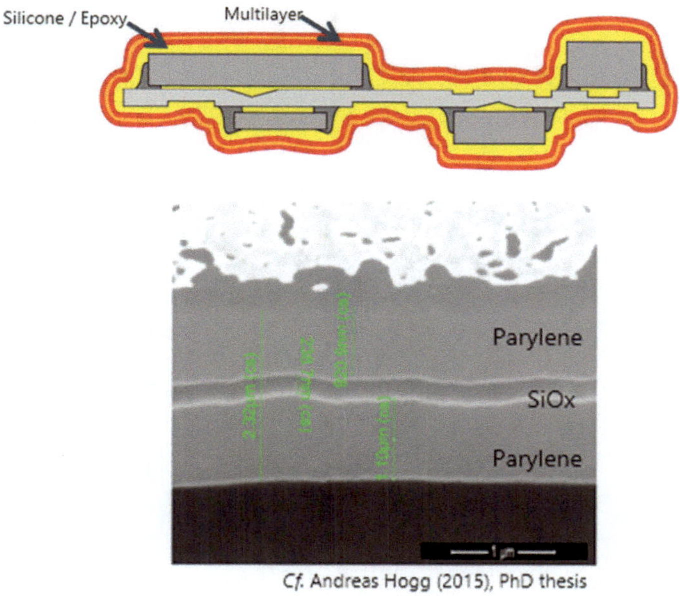

Fig. 4.46 Multilayer coating applied to a rigid PCB. (Courtesy of Coat-X SA)

- Encapsulated electronics: in a hermetic housing with feedthroughs or potted/coated in a near-hermetic configuration
- Header: connecting subassembly, attached to the main housing, near-hermetic or simply waterproof
- Extension: additional connector linked to the header connector by a flexible cable and to the body interface (electrode) by another cable, near-hermetic or simply waterproof
- Cables or wires (often called leads): flexible electrical conductors in an insulating sleeve, connecting two elements of the implanted system
- Electrodes: the interface with body tissues, as MEA, ECoG, DBS electrodes, wire electrodes, paddle electrodes, nerve cuffs, etc.
- Deported components linked to the encapsulated electronics, near-hermetic or simply waterproof:
  - Induction coil for power transfer
  - RF antenna
  - Sensors

Electrical currents are flowing between the various subassemblies attached to the encapsulated electronics. We have seen above that the electronics must be properly protected against moisture. Elements which are outside the encapsulated electronics must also be as much as possible protected from moisture ingress. Penetration of water in these external components may have the following impacts:

- Promotion of corrosion, especially in presence of DC voltage and ionic contamination (see Sect. 4.5)

a

b

Fig. 4.47 (**a**) and (**b**) Multilayer coating applied to a flexible circuit. (Courtesy of Coat-X SA)

- Leakage of current between two channels, reduced sensitivity in reading devices and reduced stimulation voltage in writing devices
- Increased cross talk between channels

Moisture will penetrate and diffuse at various points. For example, a paddle electrode with a flexible cable attached to an in-line connector will present potential routes for water ingress as shown on Fig. 4.49. The potential paths of liquid

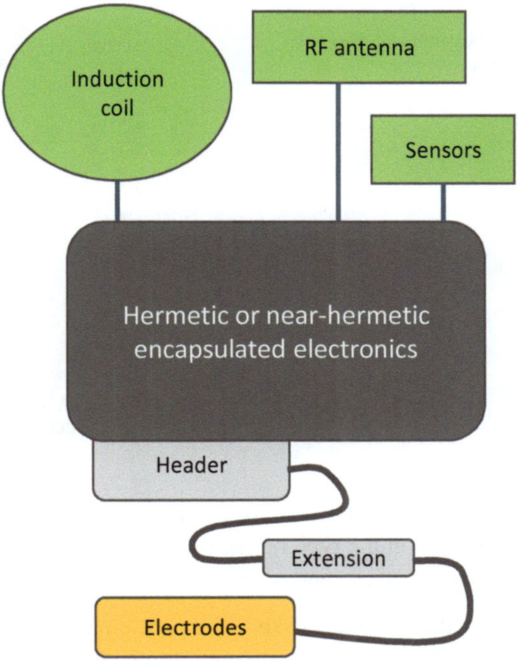

Fig. 4.48 Components of an implanted BCI

penetration are indicated in red. These are the weakest points in the encapsulation of the electrode and its connector. Moisture penetration along these routes is rather quick, in the range of days to months. The empty spaces, cavities, and other interfaces which are intended to be protected by the encapsulation will therefore be invaded by body fluids, constituted mainly of water. We have seen in Sect. 4.5 on corrosion that pure water will not trigger corrosion. Unfortunately, pure water is never found in the human body. Implanted devices are surrounded by liquids ionically loaded. In addition, even if properly cleaned during the manufacturing process, cavities of the device always present some residual ionic contamination which will add to the already ionically charged penetrating fluids.

The design of implantable connectors and extensions is made in a way to minimize moisture penetration and delay it as much as possible. The primary protection is based on silicone rubber gaskets:

- Double O-rings on the male connector: when introduced in the female connector cavity, the O-rings prevent or at least postpone body fluids to penetrate in the contact area.
- Septum on the set screws: the screwdriver is punched through the septum for fastening the screw. When the screwdriver is retracted, the opening in the septum seals and minimizes body fluids penetration.

These silicone barriers are at best waterproof but probably not hermetic. Moisture will leak through these gaskets sooner or later. At a lesser rate, moisture will also diffuse through the main body of the connector, either in a header configuration or in an extension.

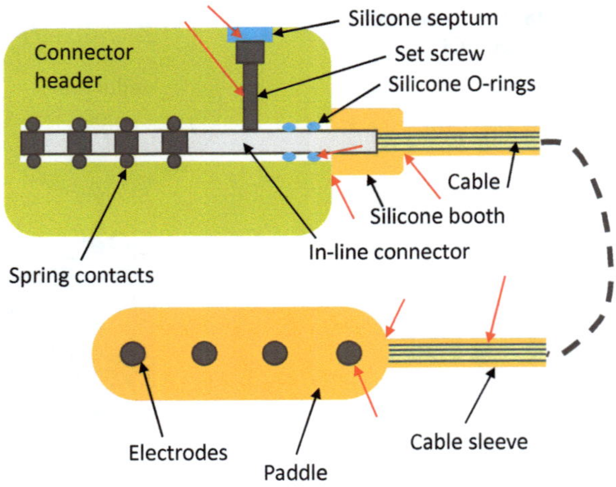

Fig. 4.49 Potential routes of moisture ingress (red arrows)

Consequences of moisture ingress in connectors depend on the geometry of the contacts, on the type of electrical signals transferred at the level of the connection, on the materials, and on the cleanliness of the various parts:

- Distant contacts (like in the pacemaker standard IS-1 [54]) are moderately sensitive to moisture, as channel-to-channel insulation resistance remains in the MΩ range, maybe decreasing to hundredths of kΩ after several years of implantation.
- In stimulation situations, the voltage transferred at the connector level is in the range of 1–20 V. Tiny leakage currents to adjacent channels have a minor effect on the stimulation performances.
- In sensing situations, decrease of insulation due to moisture may have more drastic consequences regarding sensed signal amplitude, noise, and cross talk.

The IS-1 [54], DF-1 [55], and IS-4 standards have originally been set for stimulation devices (pacemakers, defibrillators, SCS, DBS, SNS, etc.). Outside of this range of standards applied to rather big connectors with low channel counts, there is not much guidance about how to design and characterize connectors for neuro-applications. Collecting extremely low energy signals in or at the surface of the brain, in the range of μV and nA, requires high levels of insulation and minimal contact impedance.

In consequence, all the above-described elements outside the hermetic housing must be encapsulated in the best possible way to minimize moisture ingress. Connectors, cables, and electrodes cannot be hermetically sealed in metal or ceramics. The only ways to protect them are encapsulation or coating with polymers, which have limited moisture resistance as discussed in Sect. 4.9.6. The adoption of multilayer coating will provide a much better protection of electrodes and leads in a near future. But the weakest point of AIMDs remains the gaskets protecting detachable connectors. Systems consisting in hermetically encapsulated electronic and near-hermetic multilayer coating of leads and cables may fail due to moisture passing the barriers of silicone O-rings in connector cavities.

Fig. 4.50 Representation of the impedance at the interface between a metal electrode and body tissues/fluids

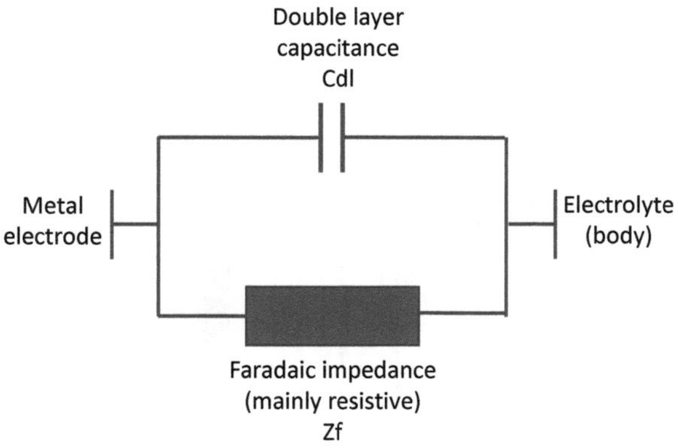

Numerous scientific papers or books [62, 63] have provided models for the complex impedance of leads. We will not develop any deep theory of how electrical signal travel from the body to the electronic circuit or the reverse. A full simulation of a multichannel electrode and the evolution of the electrical characteristics over time are out of the scope of this book. In a pragmatic approach, we prefer to recommend in vitro bench tests in an environment representative of the body and actual measurements of prototype leads and connectors, in accelerated aging setup.

Modelling the interface between the metallic conductive part of the electrode and body tissues is already a difficult task. The free negative electrons circulating in the metal are exchanged with ions floating in the body fluids, in a dynamic and no obvious way. The characteristics of this exchange are represented by a complex impedance (see Fig. 4.50) with two components:

- The Faradaic impedance illustrating the actual exchange of electrons and ions
- The non-Faradaic impedance, a capacitance representing the double layer, with no exchange of electrons/ions

This simple representation (see Fig. 4.50) with two elements hides an enormous complexity related to the variation of the impedances with time, due to the modification of the surface, growth of tissues on the surface of the electrode, modifications in the circulation of fluids, and other dynamic factors.

A simple basic model (see Fig. 4.51) of a two wires cable linking an electrode and a connector shows that a global simulation may turn out to be difficult. At each point where moisture penetrates (see Fig. 4.49), there is an ions-electrons exchange and a double-layer capacitance, with a different evolution in time. Reliable models are therefore a challenge to establish.

Each system intended to long-term implantation in the human body and including elements (electrodes, cables, connectors, extensions, sensors, etc.) conducting electricity must be evaluated and validated prior approval for commercialization. Artificial aging procedures are especially important regarding the evolution over time of impedance, current leakage, loss of insulation, and cross talk between channels. Simple theoretical calculation of current leakage and impedance will not be enough to assure a proper functional sensing or stimulation capability of the electrodes cable-connector subassembly.

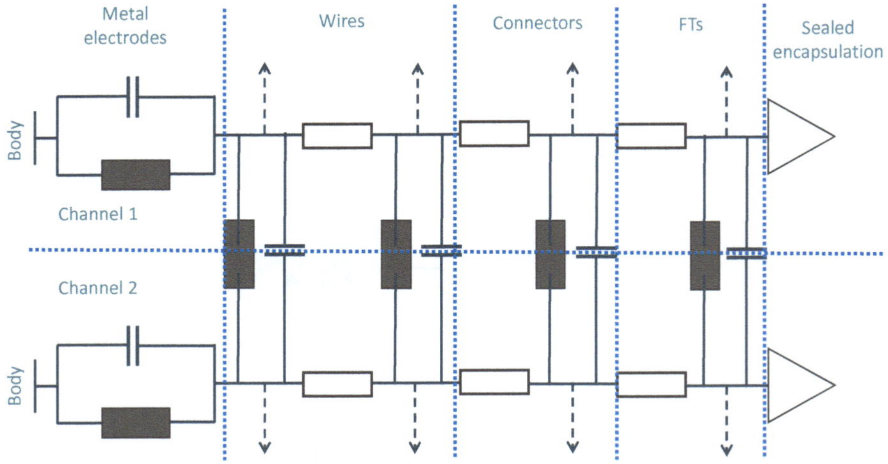

Fig. 4.51 Simplified representation of an electrode and a two wires cable

4.10 Mechanical Robustness

AIMDs are too often considered as purely electronic systems. Their important functionalities are obviously linked to electricity, like reading tiny voltages, stimulating tissues, signal handling, data exchange, or RF transmission. But nothing of this is possible if the encapsulation, cables, and other flexible items are not robust enough to protect the electronics and fulfill connectivity tasks on the long term. A quick look on the FDA database [65] gathering all adverse events occurring in AIMDs shows that a vast majority of them are of mechanical nature. Failures often are related to loss of contact, broken leads, corrosion, erosion, displacement, or other mechanical issues. Electronically sophisticated devices regularly fail for trivial non-electronic reasons. As an example, implantable programmable drug delivery pumps have their major source of adverse events at the level of the intrathecal catheter. Likely, DBS systems mainly fail due to problems with the leads, cables, extension, and connectors, rarely due to the hermetically sealed electronics.

AIMDs and BCIs encounter mechanical failures of two natures:

- *Failures to protect the electronics*:
  - Loss of hermeticity will lead to electrical failures and possibly leakage of non-biocompatible compounds.
  - Rupture or deformation of the housing: for above-the-neck BCIs placed above the skull or inserted in a partial or full craniotomy, the main reason for such failures is external impact. Following a dramatic occurrence of a broken ceramic housing due to a baseball accident, manufacturers of cochlear implants were first to establish a standard, so-called hammer test, to assess resistance of cranial devices to impacts [64]. Recent developments of BCIs with skull inserted enclosures have raised intense discussions regarding

resistance to impacts. Unlike cochlear implants implanted in the head of children playing hard games, riding bicycles, and practicing sports, BCIs are usually intended to patients with a much quieter life. Consequently, there is a rational for less stringent impact tests aimed to BCIs. As there is no standard dealing with impact test of BCIs, validation of newly developed BCIs must fix impact resistance criteria based on a rational risk analysis.

- *Loss of connectivity*:
  - Rupture of a wire in a cable, due to fatigue, excessive pull stress or torsion
  - Rupture, erosion, or wear of the protecting and insulating sleeve around a cable or electrode
  - Unintended migration of the connecting pin out of the connector block
  - Degradation of the quality of contact due to corrosion.

As mentioned earlier, the development of implanted BCI systems is a rather recent endeavor, made possible by the availability of new disruptive technologies. Consequently, BCI systems are mechanically substantially different from other more conventional AIMDs. New materials, innovative geometries, miniaturization, and above-the-neck location place BCI implants in a "no man's land" regarding standards and guidelines. There are only very few predicates on which to get inspiration. Designers of BCI devices must therefore be very careful in the definition of the mechanical characteristics and specifications. New assembly processes may also be needed. A new set of test methods and validation procedures must be studied and implemented. They will, one day, be the base of new standards.

From experience we learned that entering in a new field or moving from a known environment to a new one has the best chances of success if it follows a prudent methodology. This is especially true for mechanical aspects of active implants. Our recommendations for building performant and robust future BCIs are:

- *Learn from the past*: In previous chapters, we have pointed out failures, successes, clever designs, and sources of field actions. Mechanical failures are often very difficult to predict and happen in unexpected modes and location. Lessons learned from real cases have a great value for minimizing risks in future designs.
- *Understand the environment*: Moving from a well-known location in the body (e.g., the chest) to the head induces a lot of changes in the tissues surrounding the implant but also on the movements, availability of space, physics, surgery, aesthetics, etc. A clear understanding of the new environment will prevent the use of technologies not adapted to it.
- *Apply "reduction laws"*: Miniaturizing a device cannot be done by a simple homothetic down scaling. For example, the reduction by a factor of two of all the dimensions of a ceramic housing, including the thickness of the ceramic wall, will make the device too fragile to external impacts. Wall thickness must probably remain unchanged, to the expense of the volume of the inner cavity. In a similar way, increasing the density of FTs and reducing the insulation gap in proportion will lead to a deterioration of the insulation properties, therefore increasing electrical leakage between channels.

- *Develop adapted test methods*: Mechanical tests coming from the industry of cardiac implants may not translate well to the specificities of above-the-neck implants. For example, pacemaker leads flex over a large amplitude, about 80 times per minutes for decades. Their resistance to fatigue is critical, and special tests have been standardized to assess robustness. Fatigue tests for brain-implanted electrodes will be governed by other parameters. Besides the already discussed impact test, developed for the specific constrains of cochlear implants, there are not many standards directly adapted to BCI systems. It is the responsibility of BCI designers to develop appropriate mechanical test methods, based on a full understanding of the environment, the dynamics, and the human head.

4.11 Electrical Robustness

AIMDs include electrical components which must be protected from external influences. Under the Sects. 4.9 and 4.10, we have already discussed the mechanical protection provided by the encapsulation housing the electronics. Failure to mechanically protect the electronics will result in moisture penetration, corrosion, or short circuits. These are electrical misfunctions or disruptions with mechanical root causes.

In this chapter, we will cover failures of fulfilling an electrical function due to external electric and/or magnetic fields. In addition to create damages to the implanted electronics, external electromagnetic fields may also generate undesirable effect on the implant, mainly related to heating due to induced eddy current circulating in conductive materials. A special attention is given to MRI compatibility, an increasing factor in neuro-technologies.

4.11.1 Electrostatic Compatibility (ESC)

High-voltage (several thousand volts) charges may accumulate between two conductive elements well insulated from each other. The origins of these electrostatic charges have various origins, like friction and relative displacement of plastic materials. Static electricity occurs in everyday life, and everybody remembers getting an unpleasant electrical shock when touching an electrically grounded element. Electrostatic discharges (ESD) are of high voltage but very low energy. For this reason, it has only minor consequences for human being.

FDA and the International Electrotechnical Commission (IEC) have issued standards and guidance related to the protection of medical devices against static electricity (see, e.g., [66] or [67]).

Due to its high voltage, if directly applied to an electronic circuit, electrostatic discharges may irreversibly damage transistors and diodes. During the entire manufacturing cycle of active implants, from storage of components to packaging of the sterile device, protective measures are taken to avoid the generation of static electricity and to reduce potential exposure of sensitive parts:

- Antistatic bags for storage of components
- Electrical grounding of equipment and workbenches
- Conductive floor and operators wearing conductive shoes
- Operators wearing wrist bands electrically connected to ground
- Ionizers above the workbenches

These appropriate methods assure that the finished device, sterile and ready for implantation, has not been damaged by static electricity. Additionally, the electronics may include protective elements against high voltage applied to the inputs [68].

When the device is implanted in the body, it is surrounded by conductive fluids. Therefore, there is no static electricity in the body.

AIMDs are at risk of being damaged by static electricity for only a few seconds in their existence: *when opening the blisters in the operation room and holding the device for implantation.*

- Before opening the blister: the device has been protected during the manufacture. As soon as packaged, the device is electrically insulated from its surrounding, and no static electricity can reach the electrical inputs.
- When the device is extracted from the its sterile packaging, it may briefly be handled by nurses or surgeons without appropriate grounding. Currently developed BCIs, like wireless cortical implants (see Sect. 7.3), may even go through an electrical test (measurement of the impedance of the channels) inside the operation room, right before implantation. Handling the implant with unprotected electrodes in direct contact with sensitive electronics is a major challenge for BCIs. Even if it is for a very short time, exposure to static electricity during this critical surgical implantation phase may have serious consequences. If any channel gets damaged, it will be lost forever. This is a situation which is unique to BCI with undetachable electrodes (an example of the importance of "understanding the environment"). If the electrodes could be disconnected from the electronics, then their impedance could be measured separately, the electronics remaining safely protected from ESD. When the electrodes will be attached to the housing connector, all the parts will be grounded by conductive body fluid. Unfortunately, multichannel miniature implantable connectors are not yet available.
- After implantation, there is no risk of ESD exposure.

4.11.2 Electromagnetic Compatibility (EMC)

The space around us is crowded with electromagnetic waves over a very large range of frequencies and intensity. At low frequencies, the origins of the electromagnetic fields are essentially due to industrial activities, electrical power networks, motors, and other equipment related to energy. At higher frequencies, wireless communications are the main contributor to a dense spectrum of waves. The emergence of mobile communication, cell phones, the Internet, Wi-Fi, Bluetooth, Internet of Things (IoT), wearable devices, Global Positioning System (GPS), and other wireless applications have led to an exponential growth of the electromagnetic power density. We can now qualify the phenomenon as "electromagnetic smog," a kind of pollution constantly affecting everybody everywhere.

Wireless medical devices are covered by a global guidance document from the FDA: "Radio Frequency Wireless Technology in Medical Devices – Guidance for Industry and Food and Drug Administration Staff" [69].

The main standards governing EMC in general are IEC-60601-1 [71, 72] and ISO-14708-3 for active implantable medical neurostimulators [73].

As already exposed in Sect. 2.2.1, frequency bands are regulated by national authorities. Some bands are restricted or submitted to licenses; some are free. A small number of frequency bands are dedicated to medical devices. One must be aware of slight differences from country to country regarding the allocation of frequency bands. As medical devices are ruled by more international standards, like the pan-European CE marking, it is strongly recommended to avoid bands subject to national licenses and remain in the space of non-licensed medical bands. Unfortunately, the medical bands are few and rather narrow; meaning they also are crowded and noisy.

Two categories of EMC must be clearly distinguished (see Fig. 4.52):

- "Inward EMC": Resistance and immunity of the implanted system with regard to incident EMDs coming from the environment
- "Outward EMC": Limitation and control of the EMDs generated by the implant and susceptible to disturb other systems

During the development phase of an AIMD, great care must be taken to anticipate and foresee the evolution of the electromagnetic smog on the long term. An implant may remain in the body for several decades and is prone to be exposed in

Fig. 4.52 Bidirectional electromagnetic compatibility

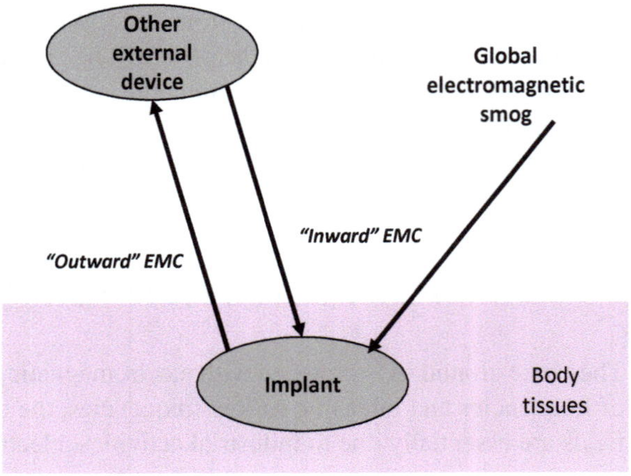

the future to an electromagnetic environment quite different from the conditions existing at the time of its design. There are examples of older AIMDs which present limited EMC, especially with cellular phones. Compliance with EMC standards and FDA guidance is necessary but not enough. See for reference "Information to Support a Claim of Electromagnetic Compatibility (EMC) of Electrically-Powered Medical Devices" [74], "Design Considerations for Devices Intended for Home Use" [75], and "Radio Frequency Wireless Technology in Medical Devices" [76].

Standards lag behind the fast evolution of wireless communication and do not anticipate changes. EMC standards cover the "normal" situation, in the case of AIMDs, electromagnetic compatibility of the device in the human body. Even if a device passed all the required EMC tests, it may fail in "abnormal" situations, for example, when exposed to extreme fields like during an MRI examination, at airport security checks, or during surgery. I have seen an EMC tested AIMD damaged during surgery by inappropriate use of electrocauterization tools.

Other abnormal situations may come from the therapeutic use of TDCS and TACS (Transcranial Stimulation) quickly described in Sect. 3.2.1. These external devices have powerful coils producing high-density direct current (DC), respectively, alternating current (AC), and electromagnetic fields interacting with the brain. The intensity of these fields is far above the limits of EMC standards. Therefore, TDCS and TACS should be contraindicated for patients having an implanted BCI. To a lesser extent, transcutaneous electrical nerve stimulation (TENS) may induce currents and fields with unpredictable impact on sensing BCIs.

I recommend to overprotect BCI systems against EMD and anticipate a much higher electromagnetic smog soon. Even if frequency bands are allocated and emission power is restricted, nobody has or will have control on the number of wireless devices on the planet. Controlling the electromagnetic pollution is going to be as complex as trying to manage the pollution of the atmosphere. Unlike other electronic devices or consumer products, AIMDs have a very long life span and must be tolerant to future more aggressive situations.

Before reaching the implanted electronics, incident EMDs are attenuated by three factors as illustrated in Fig. 4.53:

- Attenuation in the various layers of body tissues. We will see in Sect. 4.12 that electromagnetic waves are absorbed and diffracted when penetrated in the human body. Globally, low frequencies (Hz…kHz) pass through the body with minimal attenuation. Higher frequencies (MHz…GHz) get absorbed almost in proportionality with the frequencies. Absorption also depends on the type of tissue (skin, fat, muscle, bone, etc.). As a rule of thumb, the widely use 2.45 GHz band (Wi-Fi, Bluetooth, cell phone) is attenuated by a factor of 10 over 3 cm of human tissues. It means that an implant 9 cm deep in the body gets only one thousandth of the incident EMD. On the opposite, human tissues are almost transparent to low frequencies, for example, 135 kHz used for radio-frequency identification (RFID).

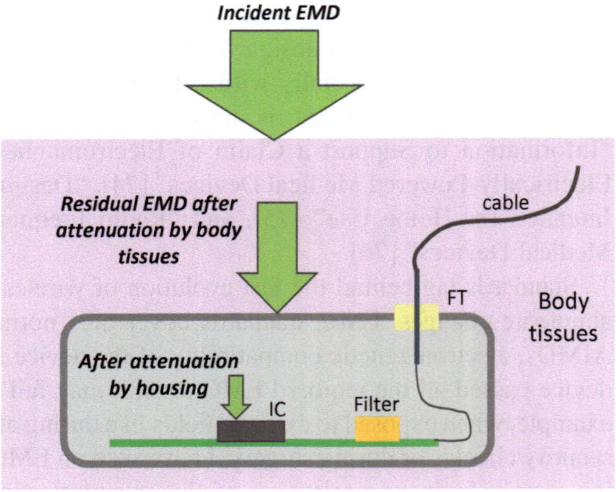

Fig. 4.53 Attenuation of incident electromagnetic disturbances

- Shielding effect of the implant housing. Conventional Ti housings are almost perfect shields for frequencies above 1 MHz. Depending on the electrical conductivity of the housing material and the thickness of the walls, a part of the lower frequency waves enters the housing shield. Ancient pacemakers were using this partial transparency at low frequencies (20–100 kHz) to place the communication coil inside the Ti can. It must be noted that a Ti encapsulation offers no protection about industrial EMDs, like the 50–60 Hz fields (and harmonics) generated by power lines, electrical motors, and transformers. Nonmetallic, non-conductive implant encapsulations like ceramics, glass, sapphire, or multilayer coating are not shielding the electronics from incident EMD at any frequency.
- Protective circuitry is usually added on the input channels of AIMDs. Sensing channels have high-gain amplifiers which are easy way for EMDs to reach the electronic chips and create damages. Electrodes, cables, induction coils, and other conductive elements located outside the encapsulation and connected to the electronics via hermetic FTs are prone to behave like receiving antennas, pick some of the energy of incident EMDs, and carry them to the electronics. Electromagnetically speaking, FTs are transparent windows in Ti shields. Electrical perturbances coming from EMDs over the FTs may be partially eliminated by adding filters on the input channels, at the level of the FT or on the PCB.

There are only few documents publicly available about filtered FTs [87]. Why are Integer (former Greatbatch) and Medtronic (MDT) investing much in the development of filters integrated in the FT instead of having simply the filter included in the electronics? I don't know for sure. Apparently, they prefer to filter away the high frequencies before they even enter in the can. But we can argue around it. A filter in the FT is a poor filter (first-order capacitive filter, with no clear cutoff and poor rejection), but we can have more sophisticated filters on the PCB or in the chips.

As it is the case for other aspects of the development of BCIs, it is very difficult to design the optimal protection against EMD. The best approach is to study carefully the existing solutions, improve them, build prototypes, and test them in an environment representative of body implantation. Companies like Zurich MedTech [77] have developed phantoms for practical tests and evaluations of prototypes but also advanced simulation software [78, 79] for the assessment of the propagation of electromagnetic waves outside and inside of the body, including the behavior around and in the implant itself. Simulations of Wireless Body Area Networks (WBANs) are now possible.

Stimulating BCI may also generate EMD which might impact other devices inside the body (see Sect. 4.11.4) or outside the body. It is important to consider active implants not only as potential victims of incident EMD but also as a source of electromagnetic perturbations.

4.11.3 MRI

There is an abundant literature on MRI, compatibility with implants, adverse events, and electromagnetic impacts [86]. The objective of this book is not to review or rephrase the ample documentation available on this topic but rather to put it in the perspective of BCI and MRI.

MRI is a nonionizing imaging technology which allows to visualize hydrogen molecules (abundant in human tissues and fat). MRI allows high-resolution 3D images of the head. Injection of contrast agents may facilitate the visualization of target tissues. MRI is intensively used in research, for a better understanding of the anatomy of the brain. It is also a major diagnostic tool, well adapted for a better treatment of patient suffering from neurological diseases.

Unlike MRI, which provide static pictures, functional MRI (fMRI) is intended to measure blood flows in the brain. Instead of target hydrogen, fMRI triggers resonance of oxygen. Oxygenation of the blood in the brain is a dynamic image of brain activity. In consequence, fMRI is a useful complement to conventional MRI for a better understanding of brain circuits and for improved diagnostics. FDA guidance related to MRI of neurostimulator is described in [73].

MRI tunnels for clinical use are enormous pieces of equipment, weighting up to 100 tons [80].

Both MRI and fMRI are based on the simultaneous exposure of the to-be-examined body part to three powerful fields:

(a) *Constant magnetic induction field*: generated by a supra-conductive coil, often called “magnet” (which is an inappropriate term, the coil having no magnetic material but inducing a constant magnetic field by the circulation of an intense DC current). Most MRI equipment used in hospitals have a magnetic induction of 1.5 Tesla (T), more rarely 3 T. Modern installations, mainly used by research laboratories, are at 7 T. Experimental equipment, dedicated to animal research, use powerful magnetic induction up to 24 T.

(b) *Gradient induction field*: superposed to the constant induction, it creates a distortion necessary for special localization. This gradient is in the range of 1–100 mT/m depending on the equipment and applications. A typical value for a hospital 1.5 T MRI tunnel for diagnostic is 30 mT/m.
(c) *Alternating excitation electromagnetic field*: in the range of 40–60 MHz. This high-intensity field (peak value around 35 kW, average power of 1 kW) is generated by a third coil.

These three powerful fields have different effects on various materials and components, respectively:

(a) The constant high induction may:

- Attract ferromagnetic (Fe, Ni, Co, and their alloys) parts creating important pull forces and torques (when the part is constrained to align with the field). Ferromagnetic materials are rare in AIMDs, with the exception implantable drug delivery pumps with stepper motors and reed switches for the reset of pacemakers.
- Attract permanent magnets, generating pull forces and alignment torques. Some AIMDs include permanent magnets, like cochlear implants (the magnet can be removed prior MRI exposure), DC motors in urinary incontinence devices, and magnetic rotors in hydrocephalic valves and eye pressure control mechanisms.
- Demagnetization of permanent magnets.

(b) The high induction gradient may:

- Attract and accelerate ferromagnetic parts which are free to move in the vicinity of the tunnel (projectile effect)
- Attract exercise pull force and torque on implanted ferromagnetic parts and conductive loops in short circuit
- Induce vibration of the implant

(c) The high-power high-frequency alternating field may:

- Induce eddy currents in any part conducting electricity with the following consequences:
  - Heat generation.
  - Counter electromagnetic field creating image distortion and artifacts. Typically, tissues behind the device might not be visible. Dark artifacts sometime are several times larger than the device itself.
- Generate induced voltage in conductive loops, coils, and PCB traces. This may arc, damage electronic components, polarize capacitors, and generate circulating currents. Induction coils for power transfer may see high voltages at the poles, especially if they consist of many turns.
- Interact with the implanted electronics with unpredictable consequences.

- Electrodes and cable may act as antennas if their length is equivalent to the field wavelength or half or a quarter of it. In such case, this unexpected antenna will pick large amount of energy and potentially damage the input amplifiers.

It must be noted that eddy currents will happen in any conductive part of the device, including Ti housings, batteries, PCBs, connectors, antennas, leads, and electrodes.

The above phenomena may have a serious impact on the patient health. They also may render the BCI inoperative, the patient losing therefore the benefit of the therapy.

Compatibility of implanted devices with MRI is sorted in three categories:

- MRI compatible or MRI safe (rare for AIMDs).
- MRI conditional. Several conditions may be imposed for proceeding to an MRI exposure:
  - Limit the induction field strength (e.g., to 1.5 T).
  - Limit the duration of examination.
  - Limit the exposed body part (e.g., below-the-neck only).
  - Surgically remove positioning magnets prior MRI examination (regularly done with cochlear implants).
  - Switch off the device before MRI examination.
  - Put the implanted electronics in a programmable "safe mode" before MRI examination (e.g., open disconnect induction coil, short-circuit input channels, etc.).

  Conditional exposure will preserve patient's safety. Nevertheless, artifacts may reduce the quality of the image.
- MRI incompatible: the wearer of the device should not be exposed to MRI.

Patients implanted with a BCI system suffer from diseases affecting their brain or nervous system. They are therefore candidates to regular MRI or fMRI examinations. In consequence, it is capital that BCI systems are designed to be MRI compatible or at least MRI conditional. A BCI system which is MRI incompatible will prevent MRI examinations subsequent to the implantation. It may be a serious limitation for patients and medical professionals.

4.11.4 Coexistence

By nature, wireless communication systems share common frequency bands. Specific RF bands are allocated by national regulators for various communication purposes (see Sect. 2.2.1). The global perception of the term "coexistence" applies usually to the possibility of using several systems in the same RF band, without interference. FCC Part 15 states that devices operating under this rule must accept any interference from primary users of the frequency band [81].

In this book, we limit the scope of coexistence to the possibility to have more than one active implant in the body of a single patient. It covers the conditions ruling the simultaneous operation of multiple devices. The Association for the Advancement of Medical Instrumentation (AAMI) [82] and the American National Standards Institute (ANSI) [83] discuss testing and risk management for wireless medical device coexistence.

Two to three decades ago, any new AIMD was excluding other active implants in patients during clinical trials. It was an easy way to avoid interferences between two and more devices in the same body. It made some sense (even if not very visionary), as cases of candidates for a second implant were rare.

Today, 1.5–2 million AIMDs are implanted every year. It becomes frequent that a patient with a pacemaker needs a SCS or a DBS. Are we going to explant the first device if the second one is providing a better quality of life? Do we explant a neurodevice when the patient needs a life-supporting defibrillator? In my opinion, excluding patients from a BCI because they have another AIMD is not acceptable any longer and will become unsustainable in the future. It is also questionable in terms of ethics. Do I have to choose between treating my incontinence and suffering intractable back pain?

In its Draft Guidance on Brain Computer Interfaces, dated Feb. 2019, mentioned under Sect. 2.2.1 and discussed in Annex 3, the FDA recommends exclusion of other AIMDs during Early Feasibility Studies (EFS) of BCIs.

I think FDA should defend the position that any patient has the fundamental right to multiple therapies provided by different implants in cohabitation in his/her body. Several devices should be able to coexist in a single human body. Regulating authorities should set standards to make this possible.

Coexistence is facilitated when two devices in a single patient's body do not use the same frequency band for communication with or from the external world. If both devices are in the same frequency band, dedicated communication protocols should avoid collision of information.

If the FDA or the European Commission does not impose coexistence rules for multiple AIMDs in a single patient, manufacturers should pursue the establishment of a voluntary standard. It is of the interest of the entire community of active medical devices manufacturers to avoid mutual exclusion and promote coexistence. Some communication protocols, like Bluetooth, have already solved the coexistence issue, by proper identification, addressing, and anti-collision procedures.

Another future evolution would be to impose communication and synchronization between devices within a single body. It would allow priority rules, firewalls, and other exchanges between devices in order to optimize a symbiotic functioning. For example, before firing a lifesaving high energy pulse, an implanted ICD may send command to all the other AIMDs in the same body to enter in a safe mode.

4.12 Communication Through Tissues

We have already seen in Sects. 2.2.1 and 4.11.2 that RF communication through human tissues is not only restricted by regulations but also by the laws of physics.

RF communication with AIMDs is possible according to several configurations:

- *Direction of the flow of information*:
  - From the implant to an external receiver
  - From an external emitter to the implant
  - Bi-directional
- *Communication distance*:
  - Proximal: the external unit is located on the skin, at the shortest possible distance from the implant.
  - Short range: the external unit is worn by the patient, for example, at the belt, in a wearable jacket, at the wrist, on its wheelchair, etc.
  - Home-based range: the external units are a base station located in the patient's house or inside a hospital.
  - Long distance: communication is established through cellular phone networks, remote antennas, or satellites.
- *Bandwidth*:
  - Reduced flow of information: upload or download of service data, software update, reading of data stored in implant memory, etc.
  - Real-time brain data: cortical data collected on many channels at high sampling rates
- *Security*: various forms of encryption, redundancy, and identification

By nature, the quantity of energy available to power an implanted device is reduced. Communication configurations which minimize the electrical consumption will be preferred. Backscattering is an interesting approach as the implant simply modulates incoming signals sent by an external unit, where power is less a constraint.

Propagation of RF waves from an implanted antenna to an external receiver is far from being a continuous homogenous path. Attenuation in the various layers may show important differences, and mismatch of dielectric constants at the interfaces between two layers is source of reflection and diffusion (see Fig. 4.54).

The model illustrated on Fig. 4.54 is a gross approximation of the reality. Better models should include the following parameters:

- The emission profile of the antenna is not unidirectional nor homogenous. It includes side lobes.
- The bottom of the housing or the PCB or a shield acts as a reflecting mirror with some scattering.
- The various layers of body tissues interpenetrate and overlap.

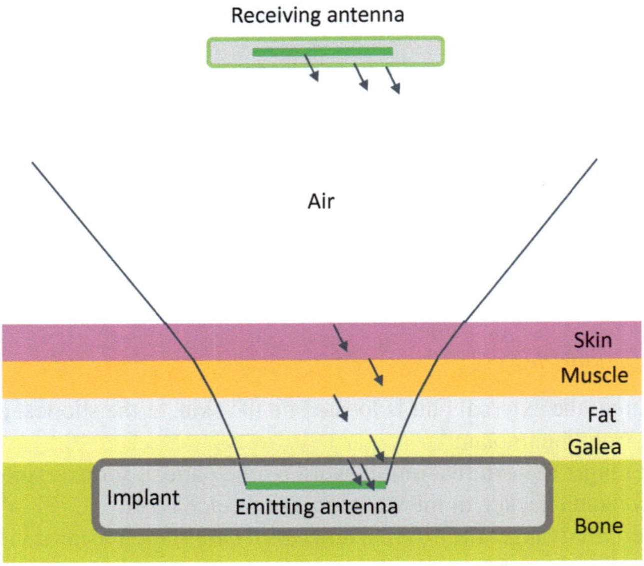

Fig. 4.54 Propagation of RF waves, from a bone inserted implant to a short-range external antenna, through several mismatched layers

- Blood vessels populate tissues in an almost random pattern.
- Hair roots and hair increase scattering.
- Sweat and hair fat add random diffusion.
- Additional layers of tissues, like fibrotic capsule on the surface of the implant, appear in the months following implantation.
- Erosion, inflammation, and necrosis may change the dielectric properties of tissues.
- The thickness of the fat layer may vary with time.
- Tissue composition and thickness have a large variability from patient to patient.
- Alignment between the emitting and receiving antenna is never perfect.
- Distance of communication may change.

Even oversimplified models are a challenge for simulation. Powerful simulation software [79] will only give a rough estimate of the communication performance. It will help engineers to make basic choices and grossly assess feasibility, but numerous iterations, prototypes, and measurements may be needed. Phantoms are used to mimic body tissues during prototype evaluations. They represent, at best, the tissue of an average patient, without blood perfusion. Engineers must be very careful in their interpretation of measurements of prototypes on a bench. At high frequencies, measurements are done in anechoic chambers to avoid reflections on flat surface. In real-life situation, the patient is surrounded by a natural environment which is far from free field. The wall and floor of the room and the presence of a headpiece, glasses, or other objects in the vicinity of the communication path may induce large variations in the performance. Preclinical evaluations may also be misleading, as no animal model is comparable to human regarding high-frequency RF communication. As described above, the evolution of tissues around the implant will change the propagation parameters with unpredictable consequences.

Variability and sensitivity to the environment increase with the frequency of RF signal. Radio communication in the hundreds of MHz range or below is not too sensitive to body parameters. Above 1 GHz, attenuation, reflection, diffusion, and scattering of radio waves in the human body become significant.

As we will see in Sects. 7.2.5 and 7.3.1, BCIs may require high-frequency RF communication channels with large bandwidth. Preliminary preclinical work at Brown University [37] has demonstrated the feasibility of short-range unidirectional (from implant to external receiver) communication at 48 Mbit/s using carrier frequencies around 3.5 GHz. Continuation of this project at the Wyss Center for Bio and Neuroengineering in Geneva, Switzerland [84], over years 2015–2018, has shown, through simulation and bench testing, the variability of performances in large bandwidth high-frequency RF communication for human use.

The nature of the human body and the law of physics impose serious limitations to the use of RF communication in the scope of large flow of information. In terms of bandwidth, BCI systems are much more demanding than any other AIMD. The most powerful commercially available RF chip for implantable medical applications, produced by MicroSemi-Zarlink [85], has a flow rate limited to 0.5 Mbit/s or 100 times lower than what is required by the BCI application described in the previous paragraph. The evolution of BCI is therefore limited by the lack of validated high-performance RF chips. Several laboratories and companies are currently developing more sophisticated RF ICs, but none of them has yet been integrated in a BCI intended for human use. It is an example of a missing block, which will be further discussed in Sect. 6.3.

4.13 Energizing Implants

AIMDs require a source of electrical energy to provide power to the implanted electronics. There are three ways to power an implant:

- Store the energy in the implanted housing (the case for more than 95% of AIMDs):
  - *Primary* (non-rechargeable) *battery*: it is the solution of choice for devices with low consumption (up to 50–100 μA), like pacemakers. When the battery is depleted, the entire device gets replaced by a new one with a fresh battery. Modern pacemakers can last up to 10 years. A drop of battery voltage indicates end-of-life (EOL), and a warning message is displayed on the physician programmer for planning an exchange of pacemaker in due times. Primary batteries are extremely reliable components, hermetically sealed in their own metallic housing. The manufacturing processes have been substantially improved since the early days of the pacemaker industry. Failures

or leakages of implantable grade primary batteries are rare, in the range of part-per-million (ppm). High-drain primary batteries are capable to provide a large current over a short period of time, as needed for implantable defibrillators.
  - *Secondary* (rechargeable) *battery*: adapted to situations where the demand in energy is high (in the range of mA), like certain neurostimulators. In the current state of the technologies, secondary batteries have a limited number of recharging cycles (500–1000). At this later stage, the capacity of the battery inexorably deteriorates and cannot any longer be recharged appropriately. Therefore, depending on the frequency of recharging cycle, after a few years, devices must be replaced by a new one with a fresh battery. Recharging a battery is a critical task, as charging current must remain under a limit to avoid overheating, and overcharging may lead to degassing and risks of explosion. In terms of risks management, rechargeable batteries represent high-impact risks of patient safety. They are mitigated by sophisticated power management circuits with redundant protection features, sometimes including temperature sensors and fuses. Secondary batteries are recharged via an induction coil magnetically coupled to an external recharging unit.
  - *Supercapacitors* (rechargeable): their energy density is much lower than in primary or secondary batteries. The sealing technology is not yet developed to a level which is compatible with long-term implants. They may leak or degas.
  - *Solid-state batteries* (rechargeable): small, thin, robust, and leak-free, but the energy density is several orders of magnitude lower than conventional batteries. Limited to nA applications, like backup memories.

- Battery-less implant with continuous energy transfer from an external headpiece (the case of cochlear implants since three decades):
  - *Inductive coupling*: a flat patch coil is attached to the implant with a magnet in the center for fixation and alignment of the headpiece. The same inductive coupling may also be used for low-bandwidth communication. Depending on distance and heat limitations, such systems may be able to transfer a few hundred mW. It is a robust and mature concept.
  - *Non-electric power transfer*: like ultrasounds or light (NIR). Still at the level of study and prototypes. None of such systems is ready for translation to human applications.

- Harvesting energy in the human body. Interesting work in this direction is conducted by several laboratories, but the power remains too low compared to conventional energy sources. It will take decades to get autonomous harvesting systems applicable to human active devices. Limitations of the harvested power will certainly exclude this alternative for neurological devices.

Wireless implantable BCI systems have unique characteristics which influence the method of energization:

- Collecting, amplifying, sampling, and wirelessly transferring large flows of information consume power in the range of tens of mW. In consequence:

  - It excludes primary batteries, which will be depleted within hours or days.
  - It is not adapted to rechargeable batteries, as frequent recharging (every day or so) will impose implanting a new device within 1 or 2 years, after around 1000 recharges.
  - Continuous power transfer through induction is the solution of choice for BCIs in a near future.

- No miniature multichannel connector is currently available. Therefore, for the next decade or so, we will not be able to detach the brain interface (MEA, ECoG) from the electronic housing. It excludes the use of rechargeable battery and confirms the choice of induction energy transfer.
- Above-the-neck implants (see Chap. 5) must be thin. There are no primary or rechargeable medical grade batteries which are thinner than 4 mm. Recent developments of thin and even flexible batteries do not meet the required moisture and leakage barriers imposed by implantable device standards. Here again, it points out induction power transfer as the best choice for BCIs.

This book focuses on searching pragmatical solutions for implementing human grade BCIs in a reasonable time frame. For this reason, *we recommend induction power transfer*. Alternative energization methods must be further developed and tested, but not at the expense of a fast and robust inductive solution. Induction energy transfer is a stable and fully validated method, which has been satisfactory for many patients over long periods of time. BCI developers should build on this solid ground and allocate their innovation skills to other capital building blocks, as the one described in the next chapter.

4.14 Implantable Connectors

AIMDs have two main components which serve different purposes and have their specific "relation" with the human body:

- *Tissue interface*: or electrodes, with the main function to electrically interact with the body, by "reading" or "writing" at specific locations. Electrodes must be in intimate contact with tissues, stay in place for a long time, be accepted, and integrate in the body. Everything is done to facilitate the symbiosis between the electrodes and the targeted tissues or organ. Insertion of the electrodes often is a delicate surgical act, which may damage the target. The better the electrode integrates in the body, the more difficult it is to remove it. Explantation of electrodes presents a surgical risk and potential tissue damage. Inserting a second electrode at the exact same location is rarely possible. Some experts assume that cortical MEA for movement restoration could be replaced by a second MEA placed beside the first one. It has not yet been clinically demonstrated. Ideally, electrodes should remain in place until death.
- *Implanted electronics*: which collects signals from the electrodes and/or sends stimulation pulses to the tissue interface. This entity is placed at the periphery

of the body, with an easy surgical access, for example, under the skin. For different reasons (depleted battery, hardware issue, technology upgrade, miniaturization), the housing containing the electronics may be exchanged and replaced by a new one.

Exchangeability requires the possibility to disconnect the cable/electrodes assembly from the implanted electronics. This function is assured by an implantable connector. Disconnection and reconnection of a new implanted electronics will be a rare event, once, twice, or a very few times during a lifetime.

Millions of pacemakers are or have been used in this context of detachability. The normal procedure to implant and replace a detachable AIMD follows a simple sequence:

- *First implantation*:
  - Incision of the skin and surgical opening of a route to the targeted tissues.
  - Introduction of the electrodes and fixation to the targeted tissues.
  - The proximal end of the cable and the male connector stick out of the skin opening.
  - A "pocket" is prepared for the implant.
  - The male pins of the cable are inserted in the connector of the electronics and secured in place (set screws).
  - The electronic housing in placed in the pocket which is then sutured.
- *Replacement of the implanted electronics* (usually several years after first implantation):
  - Incision of the skin to open the pocket
  - Extraction of the implant
  - De-connection of the leads
  - Connection of a new implant
  - The new electronics housing in placed in the same pocket which is then sutured

This strategy is well adapted to pacemakers, as the connector has only two or four channels. In addition, pacemakers are in the pectoral area where space is available. Since the 1990s and the standardization of connectors (IS-1), there has been no incentive in the industry to further miniaturize implanted devices. About 10 years ago, increasing number of channels in neurostimulators, from 4 to 8, 16, and 32, has pushed manufacturer to miniaturize further, and denser in-line connectors were developed. But, as neurostimulators remained located below the neck, detachable connectors continued to be quite bulky.

Cochlear implants, in the 1990s, were the first electronic implants to be placed above the neck (see Sects. 1.4.1.3 and 3.4.1). At that time, making a detachable connector with 22 channels was not possible; the size would have been far too large to be placed under the skin behind the ear. Consequently, the only possibility was to attach the 22-wire lead permanently to the titanium housing, *without possibility to disconnect it later*. This has fundamentally changed the configuration of the implanted system, compared to a pacemaker:

- As it would not be possible to exchange the electronics for several decades after implantation, the implanted electronics must be as simple as possible, without components prone to fail.
- The implant must be battery less forcing the development of inductive energy transfer and proximal inductive communication.

Implanted BCIs of today are in the same situation than CIs in the 1990s. We cannot manufacture reliable detachable connectors for 100 channels or more, which are small enough to be placed in the head and remain reliable for on the long term. Therefore, current developments of large channel counts BCIs are still based on a permanent connection of the electrodes to the electronics. It has major consequences regarding long-term applications, as no exchange of the electronic will be possible.

As for CIs, the lack of miniaturized implantable connectors will, for the next 5–10 years, force designer of BCIs to:

- Keep the electronics simple and reliable. It excludes possibilities to include some signal processing, compression, or decoding capabilities in the implant. All the complexity and intelligence must to be kept in the external unit.
- Transmit unprocessed data from the implant to the outside, at the cost of large bandwidth communication.
- Take the risk of patients falling back to their previous condition in case of failure in the electronics housing.

As a conclusion of this section, we can say that detachable miniaturized connectors are the main missing building block of the BCI industry.

References

1. Wolf P (2008) Chapter 3: Thermal considerations for the design of an implanted cortical brain-machine interface (BMI). In: Reichert WM (ed) Indwelling neural implants: strategies for contending with the *in vivo* environment. CRC Press/Taylor & Francis, Boca Raton
2. Becker EL, Landau SI, Manuila A (1986) International dictionary of medicine and biology. Wiley, New York
3. Schaubroek D et al (2017) Polyimide-ALD-polyimide layers as hermetic encapsulant for implants. In: XXXI International Conference on Surface Modification Technologies (SMT31), Mons, Belgium, July 5–7, 2017
4. Op de Beeck M et al (2017) Ultra-thin biocompatible implantable chip for bidirectional communication with peripheral nerves. In: 2017 IEEE Biomedical Circuits and Systems Conference (BioCAS)
5. Op de Beeck M et al (2013) Improved chip & component encapsulation by dedicated diffusion barriers to reduce corrosion sensitivity in biological and humid environments. In: 2013 Eurpoean Microelectronics Packaging Conference, Grenoble, France, Sept 9–12, 2013
6. Xie X et al (2014) Long-term reliability of Al_2O_3 and Parylene C bilayer encapsulated Utah electrode array based neural interfaces for chronic implantation. J Neural Eng 11(2):02016
7. Barrese JC et al (2013) Failure mode analysis of silicon-based intracortical microelectrode arrays in non-human primates. J Neural Eng 10(6):066014
8. https://en.wikipedia.org/wiki/Tyvek

9. https://en.wikipedia.org/wiki/Arrhenius_equation
10. https://en.wikipedia.org/wiki/Saline_(medicine)
11. https://en.wikipedia.org/wiki/Phosphate-buffered_saline
12. Takmakov P et al (2015) Rapid evaluation of the durability of cortical neural implant using accelerated aging with reactive oxygen species. J Neural Eng 12:026003
13. https://www.medgadget.com/2007/04/easyband_remote.html
14. https://www.businesswire.com/news/home/20070222005165/en/Allergan-Announces-Acquisition-EndoArt-Leading-Swiss-Developer
15. https://finetech-medical.co.uk/
16. Thompson DL et al (1998) Filtered feedthrough assembly for implantable device. Patent US5836992, Medtronic Inc
17. Fraley M et al (2002) Leak testable capacitive filtered feedthrough for an implantable medical device. Patent US6349025, Medtronic Inc
18. https://integer.net/product/feedthroughs-filtered-feedthroughs-crm/
19. https://en.wikipedia.org/wiki/Wilson_Greatbatch
20. http://www.morgantechnicalceramics.com/en-gb/products/healthcare-products/medical-feedthroughs/
21. https://www.sct-ceramics.com/en/our-products/implantable-feedthroughs/
22. http://hermeticsolutions.com/
23. https://search2.ucl.ac.uk/s/search.html?query=anne%20vanhoest&collection=website-meta&profile=_directory&tab=directory
24. Vanhoestenberge A, Donaldson N (2013) Corrosion of silicon integrated circuits and lifetime predictions of implantable electronic devices. J Neural Eng 10:031002
25. Vanhoestenberge A et al (2014) Underwater current leakage between encapsulated NiChrome tracks: implications for strain-gauges and other implantable devices. Sens Actuators A Phys 212:1–11, Elsevier
26. https://www.schott.com/primoceler/
27. https://www.us.schott.com/english/index.html
28. http://aemf.org/
29. http://aemf.org/item/neuromuscular-disorders/
30. http://hermeticsolutions.com/
31. https://rippleneuro.com/
32. https://www.imtek.de/home-en?set_language=en
33. http://www.imtek.de/laboratories/biomedical-microtechnology/bm_home?set_language=en
34. Schuettler M et al (2011) Ensuring minimal humidity levels in hermetic implant housings. In: 2011 Annual International Conference of the IEEE Engineering in Medicine and Biology Society
35. Schuettler M et al (2012) Hermetic electronic packaging of an implantable brain-machine-interface with transcutaneous optical data communication. In: 34th Annual International Conference of the IEEE EMBS, San Diego, CA, USA, Aug 28 to Sept 1, 2012
36. https://cortec-neuro.com/
37. Yin M et al (2013) A 100-channel hermetically sealed implantable device for chronic wireless neurosensing applications. IEEE Trans Biomed Circuits Syst 7(2):115–128
38. https://en.wikipedia.org/wiki/Arto_Nurmikko
39. https://www.csem.ch/home
40. https://en.wikipedia.org/wiki/United_States_Military_Standard
41. https://en.wikipedia.org/wiki/MIL-STD-883
42. MIL-STD 883, Test method standard for microcircuits, Revision J, June 7, 2013
43. MIL-STD 750, Test methods for semiconductor devices, Revision F, Nov 30, 2016
44. MIL-STD 202, Electronic and electrical component parts, Revision G, Feb 8, 2002
45. https://en.wikipedia.org/wiki/ASTM_International
46. ASTM F/34-72T, Tentative recommended practices for determining hermeticity of electron devices with a helium mass spectrometer leak detector. Annual book of ASTM standards, Part 43, 1976

47. MIL-STD 1576, Electroexplosive subsystem safety requirements and test methods for space systems, July 31, 1984
48. https://www.accessdata.fda.gov/scripts/cdrh/cfdocs/cfRes/res.cfm?id=35401
49. Ely K (2000) Issues in hermetic sealing of medical products, medical device & diagnostic industry magazine MDDI, Jan 2000
50. https://www.alphaadvancedmaterials.com/
51. https://www.orslabs.com/services/rga-package-gas-analysis/
52. https://www.orslabs.com/services/rga-package-gas-analysis/iga/
53. Munson T What is electrochemical migration – dendrite shorting of electronic circuits? White Paper, Foresite Inc. www.foresiteinc.com
54. ISO 5841-3:2013 Implants for surgery – cardiac pacemakers – Part 3: Low-profile connectors (IS-1) for implantable pacemakers
55. ISO 11318:2002 Cardiac defibrillators – connector assembly DF-1 for implantable defibrillators – dimensions and test requirements
56. ISO 16429:2004 Implants for surgery – measurements of open-circuit potential to assess corrosion behavior of metallic implantable materials and medical devices over extended time periods
57. Op de Beeck M et al (2019) FITEP: a flexible implantable thin electronic package platform for long term implantation applications, based on polymer & ceramic ALD multilayers. IMAPS Medical Workshop, San Diego, Jan 2019
58. Hogg A et al (2014) Protective multilayer packaging for long-term implantable devices, Elsevier. Surf Coat Technol 255:124–129, SCT-19263
59. Hogg A et al (2023) Ultra-thin layer packaging for implantable electronic devices, IOP Publishing. J Micromech Microeng 23:075001
60. Hogg A (2014) Development and characterization of ultrathin layer packaging for implantable medical devices. PhD Thesis, Graduate School for Cellular and Biomedical Sciences, University of Bern, Switzerland
61. http://coat-x.com/
62. Merrill D et al (2005) Electrical stimulation of excitable tissue: design of efficacious and safe protocols, Elsevier. J Neurosci Methods 141:171–198
63. Grimnes S, Martinsen OG (2000) Bioimpedance and bioelectricity, basics. Academic Press, London, ISBN 0-12-303260-1
64. IEC-60068-2-75:2014 Environmental testing, Part 2-75: Test Eh: Hammer tests
65. https://open.fda.gov/data/faers/
66. https://www.accessdata.fda.gov/scripts/cdrh/cfdocs/cfstandards/detail.cfm?standard__identification_no=33456
67. https://blog.atltechnology.com/protect-yourself-from-esd
68. Ker M-D et al (2004) Active electrostatic discharge (ESD) device for on-chip ESD protection in sub-quarter-micron complementary metal-oxide semiconductor (CMOS) process. Jpn J Appl Phys 43(1A/B):L33–L35
69. https://www.fda.gov/MedicalDevices/DeviceRegulationandGuidance/GuidanceDocuments/UCM077272
70. https://www.fda.gov/MedicalDevices/DeviceRegulationandGuidance/GuidanceDocuments/UCM348890
71. IEC-60601-1: Medical electrical equipment – Part 1: General requirements for basic safety and essential performance
72. IEC-60601-1-2: Medical electrical equipment – Parts 1–2: General requirements for basic safety and essential performance – collateral standard: electromagnetic disturbances – requirements and tests
73. ISO-14708-3: Implants for surgery, active implantable medical devices – Part 3: Implantable neurostimulators
74. https://www.fda.gov/MedicalDevices/DeviceRegulationandGuidance/GuidanceDocuments/UCM470201

75. https://www.fda.gov/MedicalDevices/DeviceRegulationandGuidance/GuidanceDocuments/UCM331681
76. https://www.fda.gov/MedicalDevices/DeviceRegulationandGuidance/GuidanceDocuments/UCM077272
77. https://zmt.swiss/
78. https://zmt.swiss/applications/wireless-body-area-networks/
79. https://zmt.swiss/sim4life/
80. https://en.m.wikipedia.org/wiki/Physics_of_magnetic_resonance_imaging
81. Electronic Code of Federal Regulations, Title 47: Telecommunication, Chapter I, Subchapter A, Part 15 (Radio Frequency Devices). https://www.ecfr.gov/cgi-bin/text-idx?c=ecfr&SID=d9799480849898b1aaf705b67e156ade&rgn=div5&view=text&node=47:1.0.1.1.16&idno=47
82. FDA, Recognized Consensus Standards, AAIM TIR69:2017, Technical information report risk management of radio-frequency wireless co-existence for medical devices and systems. https://www.accessdata.fda.gov/scripts/cdrh/cfdocs/cfStandards/detail.cfm?standard__identification_no=35548
83. FDA, Recognized Consensus Standards, ANSI IEEE C83.27-2017, American National Standard for the Evaluation of Wireless Co-existence. https://www.accessdata.fda.gov/scripts/cdrh/cfdocs/cfStandards/detail.cfm?standard__identification_no=35809
84. https://www.wysscenter.ch/projects/
85. https://www.microsemi.com/product-directory/ultra-low-power-wireless/1312-implantable-medical-transceivers
86. Brown RW et al (2014) Magnetic resonance imaging – physical principles and sequence design, 2nd edn. Wiley Blackwell, Hoboken, ISBN 978-0-471-72085-0
87. https://www.meddeviceonline.com/doc/protecting-implantable-medical-devices-from-electromagnetic-interference-0001?vm_tId=2124770&user=b231f621-6dbd-47f9-a9b8-c271e83de85a&vm_alias=Protecting%2520Implantable%2520Medical%2520Devices%2520From%2520Electromagnetic%2520Interference&utm_source=mkt_MDOL&utm_medium=email&utm_campaign=MDOL_04-30-2019&utm_term=b231f621-6dbd-47f9-a9b8-c271e83de85a&utm_content=Protecting%2520Implantable%2520Medical%2520Devices%2520From%2520Electromagnetic%2520Interference&mkt_tok=eyJpIjoiWW1VeE9ESXpZV-EUwTW1VeCIsInQiOiJqTTdoN3hEQnR5MjViYU1WTHloVEdqb3ZZd0pvc25FbkxrTzh6SzV1OGw0T2lkZmVoVEJnTmJGTzdpUkd5MTZmUUdRXC9kOTVTK3ZlM2VLTno0dTF6WVlwMW5BWWFqMGdMT0l0MGtWMmxOTXNsSFcrbFVUZHExWldSTFR3eEo2TmYifQ%253D%253D
88. Jeong J et al (2019) Conformal hermetic sealing of wireless microelectronic implantable chiplets by multilayered atomic layer deposition (ALD). Adv Funct Mater 29:1806440

Chapter 5
Below and Above-the-Neck

5.1 Below the Neck

AIMDs for cardiac applications are naturally located not far from the heart. Pacemakers and IPGs are conveniently inserted under the skin in a pectoral pocket, and their leads follow blood channels toward the various heart chambers. Apart from DBS, the first neurostimulators were targeting the spinal cord, nerve roots, or peripheral nerves. Electrodes could be attached to their respective targets and the IPG remaining at convenient distance in the chest, trunk, belly, or back, implanted in subdermal pockets.

Inserting metal encapsulated devices in soft tissues anywhere between the hips and the shoulders is relatively easy. Pockets can be created between tissue layers, with minimal risks of damaging blood channels. These pockets expend nicely to make room for the implant. Most of the time, the implant is not or minimally visible under the skin, and patients do not feel it, or at least it does not create pain or trouble. This part of the body is very forgiving to insertion of implants in soft tissues, if the device has a volume below a few dozen cm^3. Early ICDs were large and heavy, making them less accepted by patients. Programmable implantable drug delivery pumps, still in use today, are too large (about 200 cm^3) for the patient to forget about them.

This tolerance of the human body to reasonably sized implants has been a barrier to miniaturization. Since more than 20 years, pacemakers are small enough to be accepted by patients, and additional efforts to reduce their size would just increase manufacturing costs or reduce device autonomy. Even if the electronic features have constantly progressed, AIMDs have not changed much in shape and mechanical encapsulation since the 1990s.

Because of this lack of incentive to miniaturize, the industry has not anticipated the new specific needs of neurostimulators. When more channels were needed, designers simply added more FTs, more connectors, and larger header and housing to accommodate them. Compared to pacemakers which generate very short pulses

C. Clément, *Brain-Computer Interface Technologies*,
https://doi.org/10.1007/978-3-030-27852-6_5

every second, neurostimulators usually deliver continuous burst of pulses at a rate around 100 Hz. To deliver these signals, neurostimulators require larger batteries or rechargeable batteries. There too, the industry made minimal efforts to miniaturize batteries. Consequently, commercial implantable neurostimulators are two to six times larger than pacemakers but still small enough to be accepted by patients.

SCS and DBS neurostimulators are often available in two configurations with identical electrical stimulation features:

- Larger size with primary battery.
- Smaller size with rechargeable battery.

It has been seen that patients often prefer the larger device, as they don't need to bother with regular recharging. It is one more sign indicating that size is not a critical factor for below-the-neck implants.

In cases of patients with thin skin and low body mass, the implant might be protruding slightly and be noticeable. But I have rarely seen patients bothered by that, as it can always been hidden under clothes.

By being inserted in soft tissues, below-the-neck implants are not much sensitive to external impacts.

5.2 The First Steps in the Direction of the Head

In the 1980s adjustable shunt valves have been designed to releasing excess of CSF for hydrocephalic patients. This device does not strictly fall in the category of AIMDs as it has no implanted electronics nor batteries. It is nevertheless interesting on many aspects because it includes an implanted magnetic rotor which can be rotated by an externally applied rotating field. Aspects related to the protection of the magnets for assuring long-term biostability or to miniaturization for insertion in the skull make valves for hydrocephalus pioneers in the field of neurological implants.

At the beginning of the 1990s, two clinical indications targeting the head have been approved:

- *DBS for Parkinson's disease*: if electrodes are placed in the brain, the IPG remains in the chest, simply because it is too big to be implanted above-the-neck. This device has been discussed in detail previously (see Sect. 3.4.2). Cables must be tunneled under skin, from the chest to the top of the head, passing the mobile area of the neck. Even if tunneling is now a well-controlled surgical procedure, the long cable crossing the neck generates potential risks:
  - Fatigue rupture due to frequent movements of large amplitude along the neck
  - Two sections of cable, one from the IPG to an intermediate connector behind the ear, and the cable leading to the DBS electrodes. The multiplication of elements and contact areas increases risks of failures.

  - To remain flexible, the cable cannot include many wires. Eight wires seem today a reachable limit. Aleva (refer to Chap. 3, [30]) is considering to tunnel along the neck cable with 12 wires. Long-term clinical assessment is still needed.

- *Cochlear implant*: (see Sects. 1.4.1.3 and 3.4.1) this device is the first purely above-the-neck AIMD. As seen previously, the specificities and location of CIs have forced the designer to totally rethink the technologies traditionally used in active implants. We can consider CI as the ancestor of BCI. For two to three decades, CI remained an atypical product, and the neurotechnology community was slow to understand it was opening the way for more sophisticated cranial implants.

5.3 Above-the-Neck Implants

The human head is very special. On many aspects, it differs significantly from the rest of the body (see Fig. 5.1).

Even if it looks trivial, let's review and discuss some of the specificities of our heads:

- Compared to the rest of the body, the volume of the head is limited. AIMDs located above-the-neck must be miniaturized.
- Inside the skull, there is no free space available for insertion of an implant. The rest of the body can expand to accommodate the addition of a device, but not the head, as it is constrained by the rigid envelop of the cranium.
- To go inside the head, we must open the skull through an invasive surgery. This is much more difficult than making a simple incision in the skin and creating a pocket in soft tissues. The rare natural openings of the skull, at the level of the eyes, ears, and spinal cord, are not accessible for the introduction of implants.

Fig. 5.1 Specificities of the human head

- Under the scalp, only thin devices can be introduced. In addition, subcutaneous devices in the head must either be flexible or of rounded shape to accommodate the natural curvature of the skull. It is generally agreed that a subcutaneous device should have a maximum thickness of 4 mm and rounded edges. Thicker devices will be too visible and may lead to tissue erosion. Alternatively, partial or total craniotomies may be done to accommodate thicker housings.
- The brain is a very soft material. Unlike muscle or fat, brain tissue cannot hold a device in place. Device inserted inside the skull must be attached to the bone and cannot be left floating around. Relative movements of the brain may induce irreversible damage in case of excessive pressure on foreign elements (electrodes or housing) fixed to the skull.
- The head is visible and reflects an important part of our personality. Implants which remain apparent on the head are not well accepted by patients. For example, the transdermal pedestal for the connection of the Utah array (see Sect. 3.3.1) is perceived as being too visible. Aesthetics is capital for head implants. Other parts of the body can easily be hidden by appropriate clothes.
- The head is placed on the neck. It looks like a triviality, but the consequences of the large mobility of the neck are important. It is very difficult to have wire connections between the head and the rest of the body.
- We may also add that the head is exposed to impacts. Implants placed under the skin or inserted in a craniotomy are only protected by a thin layer of tissues. In addition, they are backed by solid bone or attached to the edges of the craniotomy. In case of external impact, most of the incident energy will be taken by the implant.

The above described specificities of the head having a large influence on the design of cranial active implants. Developers of BCIs must:

- Forget about traditional design rules and principles coming from below-the-neck AIMDs.
- Rethink fundamental encapsulation concepts in the perspective of the specificities of the human head.
- Partner with neurosurgeons and plastic surgeons to adapt BCI devices to surgical technics specific to the skull.
- And understand the special relation that patients have with their head.

The first above-the-neck AIMD including a battery is the RNS device from NeuroPace, extensively discussed under Sect. 3.4.7. Unique features like the curved shape, the fixation in a ferrule inserted in a craniotomy, and the proprietary detachable connector make this device a great source of inspiration for future BCIs and a pioneer concept.

Some neurosurgeons and plastic surgeons, experimented in post-traumatic reconstruction of the skull, have innovative views regarding above-the-neck implants. Work initiated by Paul Manson [1] and continued by Chad Gorgon [2], both plastic and reconstructive surgeons at Johns Hopkins University, in collaboration with a Baltimore start-up named Longeviti [3], consists in removing a substantial portion of the skull bone and replacing it by a patient-specific skull insert in polymethyl methacrylate (PMMA) [5]. PMMA is long-term biocompatible and biostable [6]. The dimensions and shape of the insert are determined by patient's CT scan. 3D printing allows to recreate in PMMA the exact replacement of the surgically removed portion of the skull. Specific medical devices could be embedded in the PMMA insert. For example, existing devices like CIs or RNS from NeuroPace could be integrated in a PMMA insert perfectly adapted to the patient's anatomy.

The first application of the concept of removing a portion of the skull and replacing it by an insert with an implant is specific to the integration of a shunt valve for hydrocephalus (see Fig. 5.2). In this execution, the insert is made of high-density polyethylene (PE), and the shunt valve is an approved product. The insert greatly facilitates surgery and fixation of the valve. Patient's comfort and aesthetics are substantially improved. The concept could be extended to the integration of BCIs.

Figure 5.3 shows a patient-specific 3D laser-printed PMMA skull insert for bone reconstruction. The concept has been further extended by merging both technologies leading to the integration of active devices in 3D laser-printed PMMA inserts. The active device seen on Fig. 5.4 is a prototype of BCI connected to an ECoG. As discussed in Sect. 4.9.6, laser-printed PMMA inserts only provide a near-hermetic encapsulation. PMMA absorbs water in the range of 0.3–0.4% in weight. Moisture will slowly diffuse through the PMMA and reach electronics and battery. Complete validation and long-term accelerated aging tests must be done to assess the degree of moisture protection provides by PMMA. This project is still far from getting approval, but the concept is promising.

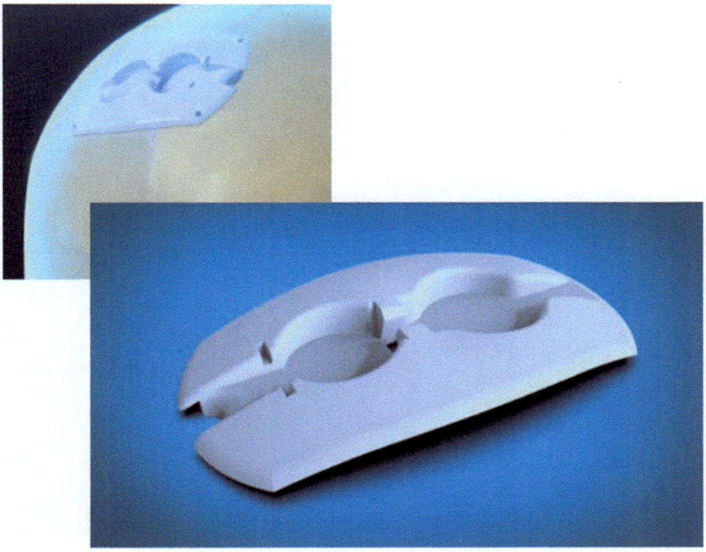

Fig. 5.2 Skull insert for hydrocephalus shunt valve, InvisiShunt®, Model OP1000. *(These figures are reprinted with the permission of Longeviti Neuro Solutions LLC; InvisiShunt® is a registered trademark of Longeviti Neuro Solutions LLC)*

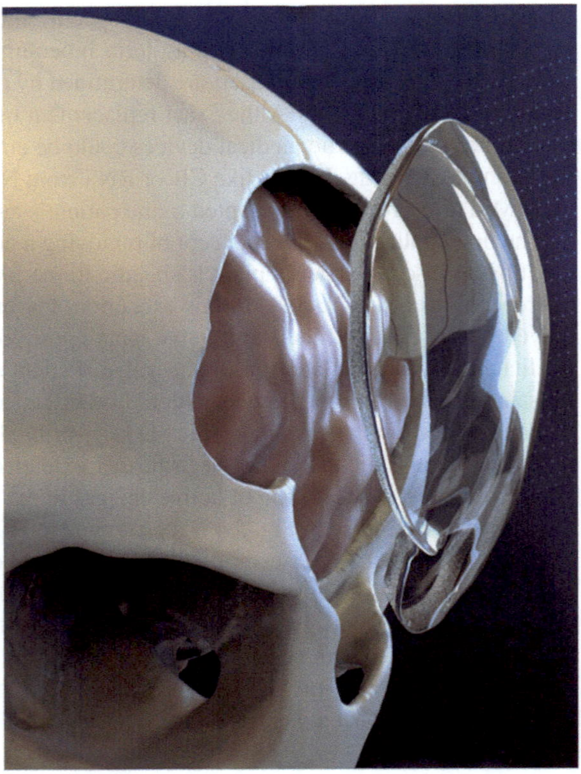

Fig. 5.3 3D laser-printed skull insert in PMMA, ClearFit™. *(This figure is reprinted with the permission of Longeviti Neuro Solutions LLC; ClearFit™ is a registered trademark of Longeviti Neuro Solutions LLC)*

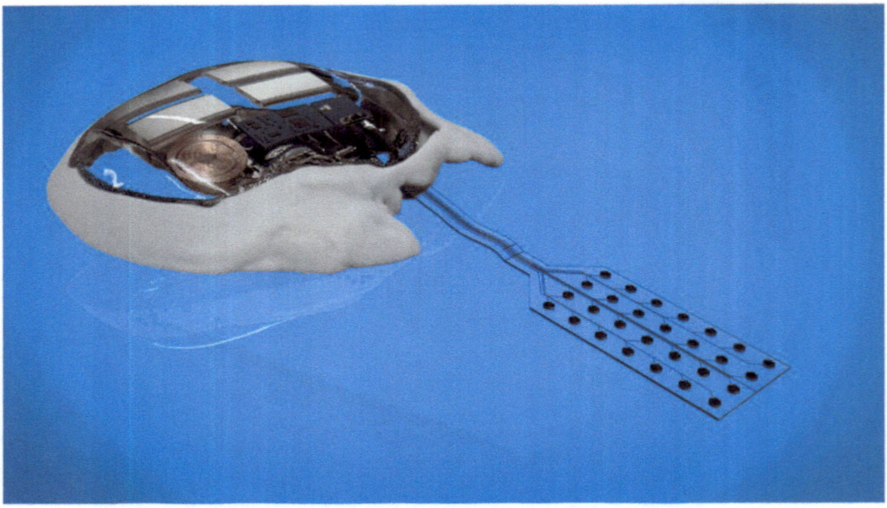

Fig. 5.4 Prototype of BCI integrated in a 3D laser-printed skull insert in PMMA. *(This figure is reprinted with the permission of Longeviti Neuro Solutions LLC)*

We saw under Sect. 4.9.7 that the weak point of devices in a near-hermetic configuration is the junction between the cable and the electronics. Moisture will creep along the cable and insulator sleeve and reach the electronics. This is a risk which may show up in the non-hermetically sealed electronics of Fig. 5.4. For this reason, the best application of the PMMA patient-specific insert concept is to integrate in the bone replacement plate a hermetically sealed implant, like a CI or the RNS system from NeuroPace.

Another remarkable achievement in the field of above-the-neck BCI implants is the Wimagine® project [4], conducted by Clinatec [5] in Grenoble, France, under the guidance of Prof. A.-L. Benabid (see Sect. 3.4.2 and reference [27] of Chap. 3), the DBS pioneer. Wimagine includes an 8x8 ECoG electrode placed under a wireless BCI encapsulated in a titanium can (see Fig. 5.5) inserted in a craniotomy (see Fig. 5.6). The device covers both the arm and leg areas of the motor cortex. In the first ongoing human trial, two devices have been implanted, one on each side of the brain and connected to an exoskeleton with the goal of restoring some movement in the four limbs of a paralyzed patient (see Fig. 5.7a–c). Even if the Wimagine system needs further improvement and validation, it is a precursor device in the field of above-the-neck BCIs and a source of inspiration (see Fig. 5.8). Part of the originality of the concept resides in the placement of the electrodes directly under the housing. Therefore, there is no cable to connect these two entities. It is step in the direction of the "brain button" (see Sect. 7.4.1), a vision to integrate the entire system (electronics, wireless communication, power management) on the back of the electrodes. The Wimagine project is also innovative in its ambition to decode movement intentions without penetrating electrodes.

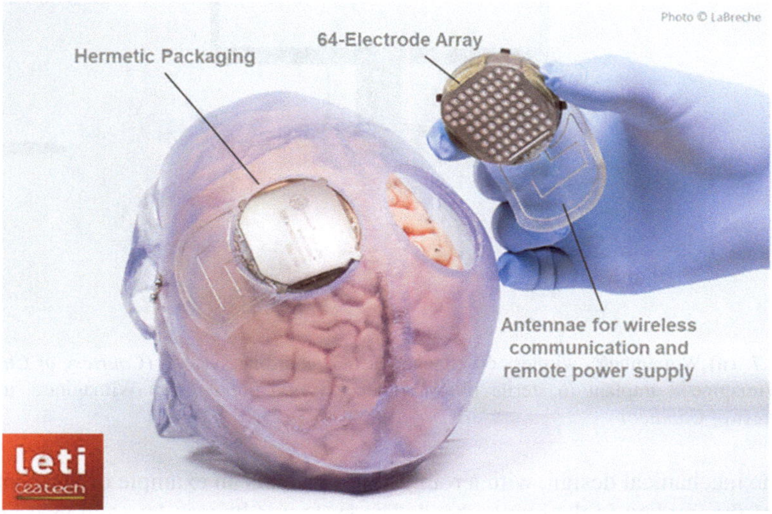

Fig. 5.5 Wimagine®, wireless 64-channel ECoG recording implant, insertion in a craniotomy. *(Courtesy of Clinatec)*

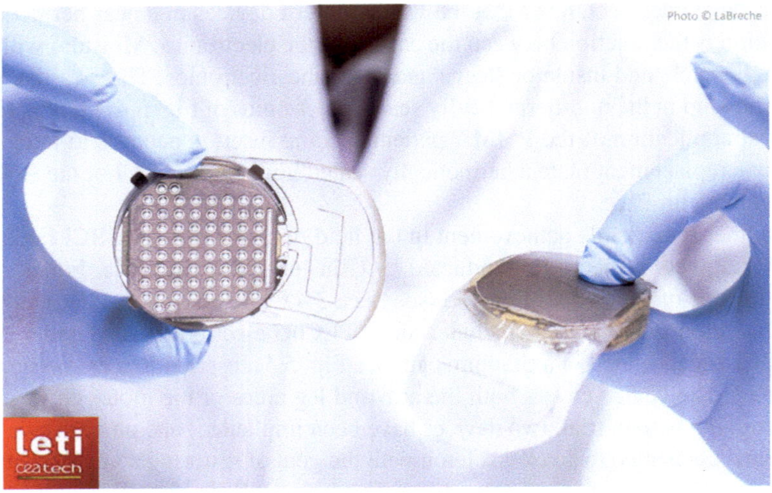

Fig. 5.6 Wimagine®, wireless 64-channel ECoG recording implant, top and bottom views. *(Courtesy of Clinatec)*

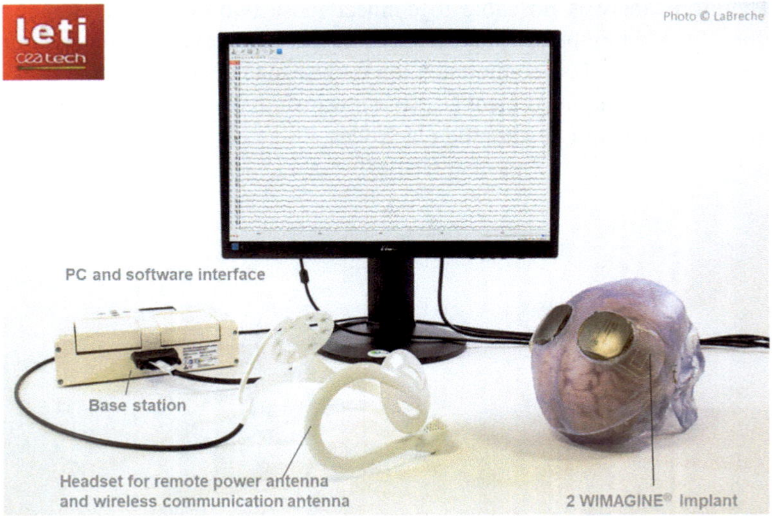

Fig. 5.7 (**a**) Wimagine®, wireless 64-channel ECoG recording system. *(Courtesy of Clinatec).* (**b**) Wimagine®, implant in sterile blister. *(Courtesy of Clinatec).* (**c**) Wimagine® implant. *(Courtesy of Clinatec)*

The mechanical design, with a rounded top plate, is an example of good integration of the implant in the human head. The system is battery-less and uses and RF communication link.

However, some difficulties to read the motor cortex corresponding to the lower limbs may be expected. Unlike the area of the cortex commanding the upper limbs,

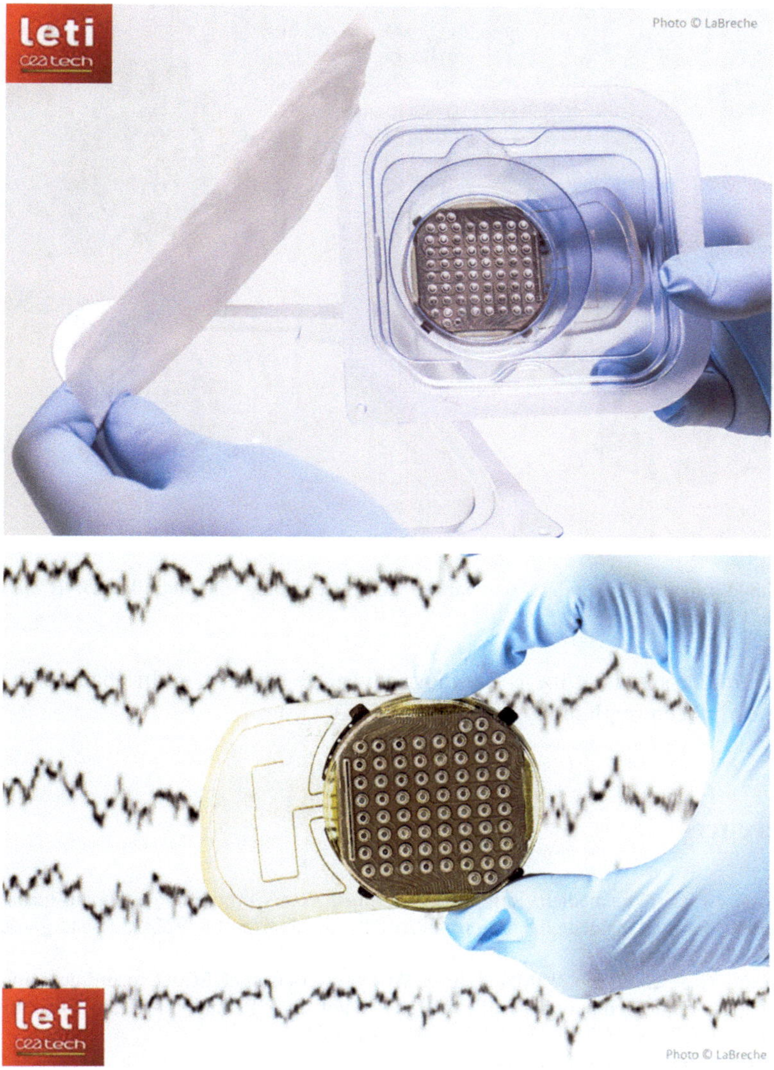

Fig. 5.7 (continued)

located in the surface on the brain, the motor area of the legs is mainly folded in a convolution. ECoG electrodes are likely to collect some useful information related to the movement of the arms but will be less appropriate to gather neuronal signals from the intentions of moving the legs. For this specific region of the motor cortex, penetrating electrodes are likely to be needed. Even the Utah array might be inappropriate as it cannot be easily inserted in the narrow grooves of the convoluted surface of the brain.

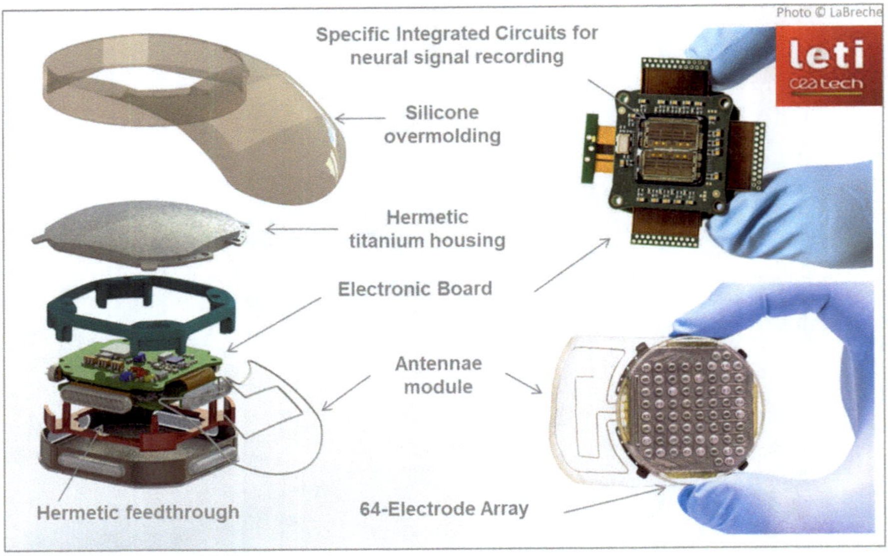

Fig. 5.8 Wimagine®, wireless 64-channel ECoG recording implant, details. *(Courtesy of Clinatec)*

Nevertheless, Wimagine is the most advanced wireless multi-channel BCI that has already been implanted in the human head.

References

1. https://www.hopkinsmedicine.org/profiles/results/directory/profile/0001194/paul-manson
2. https://www.hopkinsmedicine.org/profiles/results/directory/profile/9251953/chad-gordon
3. https://longeviti.com/
4. Mestais C et al (2015) WIMAGINE®: Wireless 64-channel ECoG recording implant for long term clinical applications. IEEE Trans Neural Syst Rehabil Eng 23(1):10. https://doi.org/10.1109/TNSRE.2014.233354, TNSRE-2103-00171.
5. http://www.clinatec.fr/en/clinatec/
6. https://en.wikipedia.org/wiki/Poly(methyl_methacrylate)

Chapter 6
Pioneers

6.1 History of Active Implantable Medical Devices (AIMD)

On purpose, we spent much time in the previous chapters to analyze what has been done in the past. This is a mind-set. As stated before, good innovation rarely finds its grounds from pure genius. Most bright ideas come from understanding our environment (see Sect. 2.2), in a wide sense of the word. This methodology applies to the creation of future BCIs. History is not a negative word. Often, scientists and engineers ignore or disregard the lessons of the past. In my opinion, the most productive approaches in terms of innovation are based on experience. Debating with "gray haired" people may be a good way to find the routes of innovation. Young engineers following this wise guidance will, one day, become the leaders in innovation.

As already described in Sect. 1.4.1.1, the development of cardiac pacemaker is the root of the current successes of AIMDs. It took a lot of energy and faith to develop the first pacemakers in the 1960s. At that time, a lot of people were dying from well-identified and diagnosed cardiac disorders, but without cure or therapy. Pioneers in this field understood that electrical stimulation could be used to accelerate hearts beating too slowly, to re-synchronize some of the four chambers of the heart, to stop fibrillation, or to restart hearts which suddenly ceased to beat. Most researchers in the cardiac field knew it was possible to do marvels with electricity, but technologies were not ready for a widespread solution.

In this context, engineers took risks, but reasonable ones. At the beginning of this industry, we had no reliable batteries and no way to encapsulate electronic circuits well enough to last for decades. Anyway, in a pragmatical approach, pacemakers with mercury batteries and basic electronics in non-hermetic encapsulations have been implanted in patient in serious conditions. The alternative was death. Risk-benefit balance is a capital concept in AIMDs. It also applies to BCI.

A good lesson should be taken from Arne Larsson [1], who got the first implantable pacemaker in 1958 at the age of 43. This patient was in critical condition with

C. Clément, *Brain-Computer Interface Technologies*,
https://doi.org/10.1007/978-3-030-27852-6_6

life expectation limited to months, at best. The device, very simple and far from current pacemakers, had been designed and built by Rune Elmqvist [2] and implanted by Dr. Ake Senning [3] at Karolinska Sjukhuset in Stockholm, Sweden. Arne's device did not last long and had to rapidly be replaced. In total, he received 26 implants, but had his life saved. He had a full active life until 2001, 43 years of life extension due to some "reasonable" risk takings. I met both Arne Larsson and his surgeon Dr. Senning before they passed away. It was a great lesson and a motivation to go on designing AIMDs.

The cardiac industry went on in the development of better, more reliable devices, applicable to life-supporting but also to quality of life improvement devices. A major step was achieved in the 1980–1990s with implantable defibrillators [4]. New technical challenges were in front of us, especially linked to the use of high voltage components in the human body. There too, important lessons must be retained on how to integrate new technologies in AIMDs. It must be put in the perspective of risk management and risk-benefit balance.

If the first cardiac active implants were mainly life-supporting devices, the bulk of cardiac indications are today focused on the improvement of quality of life. Patients suffering from cardiac disorders linked to failing synchronization between the four chambers of the heart, to atrial fibrillation, or to bradycardia find today acceptable solutions by implantation of appropriate active devices.

Cardiac AIMDs are the roots of all developments in the field explored in this book. Several key technologies discussed previously, like hermetic encapsulation, wireless communication, and detachable electrodes, would never have become available without the efforts of the cardiac industry. Similar technologies and hardware opened the road for the first neurological applications like CI, DBS, SCS, and SNS, as described in Chaps. 1 and 3.

The above discussed successes in neuro-applications are based on the pragmatical approach consisting of solving problems step by step, clever balanced risk taking, and integration of new technologies when they become available. We recommend the same wisdom in the development of current and future BCIs.

6.2 New Fields: It Works!

Even if progresses are slow, development durations very long, and costs of putting a product on the market shockingly high, new neurological applications are becoming available to patients in needs. It works!

Kids born with serious hearing disorders can nowadays get a CI early in their childhood, have a normal life, go to school, learn a profession, and feel happy. People with intractable chronic back pain may benefit from SCS to continue a full active life. If Parkinsonians are not cured by a DBS system, their quality of life is improved to such a level that they feel like not being any longer victims of the disease. Likely, SNS provides thousands of patients with a substantial improvement of their bladder control.

What has been achieved so far in the field of neuroactive devices is amazing. We can estimate that more than half a million people had their life changed by neurological implants. Ongoing efforts in the existing therapies will dramatically improve the efficacy of the treatment but also will allow to address the solutions to a vast majority of still untreated or undertreated patients. Our efforts should be deployed, in parallel, at several levels:

- Improve existing neuro-therapies for better performances, lower costs, lower risks, and larger population.
- Address poorly met medical needs with better neuro-technologies.
- Apply neuro-technologies to unmet medical needs.

6.3 Missing Technology Blocks

Since the early developments of first generation of AIMDs, engineers have been faced to unexpected challenges. The ideas were arriving quicker than the technologies needed to turn an innovative concept in a therapy. At first, we did not know much about how to marry body tissues with foreign materials. It took a couple of decades for the pioneers to understand how reactive to intrusion our body was. We were not expecting to have such a hard time to move from existing concepts to implanted devices. We have seen earlier in this book that we found some methods and technics to translate ideas in human grade implants, but we would like to do more and to do it faster. There is a constant frustration of not being able to be better. The main limiting factors to further development of AIMDs and BICs are:

- Missing technology blocks.
- Technical barriers.
- Laws of physics.

A neurostimulator has a relatively simple electronic circuit under the hood. We could dream of much more complex electronics, but the human body imposes major constrains (see Fig. 6.1) already discussed in detail in the previous chapters.

Section 7.2 will recapitulate the various barriers and challenges already identified in this book. Beside technical hurdles or limitations due to physics, there are several so-called missing blocks which have been harnessing our ambitions in the development of BCIs. The main elements missing today are:

- High-power density primary batteries.
- Thin batteries (hermetically sealed).
- Secondary batteries with >10′000 recharging cycles.
- Secondary batteries with fast recharging (>10C).
- Multichannel (>50) miniature connectors for long term implantation.
- Flat flexible implanted ribbon cables with more than 100 traces.
- Thin shock-resistant housing materials with electromagnetic transparency.
- Low-power high-bandwidth RF chips.

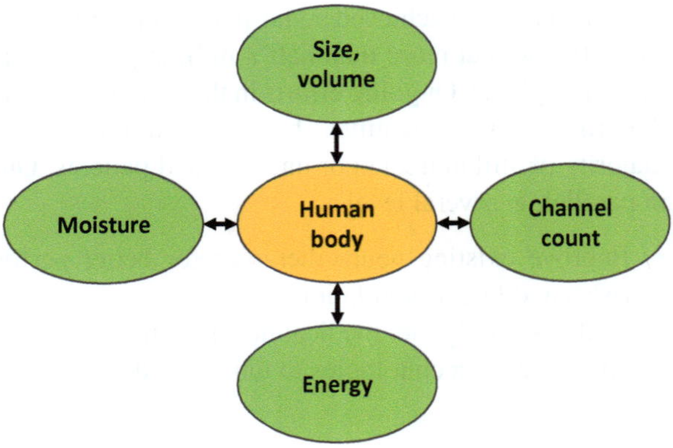

Fig. 6.1 Constrains imposed by the human body on implanted electronics

References

1. https://www.medmuseum.siemens.com/en/stories-from-the-museum/herzschrittmacher
2. https://en.wikipedia.org/wiki/Rune_Elmqvist
3. https://en.wikipedia.org/wiki/%C3%85ke_Senning
4. https://en.wikipedia.org/wiki/Implantable_cardioverter-defibrillator

Chapter 7
Doers

7.1 New Technologies

In this chapter, I illustrate some of the "real-life" situations, not only based on my experience but also my current activities and projects in which I'm involved. It is not theory, but reality.

Working with human beings forces us to minimize risk taking. The field of BCI is sometimes perceived as being overconservative. Academic groups often try to be much more innovative than industrial team. Trying to introduce new technologies too soon is a major cause of failure in neuro-technologies. Why do innovations coming from other fields, like consumer electronics, take so long time to find their way to the human brain? Here are a few clues:

- The field covered by this book is *BCI for human benefit*. The key word is *human*. We interface with the most important part of our body: the brain. Patient safety is priority number one. It impacts our choices and reduces substantially the available options. New technologies must be first proven to be safe for human before becoming part of our design options. The process of rendering a new technology safe for human grade long-term implant is long and tedious. Consequently, a new technology has the time to become an old technology until it reaches the patients!
- AIMDs are highly regulated, must comply with a lot of standards, and go through long clinical trials until the device is approved. When the approval is granted, nothing can be modified. All the specifications, components, and processes must remain unchanged until the product is retracted from the market. It is a dramatic barrier to innovation. Introducing a new technology in an existing line of products may at best require an approval supplement (PMA-S), or a full set of clinical assessments, or, in the worst case, filing for a new approval.
- Approval authorities and regulators have an intrinsic aversion for innovation, which comes from their duty to protect the population from unsafe medical products. Until proven safe, a new technology is always under the suspicion of being a new risk for patients. Pioneers who try to introduce innovative components,

C. Clément, *Brain-Computer Interface Technologies*,
https://doi.org/10.1007/978-3-030-27852-6_7

processes, or technologies must produce evidences, proofs, and rational explanations convincing enough to reassure the FDA.
- BCIs are not lifesaving devices. Regulators, developers, and even patients are not ready to take *unreasonable risks* to introduce new technologies. Benefits-to-risks ratio and benefit-to-costs ratio are two barriers to innovation. For being able to demonstrate that a new technology is providing superior benefits without increasing the risks nor the costs, innovators must do their homeworks, build prototypes, and develop and validate new processes, components, and products. Benefits might be foreseen early in the development process but will be proven only at the end of the clinicals and may be even much later in the post-market phase. New or increased risks induced by a new technology may appear very late in the design cycle. Regarding cost supplements related to the introduction of new technologies, they are usually grossly underestimated. Reality of costs is apparent at the end of a project.
- National health costs keep growing. In certain countries, the economic burden of "non-health" is becoming unbearable. Some new technologies may never reach medical applications, simply because the economic impact is too high. Even if the benefits may be attractive, the healthcare system may not be able to afford the innovation.

During the last two decades, we have seen an explosion of technologies, especially in the field of communications. Today, everybody is interconnected with many other people. Remember what we said at the beginning of this book: an individual human brain represents hundredths of billion neurons, each connected to their neighbors through about 10,000 connections. So, at the scale of the planet, our engineered communication network is not even approaching the complexity of one single brain.

Nevertheless, what is happening today on the front of new technologies is amazing. "Explosion" is not an exaggerated wording, at least in all the fields related to data, information, and communication. Some examples are:

- *Big data*: [1] We are now able to store, compare, analyze and exchange enormous quantities of information. It is a fundamental change of paradigm. We come from an environment of individual patients handled independently, with the highest respect of privacy, to a world of sharing, comparing, aggregating, and information integration. It is a revolution for the healthcare system. In the world of BCI, it is a drastic evolution in our mind-set. We start to understand that the answer to individual needs may reside in the collective database. Real-time data extracted from the brain of a patient have limited value if taken as such. Compared to similar data accumulated in the past from this very patient and put in perspective of comparable data from thousands of other patients, the value of the information gathered from an individual brain will be multiplied by several orders of magnitude.
- *Artificial intelligence and machine learning*: [2, 3] These are not only buzz words. A lot is happening in this world, covering all our everyday life, from our personal assistant to financial algorithms. Our modest efforts to interface with

the brain may get a formidable help from these new technologies. BCI patients do not lack real natural intelligence, but finding tricks to compensate for their disabilities may need support from artificial intelligence and machine learning.

- *Wireless communication*: [4] We see an amazing evolution of communication, in terms of volume, distance, frequency bands, encryption, and costs. From Internet-of-Things (IoT) [5], which was already a revolutionary concept, we are rapidly moving to the next step, Internet-of-Everything (IoE) [6].
  - *Internet-of-Things*: Consists in all the imaginable connections between:
    - People: like social networks.
    - Things: sensors, devices, actuators.
    - Data: raw and processed information.
    - Processes: added value algorithms.
  - *Internet-of-Everything*:
    - This term describes a superset of IoT including machine-to-machine (M2M) [7] communication and an extension to a more global communication concept.
    - As an analogy [6], we can say that if IoT is an equivalent of a railroad line, including the tracks and the connections, IoE also includes the ticket machines, the staff, the travelers, the weather conditions, and all the other factors related to a train ride.
  - *Internet-of-Medical-Things (IoMT)*:
    - This neologism qualifies a subset of IoT limited to health applications and communication between various medical devices.
    - The main additional constrains of IoMT are patient integrity, privacy, safety, and security.

As discuss thoroughly in this book, the environment and regulations linked to BCI are not yet ready to include the above revolutions. Internet-of-Brains will remain utopia for several decades. Some initiatives are taken in this direction (see Sect. 7.4.3), but there is still a very long way to integrate a total global interconnectivity between our brains and the networks.

7.2 Opportunities

This section will give a quick overview of the current and foreseeable evolution of technologies which may impact and facilitate new BCI developments. Following the overall philosophy of this book, we will remain pragmatical and keep the focus on technologies which are *ready for translation*. Our goal is to design devices for human beings, excluding ideas which may be attractive but are too far away from being translatable to a real human therapy.

7.2.1 Energy

We have seen in the previous chapters that energy is the main hurdle to active devices implanted in the human body. In other fields, like consumer products, exciting new technologies are emerging, for example, wireless communication features. These developments prioritize performance before energy consumption. Autonomous devices (not connected to the main power supply), like cell phones or wearable medical devices, can be easily recharged or have a quick change of battery. Cell phones of the two first generations could last more than a week between recharging sessions. Modern smartphones, with much higher performances and new features, do need to be recharged every couple of days. In this case, performance costs energy. It is also a confirmation that electronics and batteries are not progressing at the same pace.

AIMDs and BCIs do not share the privilege of connectivity to the power network or easy recharge. Implantable primary and secondary batteries have poor energy density values, and substantial improvements are not expected any soon. Regularly, research papers announce better battery chemistry, higher energy density, and faster recharge. Unfortunately, this promising work is still far remote from implantable applications.

Inductive power transmission has also limitations, mainly due to heat but also to the reluctance of patients to wear a headpiece. As discussed in Sect. 4.13, the energy available in an implant is limited. This constrain, not negotiable, has the following consequences:

- Current available of-the-shelf components (COTS) like microprocessors, controllers, field-programmable gate arrays (FPGA), RF communication chips, and others are not adapted to implantable BCIs. They all consume too much power. It is a major barrier to innovation. For the sake of speed to market, designers may be willing to use only COTS, with the drawback of poor energy performances. The alternative is the creation of low power application specific integrated circuits (ASICs), at a very high cost and a long development time. How many great BCI projects failed for this reason? They had either poor performances because of the choice of COTS components, or they could not afford the development time and money to design energy sustainable ASICs.
- The number of channels must sometimes be reduced relatively to the first intentions. Increasing the number of channels impacts directly, almost proportionally, the energy consumption.
- Sampling rate, time resolution, and communication bandwidth are limited to the available energy.
- RF communication distance is also restrained by energy. Today, long distance, large bandwidth communication is not possible within the constraints of the energy available in the implant.
- Integrating computing capabilities (decoding, signal processing, etc.) in the implant consumes too much energy. A careful balance must be done between the

energy requirements for doing "in-implant processing" and transmitting unprocessed data out of the body.

Before even starting the development of a BCI, a clear energy budget is needed. A lot of key decisions regarding the performances, the user's comfort, and the attractivity of the device will result from the choices of components depending on consumption.

It has been shown that batteries, either primary or secondary, are not well adapted to BCI for the following reasons:

- Size: the energy density of chemical batteries is poor with regard to the high energy needs of BCIs.
- Thickness: no thin batteries are available for slick skull implants.
- Limited recharging performances: maximum of 1000 recharge cycles, long recharging duration.
- Heat dissipation during secondary battery recharging.
- Risks: batteries, especially secondary ones, remain the most critical components in active implants.

In consequence, the solution of choice for energizing above-the-neck BCIs is magnetic induction [8]. Having two proximal coils, one implanted and the other external, separated by skin and a thin layer of tissues, inductively coupled, may be used for two energy-related purposes:

- Recharging a secondary battery. There is a handful of inductively rechargeable AIMDs which already got FDA approval.
- Continuously transmitting energy to a battery-less implant. A lot of experience in this concept has been accumulated by the designers and manufacturers of CIs.

A rich literature [9] covers the various executions of inductive energy transfer systems in active implants. As many solid patents exist in this field, FtO might be a hurdle. The objective of this book is not to enter in the details and the theory of the concept. Even if looking straightforward, designing an optimal inductive coupling system for a human implant is a challenging task. The topic could fill a book on its own. Some further description of the case of CI is available at Sect. 3.4.1. Designers of BCI must first clearly identify their needs and then get inspiration from existing inductively coupled coils. The key parameters to be specified are:

- Minimum power to be received by the implanted coil.
- Maximum distance between the two coils.
- Maximum acceptable misalignment between the two coils.
- Location of the implanted coil with regard to a metallic encapsulation.
- Admissible temperature increase.
- Acceptable energy absorbed by the body when exposed to the induction field, measured by the specific absorption rate (SAR) [10], which depends on several factors.
- Continuous (battery-less) versus recharging purpose (then, set max recharging duration, time between two recharges).

From these input specifications, developers will select the characteristics of the inductive configuration:

- Frequency.
- Diameters of the coils.
- Number of turns of each coil.
- Type of coil wires (material, diameter, Litz wires [11] (a multistrand wire used in RF to minimize skin effect)).

Note that current battery-less BCIs in development consume more energy than what CIs do. Therefore, the inductive system must be redesigned for the specific needs of BCIs.

As described in Sect. 3.4.1, the most appropriate way to position and hold the external coil with regard to the implant is to use two magnets attracting each other. Dimensioning these magnets is a difficult task due to skin thickness variations.

A newer execution of induction coupling is based on so-called resonant inductive coupling [12, 13] consisting in having four coils, two in the external system and two in the implant. It provides a better coupling factor but is a bit more difficult to fit in a tiny implant. Addition of ferromagnetic cores in the center of the coils, acting as field concentrators, is also an alternative to improve coupling, with the same disadvantage of complicating the integration of the implant.

A lot of exciting initiatives are ongoing with the goal to break the current limitations of energy in implants:

- Energy harvesting [14, 15]: collecting energy from the body or from the environment to feed implants. Unfortunately, none of these projects is even near of a viable solution for long-term BCI. Systems harvesting mechanical energy, from body movements, blood flow, or lung expansion, are in the range of μW, 3–4 orders of magnitude away of what BCI system may need. Peltier elements [61, 62] collecting energy from the temperature gradient in the body are also far from being powerful enough to drive any complex implanted electronics. Implantable systems harvesting energy from the electromagnetic smog are also out of range.
- Ultrasonic transmission [16] through the skin might be a viable alternative to inductive coupling, with possibly less constrains in terms of heat. However, it still requires a headpiece to transmit power to the implant. So far, the efficiency of the transmission remains in the range of a few percent, with a maximum collected power of a few mW on the implant. Current BCIs require ten times more power. Another difficulty with ultrasonic transmission is linked to the materials used for the receiver (usually piezoelectric ceramics), which are not biocompatible. For long-term applications like BCI, it therefore requires a hermetic encapsulation, a serious barrier to the propagation of ultrasonic waves. This technology is not yet mature. Several groups are working in this direction.

7.2.2 *Size*

If we want to implant them above-the-neck (see Chap. 5), BCIs must be small and thin. One strong trend is to move away from the concept of "electronics-in-a-box," as explained in Sect. 4.9.6. Near-hermetic solutions, like multilayer conformal coating, may provide ways to substantially reduce the size of the implant.

We have seen in Sect. 4.9.1 that for hermetic encapsulation, feedthroughs were a major barrier to miniaturization. Near-hermetic encapsulation does not require hermetic FTs and can therefore be done much smaller.

The absence of a battery, the so-called battery-less concept, is a fundamental step to miniaturization. Anyway, there are no such things as small reliable batteries for AIMDs. CI has based their success on a battery-less approach. The only exception so far is NeuroPace (see Sect. 3.4.7) managing to do a rather small cranial device with a primary battery. This was made possible because of the limited number of channels (8), the low-frequency resolution, and the limited communication bandwidth. Primary batteries are reliable and mature. They do not need a recharging coil and nor any complex power management system. The side-by-side design of NeuroPace is a good compromise to have a thin device including a battery.

Trying to place a secondary battery in a cranial implant is the wrong direction to go. Besides the risks inherent to such batteries, the addition of a sophisticated power management circuit and an induction coil has a serious impact on the size as seen on Fig. 4.32.

The paragraph above describes the direction to take to build the BCI of the future:

- *Near-hermetic encapsulation*:
  - No lost volume induced by the traditional "electronics-in-a-box" concept.
  - No FT.
- *Battery-less*.

These are two fundamental choices, which will allow us to design miniature, thin above-the-neck BCIs. As miniaturization is our priority, this is the path to follow. Nevertheless, there is a *high price to pay* compared to a hermetic, battery-driven, and much bigger implant:

- Long-term reliability and biostability of near-hermetic solutions will be never as good as hermetic ones.
- Patients must wear a headpiece external unit.

In both hermetic and non-hermetic configurations, the other impactor on size is the electronic circuit board. From the most conservative (and large) to the futuristic vision (and miniaturized), we can list various possibilities (see Fig. 7.1):

A. Individually packaged ICs placed on one side of the PCB.
B. Individually packaged ICs placed on both sides of the PCB.
C. Unpackaged ICs assembled by wire bonding (WB) process [17].

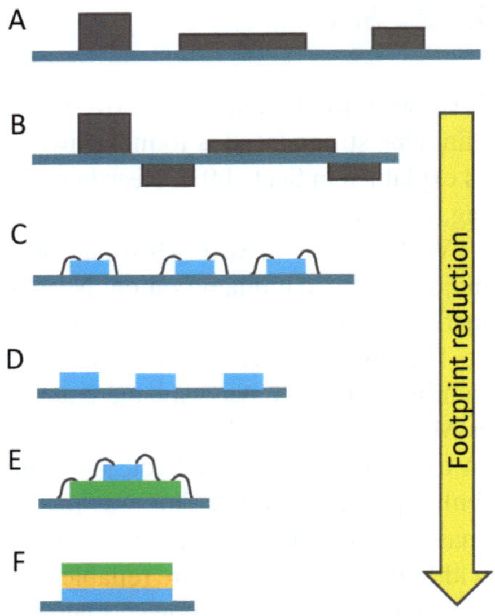

Fig. 7.1 Reducing footprint of electronic boards

D. Unpackaged ICs assembled by flip chip bonding (FCB) [18] including ball grid array (BGA) [19] setting (see Fig. 7.2).
E. Stacking chips and connect them with WB, so-called Chip-on-Chip (CoC) (see Figs. 7.3 and 7.4).
F. Stacking chips in a configuration of three-dimensional integrated circuits (3D-IC) with through-silicon vias [20].

Further miniaturization can be achieved in the "Z" dimension by using thin flexible substrates, in polyimide or LCP (see Fig. 7.4 and Sect. 7.2.3). The most advanced work in the direction of ultrathin electronic assemblies is based on the use of minced ICs, as already described in Sect. 4.9.6 and illustrated on Fig. 4.45. Note that these technologies are still far from being mature enough to be included in a human BCI.

As they may include unused features or blocks, commercial ICs do not always have the smallest footprint for a given application. We have seen before that BCI designers may consider developing their own ASICs, for optimizing performances and minimizing consumption. Another criterium in favor of ASICs could be the reduction of the footprint.

PCBs may be realized on various substrates (see Fig. 7.5):

- Rigid: most electronic boards for diverse medical and nonmedical applications are realized on multilayer substrates of FR-4, a glass-reinforced epoxy laminate material [21]. It presents the advantage of being a mature and robust technology, well adapted to AIMDs in general. For thin BCI design, the thickness of rigid boards is a disadvantage. In addition, as described in Sect. 4.9.4, the epoxy-based FR-4 is prone to absorb moisture which might be released when the board is encapsulated.

Fig. 7.2 Example of populated PCB with a mix of packaged ICs and flip chip assembled ICs for an active implantable medical device. (*Courtesy of MST AG*)

Fig. 7.3 Example Chip-on-Chip wire-bonded assembly for an active implantable medical device. (*Courtesy of MST AG*)

- Flex: thin flexible substrate in polyimide or LCP. Flexible PCBs allow non-planar configurations which are especially well adapted to thin curved skull implants.

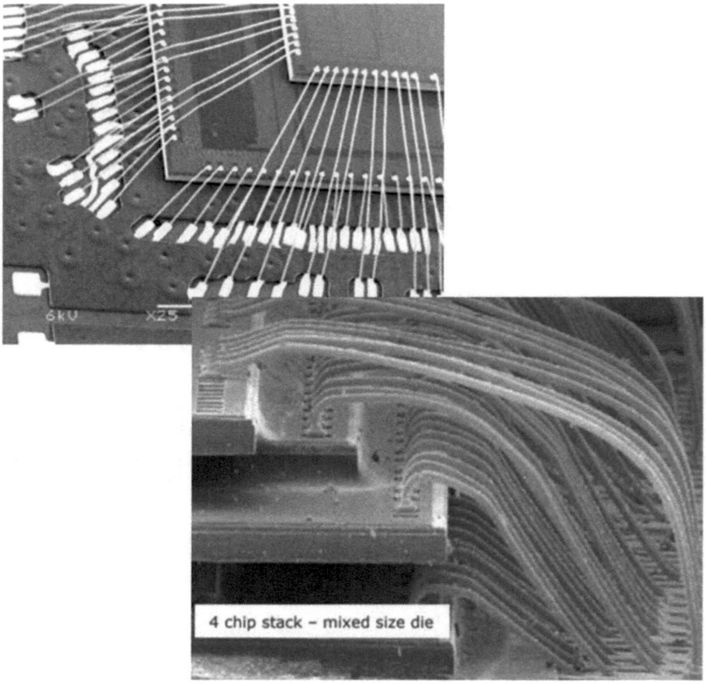

Fig. 7.4 Example Chip-on-Chip wire-bonded assembly of a stack of four chips. (*Courtesy of MST AG*)

- Rigid-flex: is a hybrid of the two above concepts, having rigid sections of PCB linked by flexible interconnecting bridges. Rigid-flex substrates open multiple opportunities for creative designs of curved or even foldable electronics.

Another approach in view of reducing the footprint of the electronics is to gather several functional blocks in a single ASICs, called System on Chip (SoC). Several teams are having ambitious goals in this direction, for example, in designing BCI SoC including sensing, stimulation, impedance spectroscopy, digitalization, compression, and even power management or RF on a single chip. The effort, in time and money, is considerable, sometime out of proportion with the objectives of a BCI system. It is a common mistake to believe that one super chip will be applicable to many different projects. Each BCI concept has its own specificities which could not be optimally covered by a generic SoC. The field of BCI is in its infancy. Designing today a SoC for BCI applications looks as an investment out of proportion.

The surface or footprint of an IC is proportional to the square of the dimension of the constitutive transistors. Following Moore's law [22], the technologies of integrated circuits have shown a constant and dramatic reduction of size. For large volume productions, like in the industries of microprocessors and smartphones, extremely small transistors (in the range of 10 nm) are now achievable, allowing to gather billions of transistors on a single chip. In parallel, silicon foundries regularly

Fig. 7.5 Rigid, flex, and rigid-flex configurations of PCBs. (*Courtesy of Dyconex-MST AG*)

increase the size of wafers. In consequence, the cost of a mask set for large wafers in small technologies has increased to very high levels, only making sense for producing millions or billions of chips.

This level of integration is out of reach for ASICs intended to AIMDs, where annual volumes remain low, in the range of 1000 to 100,000 devices per year. It means that medical ASICs must use "older" technologies, where the mask set has a cost in proportion to our ambitions. ASICs for BCI applications, in even smaller production volumes, must rely on modest technologies. Miniaturization is therefore limited by costs. During the first years of a BCI project, the cheapest way to obtain a few hundred ASICs is to join a Mixed Project Wafer (MPW) program where the surface of the wafer is shared by several project owners. The total cost of the mask set is therefore spread over several customers.

7.2.3 Connectivity

We have discussed in Sect. 4.14 the difficulties to conceive implantable connectors with multiple channels. We have also seen that it was tedious to connect BCIs with body interfaces like the Utah array. Bundles of thin gold wires are fragile, and connections at both ends of the wires must be done one by one. Under Sect. 3.3.2, high potential flexible or stretchable electrodes have been described. They certainly represent a promising solution for future BCI systems. Nevertheless, their weak point is connectivity. How do we establish a long-term reliable and biostable connection between flexible electrodes and rigid BCI housings? Not a lot of effort have been put on this critical element of the system. Missing a proper way to connect may be a serious barrier to the introduction of innovative electrodes in BCI systems.

An interesting concept is being developed by Dyconex-MST [23] in Switzerland. It is based on the realization of thin ribbon connections using liquid crystal polymer (LCP) [24, 25, 30], a thermoplastic polymer with unique physical properties. LCP is biocompatible and biostable and has a superior resistance to moisture penetration. It can be thermo-formed and thermo-sealed, allowing a lot of flexibility in the design of thin and flexible subassemblies for implanted applications. For implantable devices, LCP is superior to the conventionally used material polyimide.

LCP demonstrators have been realized for:

- Flat ribbon in a sandwich structure, sealed along the edge (see Fig. 7.6).
- Thermo-sealed one-time connection between two LCP structures (see Fig. 7.7).
- Full implanted system on a single substrate of LCP (see Fig. 7.8).

Compared to polyimide, LCP has the important characteristic of being thermo-sealable. Several thin foils of LCP can be stacked and then sealed all around the edges (see Fig. 7.8). The top and bottom layers provide a moisture-resistant encapsulation. By its biostability and near-hermeticity, LCP is very well adapted to become a key element to build the BCI of the future.

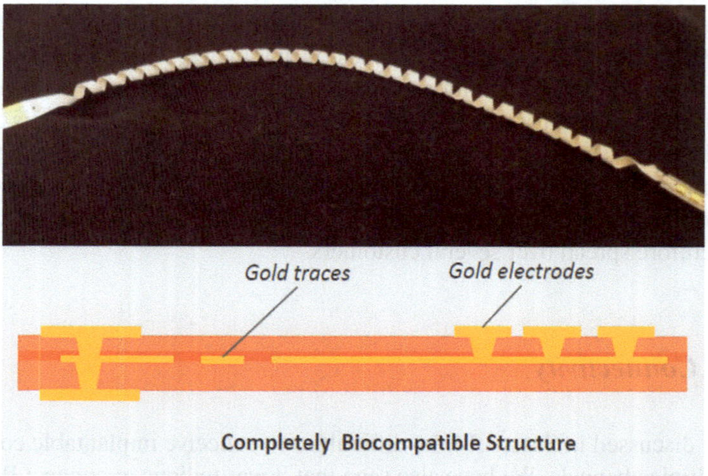

Fig. 7.6 LCP flexible ribbon with gold electrodes. (*Courtesy of Dyconex-MST AG*)

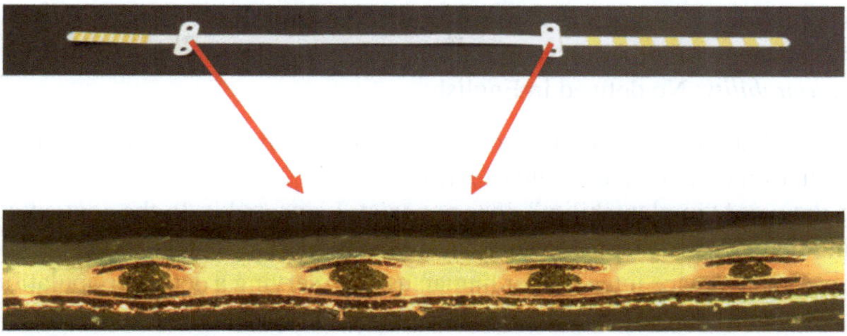

Fig. 7.7 LCP flexible ribbon with two thermo-sealed connections. (*Courtesy of Dyconex-MST AG*)

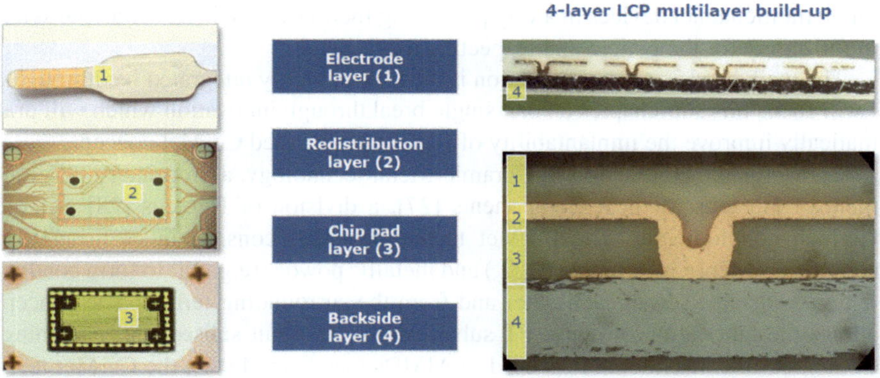

Fig. 7.8 Test chip embedded in LCP with flexible ribbon and gold electrodes. (*Courtesy of Dyconex-MST AG*)

7.2.4 Implantability

In the dictionary [26], we find the following definitions:

- *Implant* (*noun*): Any device or material, especially of an inert substance, used for repairing or replacing part of body.
- *Implant* (*verb*): To insert or graft (a tissue, organ, or inert substance) into the body.
- *Implantable* (*adjective*): Capable of being implanted. Pertaining to a device, as a micropump or porous polymer membrane, for surgical insertion under the skin for the controlled release of a drug.
- *Implantable* (*noun*): Surgery—a material, foreign to the body, that can be implanted without undue risk of rejection.
- *Implantation*: The act of implanting. The application of solid medicine underneath the skin.

- *Implantology*: The branch of dentistry dealing with the permanent implantation or attachment of artificial teeth in the jaw.
- *Implantability*: No defined in English.

We see that these definitions must be refreshed and put in the perspective of the evolution of medicine in the field of AIMD.

As the word "implantability" does not exist, I invented it. In the context of BCI, I would define it as "methodology to design, develop, validate, and get approval of an implantable device intended for human use and serving a defined population with unmet medical needs." This word covers the entire purpose of this book: how to build the BCI of the future. It has been seen in the previous chapters that the implantability of BCIs is governed by specific rules, which are not applicable to other domains of the art of implantable medical devices.

The implantability of BCIs above-the-neck is impacted by our ability to miniaturize multichannel devices in a way permitting their insertion in or within the skull, in full safety for the patient and respecting user's needs.

The main barrier to miniaturization is the lack of highly integrated feedthroughs. I will focus this sub-chapter on one single breakthrough innovation which will dramatically improve the implantability of BCIs: the so-called CerMet concept.

CerMet is the contraction of Ceramic-Metal technology, a proprietary development of Heraeus Medical Components [27], a division of Heraeus [28], a large German conglomerate. The CerMet technology [29] consists in co-firing [31] ceramic powder (e.g., alumina Al_2O_3) and metallic powder (e.g., Pt) to form conductive vias insulated from each other and from the surrounding ferrule. The concept allows the fabrication of dense FT subassemblies without subsequent machining. Compared to conventional FTs used in AIMDs (see Sect. 4.9.1), the CerMet technology allows miniaturization by a factor of 10–50. Consequently, it provides a unique opportunity to optimize the design of the BCIs of the future.

Figures 7.9, 7.10, and 7.11 illustrate the CerMet concept:

7.2.5 RF Communication

RF communication is commonly used to send information from the BCI to the outside world or in the other direction. The data transferred through the RF link may be of different nature:

- In real time:
  - Data continuously collected in the brain by appropriate sensing electrodes and transferred to an external receiver for further processing.
  - Data continuously sent by an external emitter to be transferred to the implant for stimulation purpose.
- From time to time:

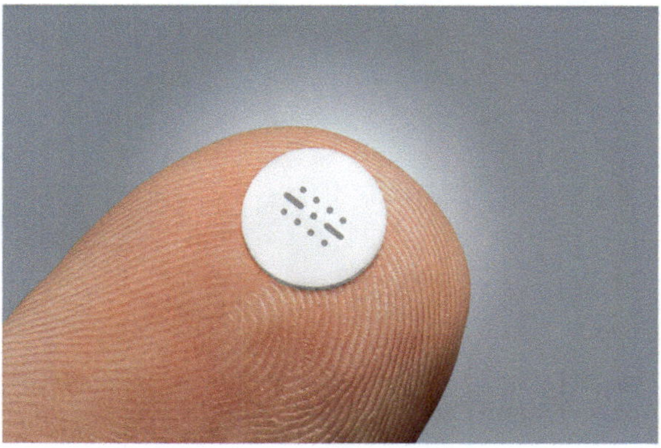

Fig. 7.9 CerMet, miniature feedthrough for implantable applications. (*Courtesy of Heraeus Medical Components*)

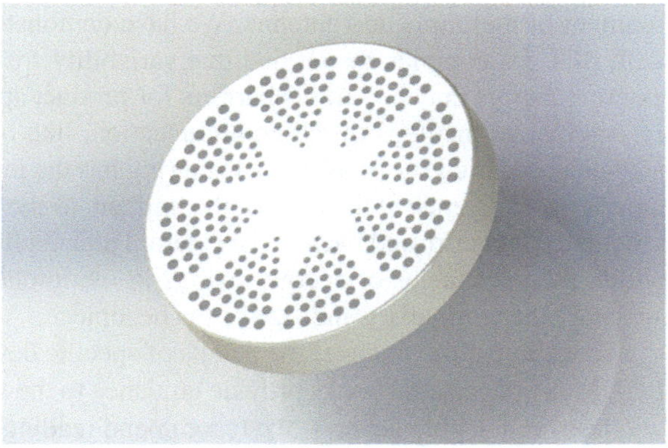

Fig. 7.10 CerMet, high-density feedthrough. (*Courtesy of Heraeus Medical Components*)

- Information from the implant to the programming interface (status of the device, battery voltage, memory content, etc.)
- Information sent to the implant (new parameters, upgrade of the firmware, etc.)

Real-time communications with implanted BCI require very large bandwidth, as already discussed in Sect. 4.11.2. Projects currently under development have real-time information flow in the range of 50 Mbit/s. This is the order of magnitude of the Wi-Fi link for computers used in home wireless local area networks (WLAN) but without the comfort of having plenty of energy from the electrical power network. As discussed in Sect. 7.2.1, energy in the implant is the main limitation for the future BCI. Large bandwidth RF links are greedy in energy. In a recent study, we have shown that sending 50 Mbit/s from a cranial implant to the external receiving antenna represents about one third to half of the entire energy budget. Section 4.12 presents more details about the limitations of propagation of RF waves though human tissues and the consequences for BCI.

Designing and optimizing large bandwidth RF communication in the human body is an under-explored field. Substantial efforts are still required to understand

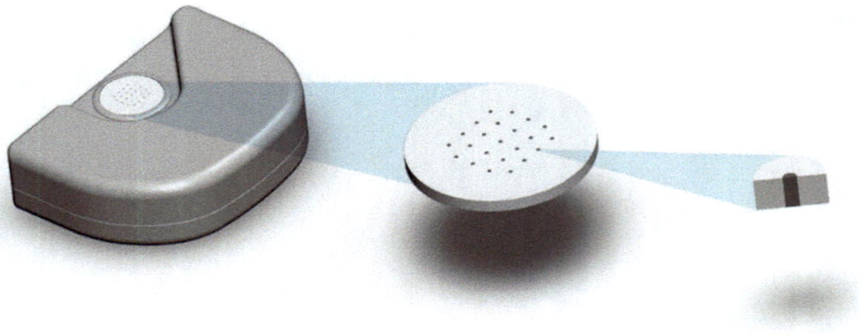

Fig. 7.11 CerMet feedthrough integrated in an implantable titanium housing. (*Courtesy of Heraeus Medical Components*)

all the transmission, reflection, scattering, and diffusion characteristics of the environment of the implanted antenna. We have demonstrated that RF communication with BCI is a complex matter and that variability from patient to patient and from external factors may be serious barriers for product approval.

Alternative methods of RF communication, like backscattering, have been and are being explored for BCIs. Backscattering has the major advantage of minimizing the energy consumed at the level of the implant, as the incident RF waves are simply modulated and mirrored back to the external unit. At high frequency and large bandwidth, the advantages of energy savings in the implant are negatively counterbalanced by the complexity and size of the headpiece.

We are not going to cover the details of specific designs of electronic circuits for BCI. Many papers and books provide guidance to the state-of-the-art schematics for wireless implantable devices. We recommend reading [70] for cortical interfaces.

7.2.6 *Optical Communication*

One promising path to avoid the limitations of RF communication described in the previous chapter is optical communication through human tissues. Feasibility has been demonstrated, under certain conditions, as, for example, bandwidth limited to 4 Mbit/s (see Reference [35], Chap. 4). An optical link, for transferring data collected by a BCI at the level of the motor cortex to a proximal headpiece (see Fig. 7.12), will have the following characteristics:

- Power consumption of the optical emitter: max 50 mW to avoid excessive heating of the body tissues.
- Emission in the near-infrared (NIR) range of 600–1000 nm (selecting the wavelength penetrating best through the human tissues).
- Data rate: min 50 Mbit/s to permit transfer of large amount of data in real time.
- Maximum distance between the emitter and the receiver: 10 mm (including hair).
- Communication should still work for up to 5 mm misalignment.

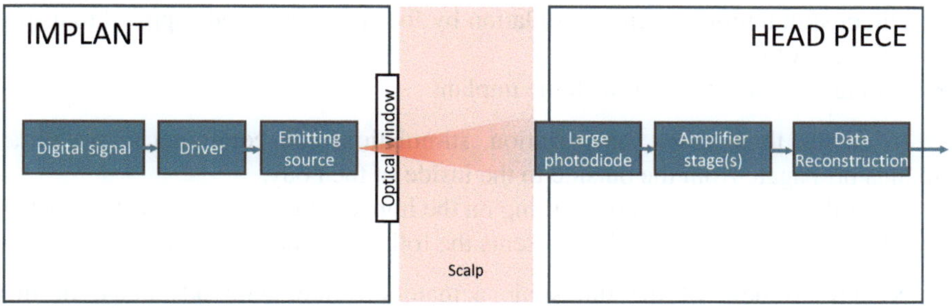

Fig. 7.12 Possible configuration of an optical link for BCI

- Should be independent from the skin color.
- Temperature increase of the device should be kept below +2 °C per ISO-14708-1.
- Respect the maximum permissible exposure (MPE) for the skin at the selected wavelength and for the duration of the transmission.

The two possible light emitters in the NIR range are the light-emitting diode (LED) and the vertical-cavity surface-emitting laser (VCSEL). Both components are available of-the-self in miniature configurations. Power budget, size, and packaging must be carefully taken in account in the selection of the components.

Proximal optical communication has two major advantages compared to RF:

- Immune from electromagnetic interferences.
- Difficult to hack.

In addition, optical links are well adapted for BCI for the following reasons:

- Possibility to transfer enormous quantity of unprocessed data.
- No "band allocation," unregulated field.
- Miniature emitter.
- Small size headpiece.
- Moderate heating.
- Low exposure of tissues.
- MRI compatibility of the emitter.

Optical links are certainly the communication system of choice for proximal unidirectional sensing BCI of the future.

7.2.7 Ultrasonic Stimulation and Communication

We have already seen (Sects. 4.13 and 7.2.1) that ultrasounds can be used to transfer energy from an external actuator to an implant. For BCI applications, ultrasound could be used for two other purposes:

- Direct stimulation or neuromodulation by focused ultrasounds applied to a specific area of the brain.
- Wireless communication with the implant.

In all the three cases (energization, stimulation, and communication), ultrasounds propagate from the outside to the inside of the body.

Several groups [32–34] are working on the impact of ultrasounds on brain activity. Ultrasonic neuromodulation presents the following characteristics:

- Noninvasiveness of the ultrasonic actuators is the main advantage of this technology.
- As ultrasounds can be focused, stimulation can be applied at cortical level but also deeper in the brain.
- Low-intensity focused ultrasounds (LIFUs) have been used to suppress epileptic seizures, initiate neural firing, and modulate behavior.
- The modes of actions of ultrasounds on the brain seem to have various origins:
  - Cavitation.
  - Influence on the ion channels.
  - Mechanical deformation of the cells' membranes.
- Focusing ultrasounds requires multiple actuators spread on the surface of the skull, making the external headpiece bulky and non-aesthetical.
- Gel must be added at the interface between the actuator and the scalp to facilitate the transfer of sound waves.
- Ultrasonic actuators, often piezoelectrical, have poor power efficiency, so the external unit cannot be easily made wearable and battery driven.

Some work [35] is also conducted on the use of focused ultrasound to stimulate peripheral nerves or the vagal nerve (see Sects. 1.3.5, 1.4.2.3, 3.3.3, and 3.4.4).

Using ultrasounds for communication between the external world to the implant is only at the research level [16, 36, 37]. For BCI stimulators, real-time data transfer by ultrasounds might be limited by bandwidth. As for energy transfer, communication requires an implanted piezoelectric receiver, with serious unresolved issues regarding encapsulation.

Because of its capability to be focused to reach deep parts of the brain and its noninvasiveness, ultrasonic neuromodulation certainly has its potential in the future BCI systems. Power transfer and communication with ultrasounds are technologies not yet mature enough to serve the field of human BCIs.

7.2.8 Integrity of Data and Security

Unlike cable or optical fibers, wireless communication, either RF, optical or ultrasonic, is by nature a method of transmission of information which has no physical support. Wireless transmission between points A and B, at a few centimeters or several kilometers distance, has the following weaknesses:

- Some information may be lost between the emitter and the receiver.
- Unwanted signals may reach the receiver and induce undesirable consequences (see Sect. 4.11.2, Electromagnetic Compatibility, and Sect. 4.11.4, Co-existence).
- Noise may degrade the quality of the transmission signal.
- The transmitted signals may interfere with other devices or equipment.
- The signal, or part of it, may be intercepted by people who have no right to access the information: hackers.
- Hackers may not only read data but send faked or malicious information to the receiver, either to fool the system, to damage it, or to take control.

RF communications are usually secured by encoding and encryption, for a better protection against hackers. We all know that no encryption is totally sure. In addition, encryption and security algorithms usually increase the volume of information to be transmitted. We have seen that BCIs already consume a lot of bandwidth, meaning it may be difficult to add complex encoding to secure BCI RF transmissions.

BCIs which only read the brain and transmit raw information in the direction of the receiver are less critical in terms of security. If hackers manage to get cortical signals, they will not be able to do much about it.

The situation is different when BCIs are for stimulation purposes. Then, hackers might be able to send inappropriate signals to the brain, having an impact on the patient. In consequence, a stimulating BCI must get a high level of protection against hackers. Several measures can be considered:

- Minimize the possibilities to "enter" in the communication channel. For this, the safest approach is *proximal communication*. When communication is between a headpiece and an implant, at 1 centimeter, there is virtually no far field. Even at 1 meter away, nobody could pick any signal.
- Minimize the power of the communication signal. In this sense, proximal configurations go in the right direction.
- Privilege optical or ultrasonic communication rather than RF, as radio waves may propagate in unexpected ways.
- Do not use standard encoding and communication protocols. For example, Bluetooth is well known in the hackers' community. They will promptly decipher any information circulating in these popular networks. Use instead proprietary communication protocols. It will secure your communication, but it will also require a lot of additional resources to design and validate the system.
- Add firewalls in the implant, with ID codes, frequent check with the external unit and other means which will prevent an intruder to do any damage.

Radio frequencies are highly regulated (see Sect. 2.2.1) and the national authorities have allocated a restricted number of frequency bands to medical device communications (some exclusively for medical use, some shared with other categories of users). If they want to use RF communication, designers of BCI systems have no freedom in the choice of the frequency. As discussed earlier, BCI is greedy in bandwidth, which excludes the use of lower frequency bands. In terms of bandwidth and availability of components (antenna, RF chips, filters), the most appropriate band for BCI is in the 2.45 GHz range. Unfortunately, this band is crowded by many other applications like Wi-Fi, Bluetooth, mobile phones, or microwave ovens. In consequence, the 2.45 GHz band is noisy, prone to get interferences between the users and, unfortunately, also the favorite hunting ground for hackers. New bands at high frequencies, above 5 GHz, are becoming available. They are still rather quiet but will soon be invaded by the Fifth Generation (5G) of cellular phones. Going up in frequency is a drawback in human tissues, as short waves get absorbed by them. Therefore, emitting at a high frequency will require higher power to pass through the same layer of tissues.

Optical communication (see Sect. 7.2.6) presents the advantage of being unregulated. Original and proprietary security algorithms and protocols can be implemented. The needs of additional bandwidth dedicated to security can easily be addressed by optical links. It is extremely difficult to hack an optical signal. Optical links are intrinsically secured. Electromagnetic perturbations do not affect optical communications. The components constituting the optical link (LED, VCSEL, photodiode) do not interfere with MRI.

Future BCIs will preferably use proximal optical links for their large bandwidth communications. It will offer superior performances and security compared to RF.

7.2.9 Costs: Reimbursement

As BCIs are intended to help patients with very serious conditions, like CLIS, paralysis, epilepsy, brain injury, spinal cord injury, amputation, or blindness, it is often thought that costs are not an issue that anyway patients or somebody else is going to pay. This is a wrong analysis. If the BCI system becomes too expensive, many patients will be excluded, even the ones suffering most.

From the very beginning of a project, engineers should be cost conscious. Overengineering is a common trend in the industry of medical devices. A modest device, at a reasonable sales cost, has more chance to reach the market than a sophisticated and expensive one.

Accelerating is the first word in the subtitle of this book. Time to market is critical for meeting our global objective: provide solutions to patients in need. We have seen before that there are several ways to accelerate a project:

- Have reasonable objectives.
- Have a clear understanding of the environment.

- Spend time to set the frame of the project. This is not wasted time but an investment for the future. The time spent for a proper preparation of the project will later induce important time savings.
- Analyze carefully the "Make or Buy" strategy (see Sect. 7.2.10).
- Choose the "best and brightest" partners.
- Do not reinvent the wheel. Integrate in the project the elements that others have already developed. This is part of the "Buy" strategy.
- Allocate appropriate resources (people and money).
- Keep project milestones. In case of problem, prefer revision and simplification of the specifications instead of delays.
- Assign the responsibility to an experienced project manager.

There is a tight relation between time and money. An accelerated time to market means a reduction of the total development costs. In its turn, it reduces the product sales price, as amortization of the investment occurs early and exposure is minimized (see Annex 2).

BCI is a complex field. Only a few groups in the world have the knowledge, the technical competencies, the networks, and the money to carry these projects to human applications. The model we have put in place at the Wyss Center for Bio and Neuroengineering in Geneva, Switzerland [38], is based on a "trilingual" concept (see Fig. 7.13):

- We speak "science."
- We speak "technology."
- We speak "human."

Accelerating neuro-technologies for human benefit is an overall concept which must be carefully balanced (see Fig. 7.14).

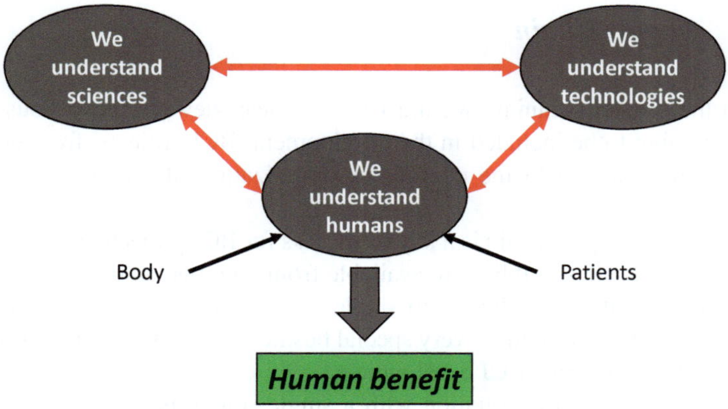

Fig. 7.13 The "trilingual" approach

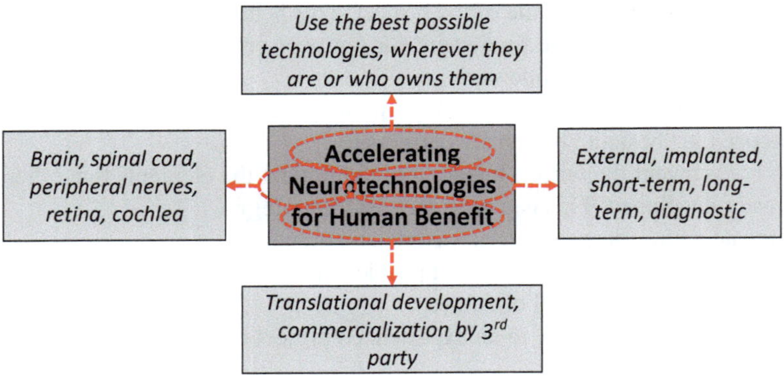

Fig. 7.14 Accelerating neuro-technologies for human benefit

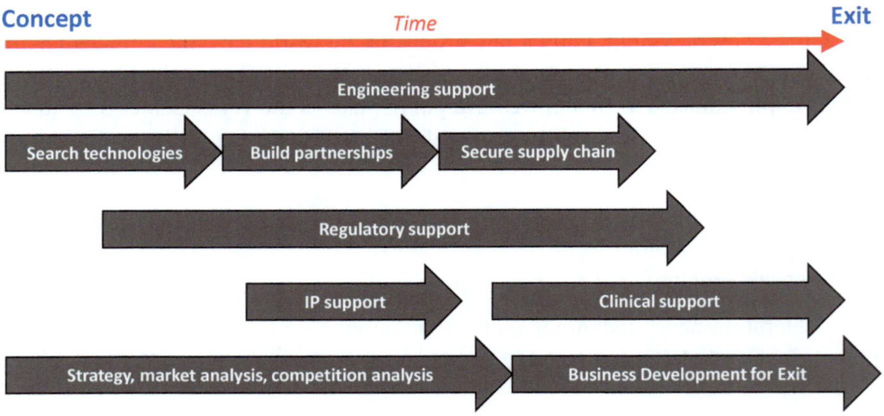

Fig. 7.15 From concept to exit

7.2.10 Supply Chain

Early in the project planning, we must have a clear view on the various partners which are going to be included in the development. BCI projects, from concept to exit, are tightly linked to the continuous optimization of the supply chain (see Fig. 7.15).

Mastering the supply chain is a key of success for BCI projects. Several strategic components and subassemblies are available from only one unique supplier. Often, there is no alternative. Dealing with single sources is not a common situation in project management. It induces very special business rules that must be fully understood before entering in a BCI development.

The way to enter in negotiation with a single source is an art which goes far beyond the usual contractual practices:

- The supplier of a unique component knows that we have no alternative.
- BCI project managers must build a solid relation with the supplier, based on mutual trust and a "win-win" approach, to an extend that the supplier becomes a partner.
- In most cases, the unique component must be slightly modified to be integrated in the BCI system. Therefore, we are not going to simply purchase an "of-the-self" component, but rather enter in a relation of co-development.
- The partner-supplier often detains the IP on the component. It means that the developers of the BCI system will not own the entire rights on the project and may be forced to pay royalties when the device is commercialized.
- The unique component is likely to also be unique for other projects in competition with ours. In consequence, there are limited possibilities to obtain exclusive rights on the use of the component.
- If we are in a single source situation, there may be several reasons for it:
  - The technologies mastered by the supplier may be very complex, the fruit of many years of development. Frequently, the partner detains secret know-how.
  - The market and number of applications are limited. This is almost always the case with BCI.

We have seen before that acceleration of complex projects relies on using existing building blocks owned by others. It is capital to have a sound "Make or Buy" strategy. In my opinion, for having a chance to reach the market with a BCI system, one must privilege the "Buy" approach (see Fig. 7.16).

Even if partnering increases the dependence on third parties, makes contractual relations complex, and probably decreases the exit value, it is the safest way to carry BCI project to success in a reasonable time frame.

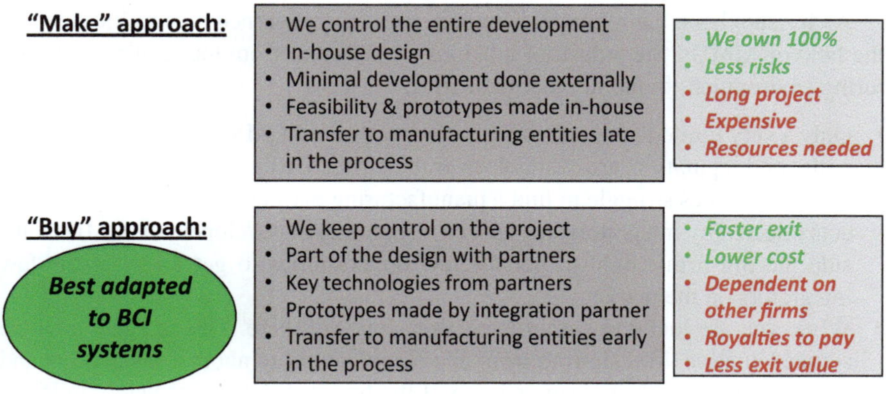

Fig. 7.16 "Make or Buy" strategy

7.2.11 *Industrialization*

Large companies in the field of AIMDs are not yet attracted by complex implanted BCI projects. Even if they master most of the technologies to enter in the BCI domain, dominant AIMD firms remain outside, for the following reasons:

- Time to market and payback are too far away.
- Risk of failure is substantial.
- Market is small (compared to their main product lines).
- BCI systems often require adaptation to each patient.

Giant firms are following carefully what is happening in the field of BCI and will certainly be interested in acquiring successful products but at a later stage. BCI project must reach demonstration of human feasibility (FIH, pilot clinical trial) before being a target for acquisition.

Beside the few large AIMD groups, there are only a few companies able to do the integration and manufacturing of BCI implants. Four or five integrators in the world have the technologies and the know-how specific to hermetic encapsulation.

Developers of BCI, from academic labs and start-ups to non-for-profit organization, are confronted to the challenge of finding an industrial partner capable to build a human grade device. The main difficulties are:

- Transfer the design to the manufacturer.
- Adapt to the processes available in the assembly plant.
- Fill the development gaps.
- Align with the quality management system (QMS) of the integrator.
- Establish supplier agreements.
- Agree on a development and industrialization contract.
- Sign a long-term manufacturing agreement.
- Find solutions to share the IP rights.

The hurdles and difficulties of the industrialization phase are always underestimated in the business plan.

As for suppliers, the relations with integrators are governed by specific rules. In the field of AIMDs, the industrial relations are different from the regular manufacturing environment found in other industries:

- Only a very limited number of integrators, with a full QMS, with experience, and with good reputation, are available over the planet. Designers and project owners must sometimes struggle to find a manufacturing partner. I have seen good projects rejected by integrators, for lack of resources. All developers of AIMDs, outside the big firms, fight to get the few opportunities to get a slot at the few integrators on the market.
- When you are lucky to get an integrator, you stay there. This is again a single source situation. Transferring from one manufacturer to another takes years and millions. It should be considered only if the first integrator completely failed to deliver. Only a few start-ups can afford such a transfer.

- BCI is characterized by very small production volumes, slow ramp-up, and high risks. It is not attractive for integrators, who will privilege easier and quicker projects. This is a recurrent challenge in BCI: nobody wants to manufacture for us.
- Dependence of the integrator is a serious business risk. If budget is not respected or in case of delays, there is not much to be done.

Seeing the difficulties described above, some developers are tempted to put in place their own manufacturing capabilities. We have seen in this book that technical challenges are huge and specific know-how is in the hand of very few people. Trying to manufacture oneself is a frequent cause of project failure. In addition, it overloads the budget and adds a couple of years to the schedule.

With regard to the complexity of BCI, it is strongly recommended to identify early in the development phase the best possible manufacturer, to secure the contractual relations, and to get them involved in the development to guarantee a smooth transfer.

7.3 Ongoing Initiatives

Several exciting projects have been described earlier in this book. Some of them are still far away from FIH and even further away from commercialization. We will now come back to already discussed initiatives and give a few instructive information on innovative projects.

7.3.1 BrainGate and Movement Restauration

An introduction to BrainGate has been done in Sect. 1.4.2.8. The consortium has not only demonstrated that it was possible to read the intentions of movement from the motor cortex but also that paralyzed patients may benefit of BCI technologies to move a cursor, activate a speller, control a robot arm, or stimulate a paralyzed limb through FES. It has also been demonstrated that even after many years of paralysis, the patients' cortex still sends motor instructions to the limbs. So far, all this has been achieved by connecting one or several Utah array(s) to a transdermal connector (pedestal), linked by a cable to powerful external computers.

The next step is to replace the cable by an external wireless transmitter, screwed on the same pedestal. Such a system has been developed by Blackrock Microsystems LLC (see Chap. 3, [8]) under the name of CerePlex-W [39]. Getting rid of the cable provides better comfort to the patient and to care people.

When available, the BrainGate project will be one of the users of fully implantable BCI systems. Such a device, called Neurocomm, is currently being developed by the Wyss Center for Bio and Neuroengineering [40]. Replacing the pedestal by a fully implantable wireless BCI will be a major step. It will avoid potential risks of infection, always possible with a transdermal connector, and greatly improve comfort and aesthetics.

The BrainGate initiative is also anticipating major progresses in the way to better decode the brain signals and to interact with advanced actuators for FES, one day fully implantable (see Sect. 7.3.4).

7.3.2 Epilepsy and Brain Circuits

The topic of epilepsy has already been introduced in Sect. 1.5.2. Two existing approved BCI systems are available to treat epileptic patients:

- LivaNova (ref [6], Chap. 1, and ref. [36], Chap. 3) is taking the route of VNS (see Sects. 1.4.2.3, 3.3.3, and 3.4.4) stimulation to address the problem with a product called SenTiva® [41]. Also note that LivaNova has a VNS system, Demipulse® [42], for treatment-resistant depression.
- NeuroPace (see Sects. 1.4.2.7 and 3.4.7, Ref. [47] Chap. 3) has an intracranial device, RNS®, described in detail above.

About 30% of the large population (about 65 million people worldwide) suffering from epilepsy are refractory to drugs. For the most severe cases, when the focal point can be localized, a resection surgical procedure is conducted.

The usual method to quantify and evaluate seizures is to measure brain activity with an EEG. For epileptic patient and their doctors, the drawbacks of EEG are:

- Examination is limited to a few hours. It is a snapshot.
- Patients must go to the hospital for EEG diagnostic. Between two EEG sessions, the status remains unknown.
- Gel must be applied on the EEG electrodes to facilitate measurement, which is not convenient for patients. In addition, putting in place an EEG cap is a time-consuming operation.
- Seizures often happen during the night or early morning and can therefore not be tracked by EEG.
- Researchers [43] have identified that the frequency and intensity of seizures vary over long periods of time, up to several week cycles. With EEG examination, neurologists ignore where patients are with regard to these cycles.

We see that EEG is not the right method to follow epileptic patients over long periods of time. Several groups are developing BCI system for continuous measurement of brain waves by placing electrodes subcutaneously, between the skull and the scalp, connected to an implanted recorder. Data are transmitted continuously from the implant to a headpiece by wireless communication. Here are examples among the ongoing developments:

- UNEEG (see Sect. 1.4.2.7, Ref. [9] Chap. 1): a simple subcutaneous datalogger with a limited number of electrodes. The device is currently in clinical evaluation.
- Wyss Center for Bio and Neuroengineering [44] with more electrodes for a better coverage of the skull surface.
- Bionic Institute [45].

BCI systems with subcutaneous electrodes may also be used for other brain circuit disorders. Some promising research is being carried for various indications, like:

- Dyslexia [46] where children, when learning to read, experience difficulties because words and sounds may show some mismatch. Prof. Anne-Lise Giraud [47] from the Neurocenter at Geneva University, Switzerland [48], has found that people with dyslexia have brains showing to be out of their normal oscillation frequencies. Using a BCI system with appropriate electrodes under the scalp may help to reorganize brain rhythms [49].
- Neurofeedback [50] for tinnitus [51]. Work [52] done at Universitätklinikum in Tübingen, Germany, using fMRI or NIR, has shown that patients with tinnitus may decrease their symptoms by their own concentration in a neurofeedback approach. The concept is now being extended to an implanted solution based on a BCI with subcutaneous electrodes placed on the skull above the auditory cortex.

7.3.3 Close-Loop Stimulation

The term "close-loop" characterizes systems where the action is dependent of a measurement of the effect. In the case of AIMDs, close-loop devices have electrodes for sensing and electrodes for stimulation. The implanted electronic will adapt the stimulation parameters according to the information collected on the sensing electrodes or other sensors. There are already some close-loop systems in the active implant field:

- Pacemakers: modern devices have sensing leads measuring electrical potentials inside the heart and tailoring the signals sent on the stimulation lead to optimize the therapy.
- Rate-responsive (RR) pacemakers are capable to adapt to stimulation rhythm to the physical activity of the patient. It consists in the combination of information coming from two sensors:
  - An accelerometer which measures the movement of the body.
  - A measurement of the electrical impedance between the tip of the lead and the housing of the pacemaker. Small variations of the impedance are induced by the dilatation and contraction of the lungs, giving an image of the pulmonary activity.

The correlation of both information will allow to discriminate situations where the heartbeat should be accelerated (e.g., climbing stairs, as the lung activity goes up) and cases where they should not (e.g., going down the stairs or sitting in a train).

- Implantable cardiac defibrillators (ICD): may also be considered as close-loop systems, as the decision to send a high-power defibrillating shock is based on signals collected on sensing leads showing that the heart either stopped or is in a fibrillation mode.
- LivaNova's VNS system for epilepsy has a heartbeat detector to eliminate stimulation in cases where the heart accelerates without seizure.
- NeuroPace RNS system has both sensing paddle electrodes on the surface of the brain and deep stimulation leads.

Several groups are presently working on close-loop DBS, with the goal of optimizing the stimulation parameters, minimizing side effects, and saving energy. Sensing is done by paddle electrodes or small ECoG grids placed on the surface of the brain.

Close-loop concepts create a lot of new issues in the field of active implants. In "normal" open-loop situation, like neuromodulation (SCS, DBS, etc.), stimulation parameters are adjusted by the physician, based on diagnostic, observation of the patient, and data collected from the memory of the device through the programming interface. In open loop, the modification of stimulation parameters is in the hand of the doctor. In close-loop system, there is a certain level of automation within the implant. The device will "decide" to change parameters according to some sensed information. In a way, the implant has some "decision-taking intelligence." This opens the door to new sorts of risks in case the implanted microprocessor makes the wrong decision. Validating close-loop system software is an unprecedented challenge. Approval authorities and regulators have not yet provided guidance on the safety of close-loop devices.

Future BCIs will certainly include close-loop features. In a way, a BCI which detects the intentions of movement linked to a FES for activating a paralyzed limb is a close-loop system.

7.3.4 Peripheral Nerves

Stimulation of peripheral nerves (see Sect. 1.4.2.5) has already been discussed above:

- SNS: see Sects. 1.3.4, 1.4.2.2, and 3.4.6.
- VNS: see Sects. 1.3.5, 1.4.2.3, 3.4.4, and 3.4.7.
- FES: see Sect. 3.2.2.

Even if PNS does not belong to the category of BCI implants, we have decided to include them in this book as an extension to the concept of interfacing with human nervous system. In some cases, there is a strong link between the brain and PNS systems:

- We have seen the vagal nerve in a kind of easy entry door to the central nervous system for the treatment of epilepsy.
- FES system in the upper limbs will one day be driven by cortical BCIs sensing intentions of movement.
- Nerves of other parts of the body may also in the future interface with the brain.
- Stimulating the nerves of an amputated limb (see Sect. 7.3.5) is also a way to carry to the brain information which is no longer available.

BCI implants are intended in priority to patients with very severe conditions like tetraplegia or CLIS (see Sect. 7.3.6). As a complement, PNS may help to treat one of the major side effects of tetraplegia, paraplegia, and spinal cord injury: incontinence, both urinary and fecal. As already discussed in Sect. 2.1.4, incontinence might be one of the highest priorities for these patients to get better care. Incontinence is a very humiliating condition. Some patients may even prefer to get a solution for their incontinence before benefiting of a BCI for motoric recovery.

CLIS patients may also loose the capacity to close their eyes. Simple PNS may provide some appropriate stimulation to force the closing of the eyelids at regular intervals, keeping the surface of the eyes wet.

PNS should therefore be considered as a provider of complementary support for patients implanted with a BCI. Coordinating the development efforts in PNS and BCI and allowing communication between the two devices may be a promising way to improve the life of patients with severe neurological conditions.

7.3.5 Haptic

Haptic [53], as an adjective, is defined as relating to the sense of touch and particularly to the perception and manipulation of objects using the senses of touch and proprioception [63].

Proprioception, often described as the sixth sense, is an important characteristic of the body enabling perception of movement and position. Proprioceptor neurons, distributed through the entire body, are sensitive to mechanical movements. These neurons carry information to the central nervous system, as a complement to other sensory receptors like the vestibular system and vision. The conjunction of these various natural sensors generates an overall perception of body movement and position.

Vision plays a capital role in the control of our movements. Haptic sensors will complement and expand the visual perception.

Haptic devices are intended to help amputees to recover some sense of touch. As PNS, haptic devices do not formally belong to the category of BCIs but should be considered as a complement, a tool to indirectly interact with the brain.

Upper limb prostheses have become sophisticated devices [54] with an active control of the artificial movements by the patient. Several groups are progressing fast in the development of advanced motorized prosthesis. Activation and control of a motorized hand prosthesis can be achieved in several ways:

- Control by the valid hand, either with a joystick or by movement sensors attached to the able hand or inserted in a glove.
- Voice control.
- Movement sensors placed on the chess or on the upper section of the amputated arm.
- EMG electrodes transdermally inserted in the muscles of the upper arm.
- Tans-fascicular or cuff electrodes transdermally attached to the nerves terminating at the amputation level.
- EMG or nerve electrodes connected to an implanted wireless device.
- External BCI, for example, EEG cap.
- Implanted BCI with electrodes placed on the motor cortex area corresponding to the amputated limb.

All the above controls are open loop, meaning that the movements of the various degrees of freedom of the prosthesis are not related to any haptic sensors and lack the "sense of touch." The preemption forces applied on the object grasped by the prosthetic hand only depend on the power of the actuators. If the command to grasp is sent to powerful motors, the object may be damaged. A first level of haptic feedback is to add force sensors on the tips of the fingers of the prosthesis to automatically impose a limitation of the applied force. This safety feature provides a way to protect the sized object from damages but does not provide any force feedback to the patient. Some feedback to the patient may be provided by vibrations, sound, light, or graphics on a screen.

The second level of haptic is to provide direct feedback to the patient by stimulation of the nerves. Several teams are working on that, but no product with active nerve feedback has yet reached approval.

For patients with spinal cord injury, the goal is to send the haptic feedback directly to the brain by placing electrodes on the sensory cortex. Such a true BCI approach is still in a conceptual phase.

7.3.6 Complete Locked-In Syndrome (CLIS)

CLIS has already been discussed under Sect. 3.4. So far, limited achievements have led to simple Yes/No answers detected by measuring brain activities by heavy tools like EEG, fMRI, and NIR. These methods are too constraining for home use and require presence of a technical staff to assist the patient.

Moving to an implanted BCI shall provide some autonomy of use for the patient and his/her family. The first step is to place a Utah array on the brain with a wired connection to an external decoding system linked, for example, to a speech generator.

The main difficulty is related to the extreme fragility of CLIS patients and the risks related to an invasive surgery.

The next step is to place an implanted wireless BCI connected to a fully autonomous system with easy to use decoder which can be operated at home, allowing the family to enter in contact with the CLIS patient. A lot of work is still needed to reach this noble objective, but it is a main incentive for the entire BCI community.

7.3.7 Others

Several promising initiatives are currently happening in the field of advanced neurological devices in relation with BCI. They illustrate the enormous potential of new technologies in this field and pave the road of successful applications of new technologies to treat patient with severe brain or spinal disorders.

7.3.7.1 Targeted Epidural Spinal Stimulation

Spinal cord injury (SCI) is a major source of disability, leading to paraplegia or tetraplegia. About a half million people are bound to a wheelchair today, with more than 25,000 new cases every year. Prof. Grégoire Courtine (see Sect. 3.3.2 and Reference [9] in Chap. 3) has been working for 14 years on neurostimulation therapies to treat persons with SCI, with the objective of having them able to walk again [55, 56, 64]. The approach is based on optimized paddle electrodes to stimulate targeted areas of the spinal cord, linked to a novel implanted electronic system with precise timing to adjust electrical pulses controlled by a remote control and sensors in the shoes (see Fig. 7.17). The system is developed by GTX Medical [57] in Holland and Switzerland. The concept of Targeted Epidural Spinal Stimulation (TESS) is based on research led by Prof. Courtine (EPFL) and Prof. Jocelyne Bloch [58] from the Centre Hospitalier Universitaire Vaudois (CHUV) in Lausanne, Switzerland.

After SCI, communication between the brain and the muscles of the legs and arms is lost or severely reduced. The injury may also affect other body functions related to bladder, bowel, sleep, and sexual organs. The therapy developed by the team consists in a surgically implanted paddle lead located on the spinal cord region that controls leg movements. The electrodes and stimulator are like SCS systems.

The first clinical trials have shown very promising results, where neurostimulation enables stepping even in people with completely paralyzed legs. The therapy triggers activity-dependent plasticity and therefore recovery of walking capabilities, even in the absence of stimulation. The therapy can be used outside hospital environment for continuous improvement and training.

On a longer term, the objective is to wirelessly connect the spinal cord stimulator to a BCI reading the walking intentions of the patient. Like upper limbs movement

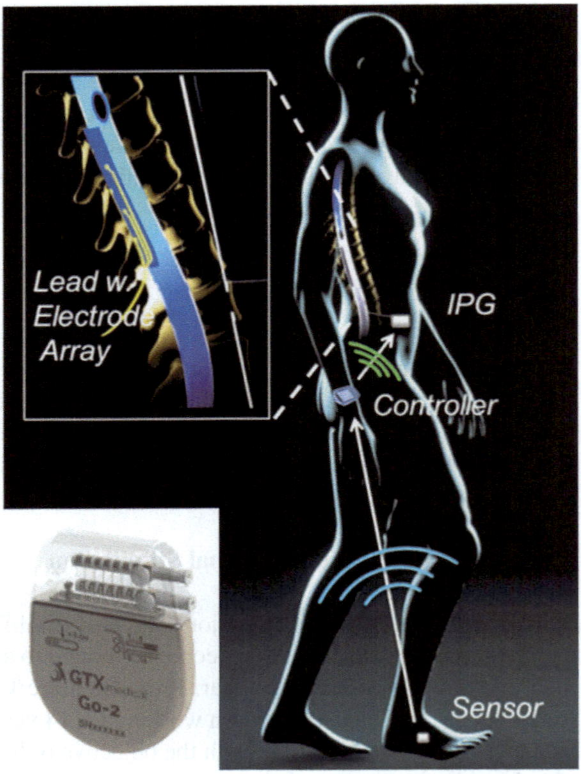

Fig. 7.17 Go-2, Targeted Epidural Spinal Stimulation system. (*Courtesy of GTX Medical and Prof. Grégoire Courtine, EPFL*)

restoration through a BCI cortical system (see Sect. 7.3.1), the future TESS systems will give patients full control on their legs.

7.3.7.2 Stentrode

Synchron [59] is developing an original concept of brain interface called Stentrode™, already described in Sect. 3.2.2. The idea is to introduce, by minimally invasive surgical technologies, a stent-like electrode (see Fig. 7.18) in the brain-blood channels. The electrodes will pick brain signals to help diagnose and treat a range of neurological disorders. Stentrode is an implanted BCI which does not require opening the skull to place the electrodes [60]. It is connected to a wireless electronic device placed in the pectoral area. At system level, the difficulties reside in the insertion of a flexible cable with 16 wires in the veins. From the pectoral location of the wireless implant to the Stentrode, the cable passes the very mobile area of the neck and will encounter flexibility and fatigue issues already described in Sects. 2.2.4 and 3.4.2 for DBS systems. In addition to fatigue due to movements, the cable will endure the regular displacements created by the blood flow and heartbeat. For long-term implantation in the blood channels, the pacemaker industry has shown

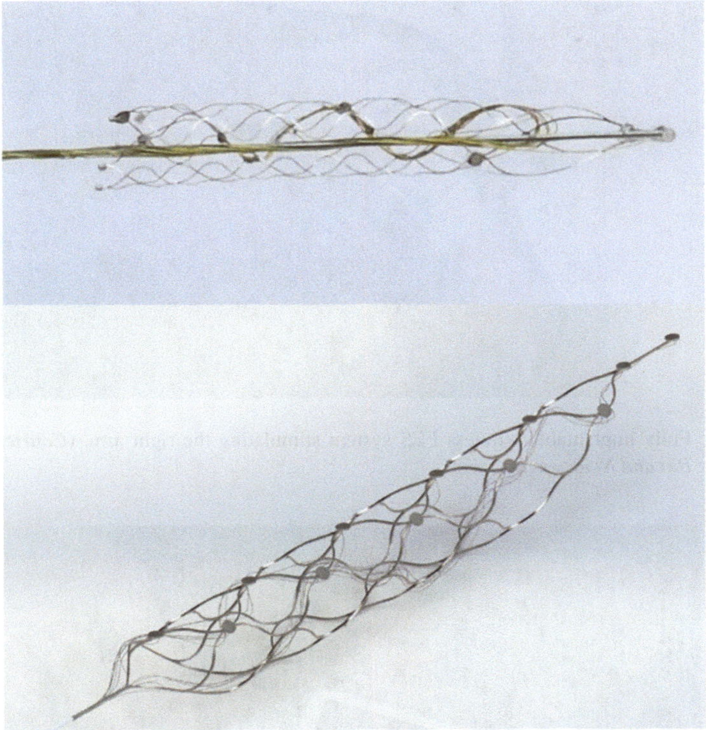

Fig. 7.18 Stentrode™ endovascular stent electrode. (*Courtesy of Synchron, Inc.*)

that the only reliable wiring is realized by coiled electrodes. Sixteen coiled wires will occupy far too much space to be placed in the brain veins. Even if a lot of technical issues remain to be solved, the concept is a promising path for future BCIs.

7.3.7.3 Implanted FES

Implanted systems for restoration of movement of paralyzed arms have already been presented in Sects. 3.2.2 and 3.4.8. The Networked Neuroprosthetic (NNP) System (see Reference [50], Chap. 3) is a fully developed project which already got approval. It has unique features that have already been described. Today, the implant is controlled by an external system which sends the stimulation patterns.

The next step is to control the implanted FES system (see Fig. 7.19), through a wireless RF link, with an external processor worn by the patient (see Fig. 7.20). The second part of the overall system consists in a wireless BCI (see Fig. 7.21) reading the movement intentions and transmitting them wirelessly to the processor. The difficulties to merge two of the most complex systems (see Fig. 7.22) described in the book are substantial, and the challenge will not be achieved before many years, probably at least a decade.

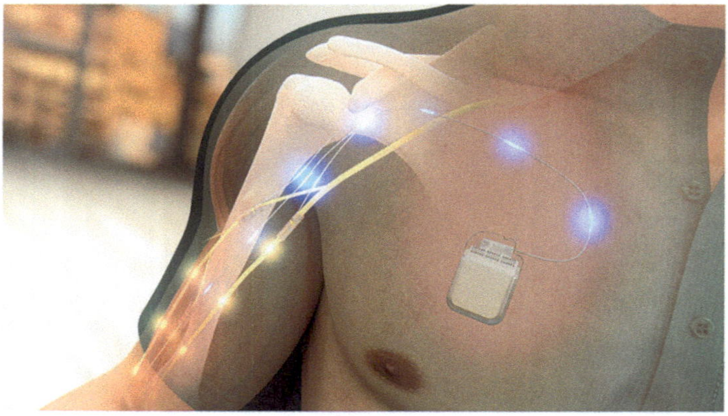

Fig. 7.19 Fully implantable wireless FES system stimulating the right arm. (*Courtesy of Wyss Center for Bio and Neuroengineering*)

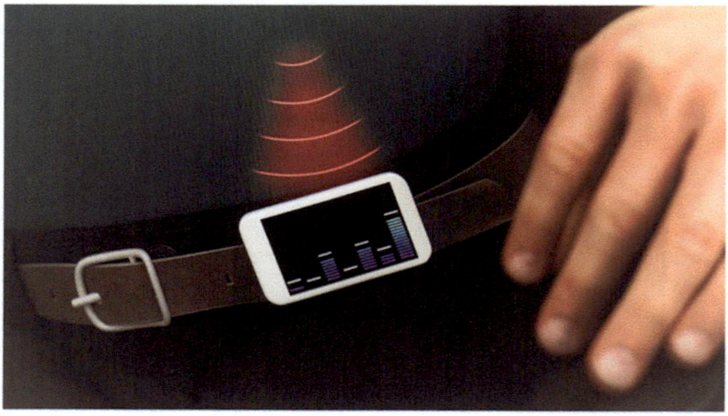

Fig. 7.20 Belt-worn processor receiving information from a wireless BCI and sending control patterns to an implanted wireless FES. (*Courtesy of Wyss Center for Bio and Neuroengineering*)

7.4 Trends

Miniaturization and energy reduction will pace the evolution trends of the future BCIs. The previous chapters have described the technical barriers and limitations of current devices and shown where substantial progresses are needed.

Not all fields of science and technology are progressing at the same speed. Since the early days of AIMDs, the main improvements came from electronics. We are adding more and more electronic features and capabilities in our implants, but energy, mechanical aspects, and material science are not improving much. After the various discussions above, we now understand that there is a huge gap between what new electronics could do to improve health and our capabilities to package and connect these implants in a way they can stay in the body for a long time. We would

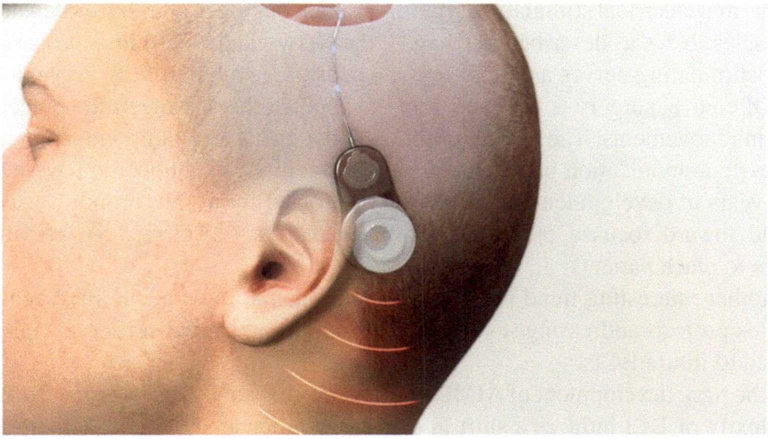

Fig. 7.21 Cortical wireless BCI collecting movement intentions. (*Courtesy of Wyss Center for Bio and Neuroengineering*)

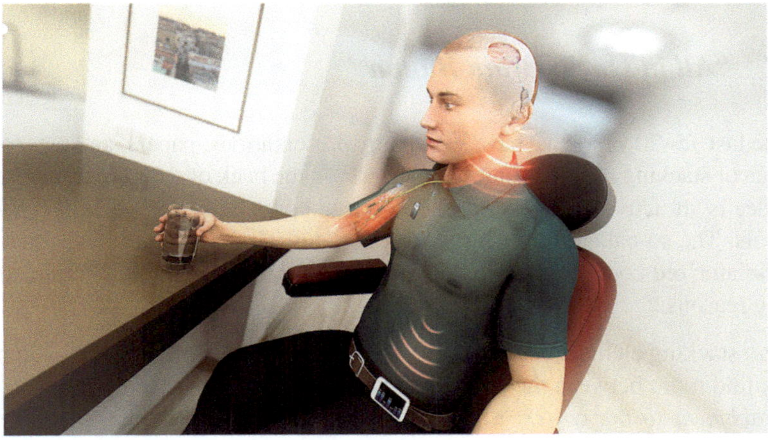

Fig. 7.22 Future vision of a fully implanted wireless system merging a FES activation of a paralyzed limb to a cortical sensing BCI with external wearable processor. (*Courtesy of Wyss Center for Bio and Neuroengineering*)

like to have more channels, but we do not know how to do the connections. We would like to transfer large volumes of data, but the RF waves are absorbed in the body. We would like to implant powerful processors, but it's getting hot.

It's time for this industry to prioritize the research and development efforts on solutions to fill these gaps. It does not make sense to have more powerful electronics if, at the end, we cannot implant the device. Success of future BCIs should be measured by their capacity to improve health. This field is not only benchmarked by an electronic technology race. We should favor step-by-step approaches. A good example is FES for activation of the arm. First, some results have been achieved by using transdermal electrical stimulation. Then, percutaneous electrodes have been inserted in muscles. Now, as described in Chap. 3, the Networked Neuro Prosthesis is capable of stimulating nerves and muscles, add sensors, in a wired fully implanted system. Of course, surgery is long and complex, but the therapy meets the objective of restoring movements. The team has not done the mistake to jump immediately to a wireless communication between master and satellites, which would have meant 5 more years of development and much larger probability to fail. I think we see now a trend toward focused projects, with more modest objectives, but providing a chance to reach patients.

Another interesting trend is seen in a more open attitude of the FDA regarding complex projects addressing severe conditions. This new mind-set is especially well adapted to future BCIs.

In the past, development of AIMDs was mainly driven by the industry. Today, the complexity of BCI induces a shift in development leadership to academia, philanthropic foundations, and large consortium supported by national and international grants. I think this trend will further develop and that development groups will pool their resources and aim to a common goal.

7.4.1 "All in One": Brain Button

For the last 15–20 years, we have seen many presentations, papers, and patents with the idea of stacking the electronics and the RF on the back of a microelectrode array. The idea is attractive as it gets rid of cables and connectors, the "Achilles' heel" of implants. The so-called brain button, autonomous wireless battery-less multichannel miniaturized cortical interface, has failed so far to reach the market for the following reasons:

- *Size*: stacking elements on the back of a MEA leads to a thick device which does not find room between the cortex and the inside wall of the skull bone.
- *Attachment to the cortex*: a too large and heavy device cannot be secured in place.
- *Moisture penetration*: a brain button cannot be encapsulated in titanium because of the number of channels and the lack of miniature FTs. Simple potting does not protect the electronics for long-term implantation.
- *Energy*: a small diameter coil on the brain button cannot capture enough energy for operating the electronics.
- *Communication*: is difficult from a miniature antenna placed on the brain button.

Some ASICs have even been developed with the same footprint as the Utah array, with the ambitious goal to be flip-chipped and stacked directly at the back of the electrodes. Without proper hermetic encapsulation, the stack of chip will fail due to moisture ingress. It was too early to go to such a level of integration. Several groups did the same mistake of designing very dense integrated electronics without having solved the issue of moisture penetration.

The concept of brain button is likely to get rejuvenated by the arrival to maturity of some of the technologies described earlier in this book:

- Near-hermetic encapsulation like Coat-X or ALD (see Sect. 4.9.6).
- Minced chips (see Sect. 4.9.6).
- CerMet high-density FTs (see Sect. 7.2.4).
- Optical communication (see Sect. 7.2.6).

It is likely that a tiny brain button, combining all the above, will become reality if properly designed in a project driven with the recommendation of this book on how to build the BCI of the future.

7.4.2 Bioelectronics

Unprotected CMOS chips slowly dissolve in CSF and other body fluids. Conventional chips do not resist on the long term in the human body. This is one of the reasons why electronics are currently encapsulated in hermetic housings, with all the difficulties described earlier in the book.

In the future, it is expected to see the emergence of new semiconductor materials which are biocompatible and biostable. Bioelectronics [65] is globally defined as the extension of electrical engineering and electronics principles to biology, medicine, behavior, or health. In this wide spectrum are included biomolecular physics, bioinspired electronics, and other general concepts.

In this chapter, we limit bioelectronics to a much narrow scope. Organic materials contain carbon and carbon chains linked with other atoms. Some conductive polymers are already used in active implants, especially to improve the contact impedance of electrodes. Some further work has been done in the direction of organic semiconductors finding today their main application in flat display screens.

If organic semiconductors can be made biocompatible and biostable, we could envision to print electronic circuits directly on flexible electrodes. For BCI applications, it will open a wide range of improvements, for example, the possibility to include preamplifiers or multiplexers directly on the electrodes, leading to a better signal-to-noise ratio and a reduction of wires and connection to the main and conventional electronics. Very high channel counts (more than a couple of hundreds) cannot be designed with one wire for each channel. Some preamplification and multiplexing capability must be introduced at the level of the body interface. Connecting many channels multiplexed on a reasonable number of wires or traces will change the landscape of complex BCIs.

Research on biocompatible organic semiconductors is progressing rapidly. Long-term biostability has not yet been demonstrated. Validation of such technologies for human grade devices and particularly BCI will take a long time, but the trend is undeniable.

7.4.3 *Networked Implants*

We have seen that cables and connectors are among the main technological barriers for the development of BCI. In consequence, several laboratories [66, 67] are working on the concept of distributing large numbers of tiny implants in the brain, communicating with each other and with the external world through a local wireless RF network. Ambitious initiatives like neural dust [68] from the Defense Advanced Research Projects Agency (DARPA) and neurograin [69] from Brown University have drawn attention in the last years. In 2017, DARPA also launched the Neural Engineering System Design (NESD) program aiming to enter in communication with one million independent neurons. All these initiatives have objectives far beyond realistic translational applications. Unmet medical needs and reasonable time to market are not part of these research projects. As such, the purpose of this book does not match the ambitions of these programs. No BCI for human use will result from this advanced research within the next decade. These projects go too far and do not participate much to the improvement of the condition of people with neurological disorders. I would be tempted to classify them as "Dreamers," covered in the next chapter. Nevertheless, the reflections around the needs to distribute the interfaces with neurons across the brain in a wireless approach have value and certainly show a trend for future BCIs.

Coming back to pragmatical views, we have seen that real life is presenting serious barriers in terms of available energy and communication within the human body. Networked implants are still part of the very long-term future.

References

1. https://en.wikipedia.org/wiki/Big_data
2. https://en.wikipedia.org/wiki/Artificial_intelligence
3. https://en.wikipedia.org/wiki/Machine_learning
4. https://en.wikipedia.org/wiki/Wireless
5. https://en.wikipedia.org/wiki/Internet_of_things
6. https://www.quora.com/What-is-the-difference-between-IoE-and-the-Internet-of-Things-IoT
7. https://en.wikipedia.org/wiki/Machine_to_machine
8. https://en.wikipedia.org/wiki/Wireless_power_transfer
9. Agarwal K et al (2017) Wireless power transfer strategies for implantable bioelectronics: methodological review. IEEE Rev Biomed Eng 10:136–161
10. https://en.wikipedia.org/wiki/Specific_absorption_rate
11. https://en.wikipedia.org/wiki/Litz_wire
12. https://en.wikipedia.org/wiki/Resonant_inductive_coupling
13. Lee J et al (2016) A review on wireless powering schemes for implantable microsystems in neural engineering applications. Biomed Eng Lett 6:205–215
14. Priya S, Inman DJ (eds) (2009) Energy harvesting technologies, Springer, ISBN 978-0-387-76463-4. http://preview.kingborn.net/942000/8fef646fa5ec49b58fb65a13fd39d404.pdf

15. Hannan M et al (2014) Energy harvesting for the implantable biomedical devices: issues and challenges. Biomed Eng Online 13:79. https://biomedical-engineering-online.biomedcentral.com/articles/10.1186/1475-925X-13-79
16. Tsai J-Y et al (2011) Ultrasonic wireless power and data communication for neural stimulation, 2011 IEEE International Ultrasonics Symposium, Orlando, FL, USA, 18–21 Oct. 2011
17. https://en.wikipedia.org/wiki/Wire_bonding
18. https://en.wikipedia.org/wiki/Flip_chip
19. https://en.wikipedia.org/wiki/Ball_grid_array
20. https://en.wikipedia.org/wiki/Three-dimensional_integrated_circuit
21. https://en.wikipedia.org/wiki/FR-4
22. https://en.wikipedia.org/wiki/Moore%27s_law
23. https://www.mst.com/dyconex/
24. https://www.mst.com/dyconex/products/lcp/index.html
25. Bagen S et al (2015) Liquid crystal polymer substrates to enable advanced RF and medical applications, IMAPS, New England 42nd Symposium & Expo, http://www.imapsne.net/2015%20presentations/F/F3.pdf
26. https://www.dictionary.com
27. https://www.heraeus.com/en/hmc/heraeus_medical_components/home_medical_components.aspx
28. https://www.heraeus.com/en/group/home/home.aspx
29. https://www.heraeus.com/en/group/technology_report_online/articles/tr_2016/03_CerMet_Tech.aspx
30. Bagen S et al (2016) Electronics packaging methods and material for implantable medical devices, IMAPS, New England 43nd Symposium & Expo, http://www.imapsne.org/virtualCDs/2016/2016%20Presentations/E/E5.pdf
31. https://en.wikipedia.org/wiki/Co-fired_ceramic
32. Formenko A et al (2018) Low-intensity ultrasound neuromodulation: An overview of mechanisms and emerging human applications, Brain Stimulation, Volume 11, issue 6, November–December 2018, pp 1209–1217, Elsevier, https://www.sciencedirect.com/science/article/pii/S1935861X18302961
33. Legon W et al (2018) Transcranial focused ultrasound neuromodulation of the human primary motor cortex, Scientific Reports 8, Article number: 10007 (2018), Nature, https://www.nature.com/articles/s41598-018-28320-1
34. Kubanek J (2018) Neuromodulation with transcranial focused ultrasound. Neurosurg Focus 44(2):E14
35. Focused Ultrasound for Peripheral Neuromodulation, Research Newsletter, EPFL, STI, CNP, TNE, 2018, https://tne.epfl.ch/page-76740-en-html/page-154687-en-html/
36. Laqua D et al (2014) Ultrasound communication for intelligent implants. Biomedizinische Technik/Biomed Eng 59(1):727–730, October 2014, https://www.researchgate.net/publication/267324432_Ultrasound_communication_for_intelligent_implants
37. Wang M et al (2017) Exploiting Spatial Degrees of Freedom for High Data Rate Ultrasound Communication with Implantable Devices, https://arxiv.org/ftp/arxiv/papers/1702/1702.05154.pdf
38. https://www.wysscenter.ch/
39. https://blackrockmicro.com/wp-content/uploads/2016/08/LB-0805-1.00-CerePlex-W-IFU.pdf
40. https://www.wysscenter.ch/project/thought-controlled-arm-to-help-people-with-paralysis-reach-and-grasp/
41. http://en.eu.livanova.cyberonics.com/healthcare-professionals/vns-education
42. https://livanova.com/en-US/Home/Products-Therapies/Neuromodulation/Healthcare-Professionals.aspx
43. Baud M et al (2018) Multi-day rhythms modulate seizure risk in epilepsy. Nat Commun 9, Article number:88

44. https://www.wysscenter.ch/project/epilepsy-monitoring-seizure-forecasts/
45. http://www.bionicsinstitute.org/new-approach-epilepsy-diagnosis/
46. https://en.wikipedia.org/wiki/Dyslexia
47. https://neurocenter-unige.ch/research-groups/anne-lise-giraud/
48. https://neurocenter-unige.ch/research-groups/#filters=&search=
49. https://www.wysscenter.ch/project/brain-stimulation-to-help-people-with-dyslexia/
50. https://en.wikipedia.org/wiki/Neurofeedback
51. https://en.wikipedia.org/wiki/Tinnitus
52. Haller S et al (2009) Real-time fMRI feedback training may improve chronic tinnitus. Eur Radiol 20(3):676–703
53. https://en.wikipedia.org/wiki/Haptics
54. https://en.wikipedia.org/wiki/Prosthesis
55. Wagner F et al (2018) Target neurotechnology restores walking in humans with spinal cord injury. Nature 563:65–71
56. Formento E et al (2018) Electrical spinal cord stimulation must preserve proprioception to enable locomotion in humans with spinal cord injury. Nat Neurosci 21:1728–1741
57. https://www.gtxmedical.com/about-us/
58. https://en.wikipedia.org/wiki/Jocelyne_Bloch
59. https://www.synchronmed.com/
60. Opie N et al (2018) Focal stimulation of the sheep motor cortex with a chronically implanted minimally invasive electrode array mounted on an endovascular stent. Nat Biomed Eng 2:907–914
61. Enescu D (2019) Thermoelectric energy harvesting: basic principles and applications, Open access peer-reviewed chapter, Published January 21st, 2019. https://doi.org/10.5772/interchopen.83495
62. https://en.wikipedia.org/wiki/Thermoelectric_generator
63. https://en.wikipedia.org/wiki/Proprioception
64. Capogrosso M et al (2018) Configuration of electrical spinal cord stimulation through real-time processing of gait kinematics. Nature Protocols, Springer Nature Experiments 13(9)
65. https://en.wikipedia.org/wiki/Bioelectronics
66. Lee J et al (2019) An implantable wireless network of distributed microscale sensors for neural applications. Accepted for IEEE EMBS Conference on Neural Engineering, March 20–23, 2019, pp. 871–874
67. Laiwalla F et al (2019) A Distributed Wireless Network of Implantable Sub-mm Cortical Microstimulators for Brain-Computer Interfaces, accepted for IEEE EMBS Conference on Neural Engineering, March 20–23, 2019
68. https://www.darpa.mil/news-events/2016-08-03
69. https://news.brown.edu/articles/2017/07/neurograins
70. Bafar VM, Schmid A (2013) Wireless cortical implantable systems. Springer Nature Switzerland. ISBN 978–1–4614-6701-4

Chapter 8
Dreamers

8.1 Interfacing with Your Brain

We saw before that our main goal is to find a way to interface with our brain in order to restore lost functions. It is a technical challenge on many aspects. The brain is such a great source of information that even getting a tiny part of the wealth of information present in this formidable network is enough to do great things. Is it worth trying to extract signals emitted by our neuron networks? Can we really interface with the brain? The answer is definitely "yes." Some valuable experiments and even some confirmed achievements like the ones described in the previous chapters do give us a motivation to continue and develop further.

As quickly covered in Sect. 1.4.3, people dream to go much further than the modest achievements we got so far. Pioneers and doers have focused their work on treating medical conditions, like paralysis, epilepsy, blindness, or deafness. We have shown we can do marvelous things by inserting electrodes in the head of suffering patients. A new generation of scientists is considering going much beyond simple therapeutic or diagnostic applications. They want to explore the field of BCI to non-therapeutic purposes, to extract information for reaching other goals, to try bridging our brain with machines and artificial intelligence, and to leverage our natural performances with the help of technologies. We all know that the current BCI system can be extended and applied to non-therapeutic applications. In view of the great difficulties we have restoring some modest movement capabilities in paralyzed patients, we may wonder if it is reasonable to go beyond medical objectives. Shouldn't we first try to do a better job for patients in need?

Dreamers are people who believe we can jump ahead and do much more. I see two categories of dreamers:

- *Dreamers in the medical field*: their projects have the same goals as BCIs described in this book. They want to restore lost functions, rehabilitate, and cure neurological disorders. Unlike the "reasonable" approach taken by genuine translational developers, taking one challenge after the others, dreamers want to

C. Clément, *Brain-Computer Interface Technologies*,
https://doi.org/10.1007/978-3-030-27852-6_8

take a giant step at once. An example has been covered in Sect. 7.4.3 where we saw proposals to spread thousands of tiny electronic particles communicating in a RF network. The concept is exciting, but the objectives are unreachable within a normal project timeline, say 10 years. This is the reason why I call them dreamers. They cannot provide answers to fundamental questions. In the example of tiny grains, nobody knows how to insert them in the tissues, how to explant them later, how to attach them to avoid migration, and how to avoid the grains damaging the brain in case of violent acceleration like in a car accident. Dreamers sometimes also deny the laws of physics. In the case of tiny grains, their energization by microwaves will heat the tissues far above the acceptable levels. Many of these issues are showstoppers. Knowing that there are no solutions, we can even wonder why dreamers persist on such projects. It may be for the sake of the advancement of science, but it cannot be justified for human health improvement. The NESD program (see Sect. 7.4.3) aiming for one million independent connections in the brain within 4 years is unrealistic in the perspective of human health. It may fuel new ideas on the long term, but is it a wise use of taxpayers money?

- *Dreamers in nonmedical applications*: their objectives are less noble than the other group described above. They want to use BCI technology for more trivial applications, like brain-to-brain communication, controlling machines or vehicles directly from the brain, merging artificial and natural intelligence, or augmentation of our capabilities. Famous wealthy entrepreneurs are investing enormous sums of money in this direction. They even hire crowds of neuroscientists to implement their dreams. The reasons why I qualify them as dreamers are:
  - Neuroscientists are involved, but only a few experienced technologists carry a reality check.
  - These groups usually ignore the natural limitations of the human body.
  - The laws of physics are not fully understood, especially regarding energy and wave propagation.
  - The surgical procedures to put these nonmedical devices in the head of people include high risks.
  - Costs related to developing and manufacturing BCI systems are grossly underestimated.
  - Even if the application is nonmedical, implants must follow safety regulations imposed by health authorities.

Are you ready to have a full craniotomy, with the associated serious clinical risks, have a long recovery time and visible scars, pay all the huge cost by yourself, and be able to drive your car directly from your brain?

I'm not in favor of the initiatives of the dreamers. Even if the young successful entrepreneurs pay from their own pockets, they are taking valuable resources, especially scientists and engineers, away from the medical field. Is it acceptable to book an operation room, a surgical team, and hospital time for placing a nonmedical implant in the head of a person? No! Especially when medical resources are in high demand.

8.2 Repaired Man

Since the very beginning of the use of implants, bone, dental, vascular, and later active implants, we have been speaking about the Repaired Man. A vast majority of implants are introduced in the body to repair it after an accident (e.g., a bone plate to join segments of a broken leg), to correct a slowly evolving impairment (e.g., opening a stenosed coronary artery with a stent), or to block symptoms (like DBS for Parkinson's disease).

A more recent evolution is related to the improvement of quality of life. Strictly speaking, active devices allowing patients to have a better life are not repairing people. They may be used for preventing degeneration, compensate for decreasing performance due to age, or attenuating chronic pain. A widely known example is pacemakers implanted in patients suffering from age-related bradycardia. Their life is not at risk, and their heart is functioning properly, but it is not able to accelerate as when they were young. Their rate-responsive pacemakers will make them feel as they were 20 years old again. Is it reparation, prevention or comfort?

Usually, patients have only one implant or two (not counting dental implants or bone repairs). We are still very far away from the day of the bionic man, a human with most of his/her body parts and organs replaced by implants. Already in 1973, a science fiction TV program "The Six Million Dollar Man" [1] was presenting a supernatural bionic man. Some dreamers believe in the concept of a bionic man or cyborg [2]. It is an illusion to think that we can prolong substantially our lives by replacing, one after the others, all the obsolete parts of our bodies. We will probably not want this type of artificial life. Anyway, our social security systems will not afford the costs of the bionic man. Maybe a few wealthy people will be tempted by a bionic prolongation of their lives, but they remain dreamers.

8.3 Augmented Man

The border between therapy or repair and augmentation is often fuzzy. We all know the cases of amputees who managed to run quicker than valid athletes by using special prosthesis. In their case, the first objective was reparation. The fact that technologies allow them to go beyond reparation and reach superhuman performances is maybe not a major concern. We can even see this as a revenge against adversity. It is anyway augmentation and must be controlled in the best possible way. Regulators have not yet defined guidelines and recommendations regarding human augmentation. It will be difficult to distinguish augmentation as a side effect of reparation and augmentation made on purpose.

Take the example of BCI for movement restoration. We have seen during the early BCI work with patients moving a cursor on a screen that they had the capability to click on the "virtual mouse" quicker than what we achieve with our finger. The brain command to click is picked by the BCI directly on the cortex, bypassing the time delay of the propagation of the signal along the nerves, from brain to finger. Is the couple of dozen milliseconds in favor of the paralyzed patient augmentation? Certainly not.

Another illustration about the fuzziness of the border between reparation and augmentation is the story of Neil Harbisson [3], who presents himself as a cyborg and transspecies activist. Born color-blind, seeing only in black and white, he decided to design a device which will give him some perception of colors. His idea was to transform the light spectrum in a sound spectrum. A given color will correspond to an auditory frequency. A photodiode is picking the dominant color of the surrounding, and dedicated electronics make the translation in the corresponding sound, which is transmitted to the skull by a bone-anchored hearing aid (BAHA) [4]. A BAHA is a percutaneous vibrating system attach to the skull bone. Vibration propagates along the skull and reach the inner ear, where they transmit to the liquid of the cochlea and induce a sound perception. BAHA are commercially available devices [5] to treat certain forms of deafness. In the case of Neil, the device has been modified in a way it does not transmit sounds picked up by a microphone but translation of the visible spectrum.

Neil did not recover the diversity of colors perceived by our retina. The single photodiode only picks the dominant color. But Neil wanted to push the experience further and extend the perceived light spectrum outside the visible field, adding ultraviolet light and infrared light. His idea, a bit provocative, was to "see" what some insects or animals could but what remain invisible to humans. This is his transspecies approach. It was also his way to take some revenge against Mother Nature who deprived him for color vision. He even pushed the analogy with insects by designing his device as a kind of antenna at the top of his head.

Even if the device is a unique case, not intended to be applicable to anybody else than himself, Neil wanted a certain recognition for his initiative and asked the approval of an ethic committee. Approval was denied because of the extension of the light spectrum to light invisible to us. The trick was categorized as augmentation. If he had restrained the device to translation of the visible spectrum, the ethic committee would have most certainly accepted it. A nice story to illustrate is how difficult it is to regulate augmentation of human beings.

Some cases of augmentation are clear and even developed on purpose. Some dreamers plan to create super BCIs to make us smarter, quicker to decide, able to communicate directly from our brain to machines and other brains, extend our memory, and many other "improvements." Some authors describe this augmentation as intelligence amplification [6]. Discussion about augmentation is of almost a philosophical nature and goes beyond the purpose of this book. Nevertheless, we must understand that our technologies aimed at medical applications open the door to less noble projects. Regulations of the use of BCI for augmentation must be put in place quickly, before being exposed to serious ethical issues.

8.4 Ethical Aspects

All along the various chapters, we have preached with enthusiasm about the availability of new technologies and the new frontiers of neuro-technologies. We have seen that we can acquire information directly from the brain but also that it is possible to influence and stimulate our brain in an artificial way. Our nervous system is an image of ourselves, part of us. Connecting to our brain is an unprecedented intrusion in our body. Authorities controlling and approving neuro-devices are taken by surprise by the exponential evolution of technologies. The impact of new technologies, their increasing invasiveness, the access to central parts of our nervous systems, and the connections with our sense are new dimensions of the already complex environment of human health.

The introduction of BCI systems raises a new set of fundamental questions about us, society, ethics, and our relations to technologies. A few examples:

- Are we going too far?
- Have we already gone too far?
- Do we have the right to interface directly with the brain?
- Are we going to lose control?
- Is artificial intelligence going to take the place of our natural thinking?
- Am I going to lose my privacy?
- Is big data a tool for improvement or a way to control us?

I do not have the answers. Probably not many people have a clear understanding of these complex notions. The objective of this book is to focus on practical and technical matters in the perspective of building the BCI of the future. But it does not mean that ethics is not important. Too often in the past, engineers have disregarded ethical matters and designed system with questionable targets and unnecessary features. They forget about the patient. As we interface with the brain, even engineers must be careful and take some distance. Ethical thoughts should be part of the entire development process, from the definition of user's need and human factors to the clinical trials and the management of risks.

The growing importance of breakthrough concepts like artificial intelligence, big data, machine learning, gene engineering, and other technologies based on information is redefining the environment of neurosciences. What happens to the human being in the middle of these revolutions? One thing is certain; we should constantly put the patient in the center. It is the best way to keep an ethical perspective. Having the patient as a priority was also the starting point of this book, when we described our translational values.

References

1. https://en.wikipedia.org/wiki/The_Six_Million_Dollar_Man
2. https://en.wikipedia.org/wiki/Cyborg
3. https://en.wikipedia.org/wiki/Neil_Harbisson
4. https://en.wikipedia.org/wiki/Bone-anchored_hearing_aid
5. https://www.cochlear.com/us/en/home/diagnosis-and-treatment/how-cochlear-solutions-work/bone-conduction-solutions
6. https://en.wikipedia.org/wiki/Intelligence_amplification

Chapter 9
Conclusions

We introduced this book by focusing on the importance of the word *build*. When we build a house, we must have a clear plan, solid foundations, a timeline, a budget, the best possible suppliers of materials, experienced contractors, and a good coordination. Building the BCI of the future requires the same approach. Rigor, discipline, and structure are the keywords for success. We have explained a methodology based on step-by-step improvements, modest and reasonable objectives, full understanding of our environment, and collaborative exchanges.

Unlike building a house or a consumer product, building a medical device has one capital additional characteristic: patients. Patients are not customers. They are much more than this. They are in the center. They are *our* patients. Meeting their expectations and understanding their needs are our duty and our priority.

Inserting electronics in the head of a patient is a technical challenge. The difficulties are mainly related to materials and mechanical aspects. Whatever progress we make in the miniaturization and computing power of the electronic circuits intended to be placed in the body for many years, it will be worthless if we do not find ways to encapsulate, protect, and connect them to body interfaces. Unless we find alternatives to titanium encapsulations with hermetic feedthroughs, we will fail to implement advanced electronics in BCI systems. More attention should be put on the practical realization of above-the-neck implants. More engineers should be trained in material sciences. “Implantology,” regulatory, and clinical affairs should be part of the university cursus. The industry of AIMDs is taken by surprise by the exponential evolution of electronics, data management, artificial intelligence, communication, accumulation, and exploitation of big data. Future successes in the field of BCI will rely on mastering the encapsulation, connectivity, and wireless communication.

BCI of the future will be different from the current brain interfaces. How to build them in a way to meet user needs and remain affordable to the healthcare systems are the keys for success. In this book, we tried to give some clues, recommendations, and guidance to designers, developers, and manufacturers of the future BCIs, hoping they will share our faith in the bright future of neuro-technologies. If wisely

C. Clément, *Brain-Computer Interface Technologies*,
https://doi.org/10.1007/978-3-030-27852-6_9

allocated, the enormous resources and experience needed for the translation of BCI systems to widespread human use will provide a tangible payback, maybe not an immediate financial payback but, certainly, a substantial return for patients suffering from neurological disorders. Helping them, their families, and the caregivers is our incentive to go on.

It is worth it!

Chapter 10
Annexes

10.1 Annex 1: Risk Management

Many books, articles, and publications related to risk management have been published. A subset of this large collection is focused on risk management in the field of medical devices. We do not intend to educate our readers on such wide and complex matters. The objective of this annex is to underline the importance of a proper methodology to manage risks in BCI systems projects. Brain interfaces deal with the most sensitive and sophisticated part of our body. We must be certain that we have taken all the possible measures to contain risks within "reasonable" limits.

Extracted from the general definition [1] of risk management, my favorite sentence is "Risk management is the identification, evaluation and prioritization of risks followed by coordinated and economical application of resources to minimize, monitor, and control the probability or impact of unfortunate events or to maximize the realization of opportunities" [2].

In the dictionary, risk is defined as "a situation involving exposure to danger" or in other words "the effect of uncertainty on objectives." Wikipedia [3] gives another definition: "Risk is the possibility of losing something of value. Values (such as physical health, […], financial wealth) can be gained or lost when taking risk resulting from a given action or inaction, foreseen or unforeseen (planned or not planned). Risk can also be defined as the intentional interaction with uncertainty [4]. Uncertainty is a potential, unpredictable, and uncontrollable outcome; risk is a consequence of action taken despite uncertainty."

Many standards and guidelines are setting the frame of risk management for medical devices. The main one is ISO-14971 [5, 6]. The various quality standards of medical devices cover many aspects of the entire life cycle of the product, from design control and supplier management to post-market surveillance. In consequence, risk management is a continuous process, the backbone of the strict methodology applied to AIMDs.

C. Clément, *Brain-Computer Interface Technologies*,
https://doi.org/10.1007/978-3-030-27852-6_10

In the conventional approach, risk analysis is first intended to make medical devices safe for human use. Patient safety is therefore the main target of risk management.

A modern approach of global project management is to extend the philosophy of risk management, including, in addition to patient safety:

- Financial risks: Identify, from the beginning of the project, risks of getting short of money during the crucial development phases. Mitigation of such risks exist, for example, reduction of the number of features, revision of specifications to more modest goals, partnerships, and cession of IP.
- Supply chain risks: In the field of BCI, some components and subassemblies come from single sources, with no alternative. The risk of losing critical suppliers must be mitigated by appropriate long-term agreements.
- Obsolescence risks: AIMDs have long approval cycles and are designed to be manufactured, without change, over long period of time. Electronic components, especially microprocessors, have short life cycles and may not be available any longer when the medical device is still on the market. Such situations are not rare. Risks of disruption of the availability of critical component must be anticipated a long time ahead the occurrence of the issue, by preparing a plan B (including notification or reapproval by the competent authority) or by accumulating a substantial inventory for covering manufacturing until the end of the product line.
- Technological risks: Without affecting the safety of the patients, some technical issues may render the product inappropriate or not able to deliver the therapy. Seen from a patient perspective, the risk of not being treated is a serious one. Plans should be made to first identify such risks and then to evaluate alternative technologies.
- Disruptive risks: We have seen in the past entire industries wrecked by the arrival of a new fully disruptive technology. Such risks should also be identified for BCI systems. What if a new drug is discovered to treat all forms of epilepsy? What if we find a way to read movement intentions with a high-resolution external system? Mitigation of such risks is difficult, but at least we should be able to identify trends and get prepared.
- Business risks: As for technological risks, business risks do not directly impact patients in terms of safety and integrity. Mergers, acquisitions, IP conflicts, arrival of new competition, change in reimbursement strategy, international conflicts, barriers to trade, new regulations, and other high-level changes in the environment may affect the availability of the therapy and therefore indirectly impact patients.

There are plenty of templates and checklists to provide support and structure to designers of medical devices. The purpose of this book is not to go in the details. I do not recommend any specific method, if the following fundamental principles are covered:

- As a priority, deal with patient safety risks.
- Then extend to more global project risks as described above.
- Identify failure modes, for all parts and sub-systems.
- Do an initial failure mode analysis:
  - Root cause(s) of the failure.
  - Impact of the failure.
  - Probability of occurrence.
  - Severity.
  - Risk is usually quantified by the product of the severity rating and of the probability rating.
- Do a first run of risk control and take mitigation measures.
- Reassess post-mitigation risk occurrence and severity.
- Decide if residual risks require a second run of mitigations.
- Constantly review the risk situation during development and after approval.
- Conduct a full review of the risks anytime a change is done in the specifications, in the supply chain, or in the manufacturing processes.

A very common mistake, and source of project failure, is to introduce changes without a full assessment of the impacts of the changes. Design or specification modifications may:

- Modify the severity and probability of occurrence of existing, identified, and already mitigated risks.
- Render previous mitigations inappropriate.
- Introduce new risks.
- Require new mitigations.

Risk management is a mind-set. It was the key of success for many projects. Working hard on risk management from the very beginning of the project is not a waste of time but rather an investment for the future. Applying a strict risk management policy ensures that the goals will be reached without late major modifications.

10.2 Annex 2: NeuroVirtual

NeuroVirtual is not a real company, but it is a representative of a likely model of a start-up developing a complex implanted BCI. As an illustration, I created the business plan of NeuroVirtual, which looks like several companies I have been involved in my professional life.

We will first establish a likely business plan for NeuroVirtual, as seen at the beginning of the project. This document will be presented to investors. Then, we will add a delay of 2 years, happening during the development phase. In a third step, another delay of 2 years will be introduced, to reflect difficulties during the clinical phase. Such deviations from the original plan are common. Business plans rarely include provisions for redesign or for unexpected problems, for example, during clinical trials. The goal of this analysis is to understand the consequences of unexpected delays on finances and returns.

10.2.1 Original Business Plan

The original business plan (see Fig. 10.1) of NeuroVirtual is based on a conventional start-up structure. The plan is the following:

- Year 1: The founder put in place the structure of the company, finds office space, hires an engineer, and files for patents. Two rounds of seed funding provide the cash needed for the first year.
- Year 2: First investors contribute 1 M$ for the development. Two more engineers are hired.
- Year 3: Second round (5 M$) for finalization of the development and building the first prototypes.
- Year 4: Animal tests and Verification and Validation (V&V) phase. Third round of 15 M$ for the preparation of the first human grade devices. The company now has 12 employees.
- Year 5: First-In-Human (FIH) and extension of clinical trials.
- Year 6: Application for CE mark and by year end new round of 15 M$ for preparing launch.
- Year 7: Product approval and first sales. First (modest) revenues 7 ½ years after start. The total exposure (bottom of the cumulated cash-flow curve) is 43.5 M$.

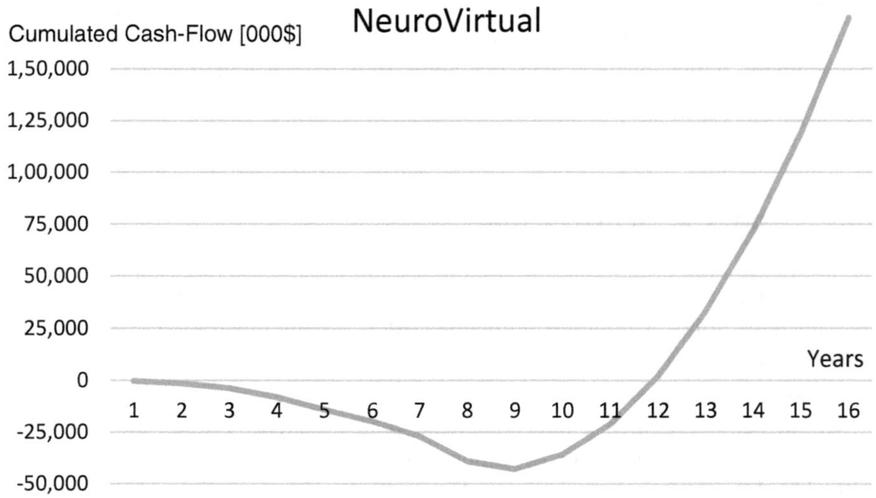

Fig. 10.1 Cumulated cash-flows according to original business plan

- Year 8: Ramp-up. A last financing round of 10 M$ is necessary to grow manufacturing capacities and distribution channels. In total, investors have gathered 46.5 M$ over 8 years. This last round provides a buffer of 3 M$ for unexpected issues.
- Year 9: By mid-year, sales are bigger than expenses. The company starts to generate cash.
- Years 10, 11, and 12: Growing positive cash-flow. At the end of the 12th year, enough cash has been produced to reimburse investors. This is the payback point, unless the company has been sold before, for a higher price.
- The slope of the cumulated cash-flow curve from year 13th indicates the profitability of NeuroVirtual (Table 10.1).

10.2.2 Business Plan with 2 Years Delay During Development

Let us simulate some problems during the development phase (see Fig. 10.2). It is not rare to find major issues which require a major redesign. Even a well-prepared development plan, with clear specifications, might have overlooked some fundamental limitations.

Consequences of such a delay are severe:

- The entire project gets delayed by 2 years. Payback will now happen in the 14th year, which will not satisfy investors.
- Redesigning the device will induce about 3 M$ additional expenses, in material, labor, and external services.
- Maximum exposure becomes 46.5 M$, exactly covered by investments. The 3 M$ buffer of the original plan has been consumed (Table 10.2).

10.2.3 Business Plan with 2 More Years Delay During Clinicals

We have seen above the dire consequences of a 2 years delay during the development phase. Sometimes, additional problems occur later in the project. It is quite common to have delays in the clinical trials, due, for example, to the difficulties to enroll patients. It is also frequent that approval authorities require complements and precision to the submission. All this may lead to 2 more years lost in the final phase of the project (see Fig. 10.3).

Consequences of a late delay are even more severe than what was discussed in the previous chapter:

- As the company is already fully staffed, salaries weight heavily on the finances. The delay of 2 years to get approval also postpones the first profit by 2 years.

Table 10.1 Details per original business plan

		Per original business plan							
Yr	Q	Expenses	Revenues	Cash-flow	Cumul	Investm.	Capital	FTE	Phase
1	1	50		−50	−50	200	200	1	Seed funding
	2	70		−70	−120		200		
	3	100		−100	−220	300	500	2	2nd seed funding
	4	150		−150	−370		500		
2	1	200		−200	−570	1′000	1′500	3	Round A
	2	250		−250	−820		1′500		
	3	300		−300	−1′120		1′500	4	
	4	350		−350	−1′470		1′500		
3	1	400		−400	−1′870	5′000	6′500	6	Round B
	2	500		−500	−2′370		6′500		
	3	700		−700	−3′070		6′500	8	Prototypes
	4	700		−700	−3′770		6′500		
4	1	1′000		−1′000	−4′770		6′500	10	Animals
	2	1′000		−1′000	−5′770		6′500		V&V
	3	1′000		−1′000	−6′770	15′000	21′500	12	Round C
	4	1′500		−1′500	−8′270		21′500		Serie 0
5	1	1′000		−1′000	−9′270		21′500	14	FIM
	2	2′000		−2′000	−11′270		21′500		Serie 1
	3	1′500		−1′500	−12′770		21′500		Clinical
	4	1′500		−1′500	−14′270		21′500		
6	1	1′500		−1′500	−15′770		21′500		
	2	1′500		−1′500	−17′270		21′500		
	3	1′500		−1′500	−18′770		21′500		Application CE
	4	1′000		−1′000	−19′770	15′000	36′500		Round D
7	1	2′000		−2′000	−21′770		36′500	18	Prepare launch
	2	1′500		−1′500	−23′270		36′500		Approval
	3	2′000	50	−1′950	−25′220		36′500	20	First sales
	4	2′000	200	−1′800	−27′020		36′500		
8	1	3′000	400	−2′600	−29′620		36′500		Ramp-up
	2	3′500	600	−2′900	−32′520		36′500		
	3	4′000	800	−3′200	−35′720		36′500		
	4	4′500	1′200	−3′300	−39′020	10′000	46′500		Round E
9	1	5′000	2′000	−3′000	−42′020		46′500		
	2	5′500	4′000	−1′500	−43′520		46′500		
	3	6′000	6′000	0	−43′520		46′500		*End loosing*
	4	6′500	7′000	500	−43′020		46′500		*First profit*
10	1	7′000	8′000	1′000	−42′020		46′500		
	2	7′500	9′000	1′500	−40′520		46′500		
	3	8′000	10′000	2′000	−38′520		46′500		
	4	8′500	11′000	2′500	−36′020		46′500		

(continued)

Table 10.1 (continued)

		Per original business plan							
Yr	Q	Expenses	Revenues	Cash-flow	Cumul	Investm.	Capital	FTE	Phase
11	1	9′000	12′000	3′000	−33′020		46′500		
	2	9′500	13′000	3′500	−29′520		46′500		
	3	10′000	14′000	4′000	−25′520		46′500		
	4	10′500	15′000	4′500	−21′020		46′500		
12	1	11′000	16′000	5′000	−16′020		46′500		
	2	11′500	17′000	5′500	−10′520		46′500		
	3	12′000	18′000	6′000	−4′520		46′500		
	4	12′500	19′000	6′500	1′980		46′500		*Payback*
13	1	13′000	20′000	7′000	8′980		46′500		
	2	13′500	21′000	7′500	16′480		46′500		
	3	14′000	22′000	8′000	24′480		46′500		
	4	14′500	23′000	8′500	32′980		46′500		
14	1	15′000	24′000	9′000	41′980		46′500		
	2	15′500	25′000	9′500	51′480		46′500		
	3	16′000	26′000	10′000	61′480		46′500		
	4	16′500	27′000	10′500	71′980		46′500		
15	1	17′000	28′000	11′000	82′980		46′500		
	2	17′500	29′000	11′500	94′480		46′500		
	3	18′000	30′000	12′000	106′480		46′500		
	4	18′500	31′000	12′500	118′980		46′500		
16	1	19′000	32′000	13′000	131′980		46′500		
	2	19′500	33′000	13′500	145′480		46′500		
	3	20′000	34′000	14′000	159′480		46′500		
	4	20′000	35′000	15′000	174′480		46′500		

- Maximum exposure is now 58.3 M$, 11.8 M$ more than before. Consequently, Round E must be increased from 10 M$ to 25 M$, leaving again a buffer of 3 M$ for unexpected issues during launch.
- At the end of the 16th year, payback is not yet reached, and investors have taken higher risks (Fig. 10.4 and Table 10.3).

10.3 Annex 3: FDA Draft Guidance on Brain Computer Interfaces

We have seen that BCI systems are not covered by specific standards and guidance. So far, designers of BCI systems have based their work on general standards applicable to AIMDs. As BCI are now becoming a fast-growing field, it is high time to fill the standard and regulation gap.

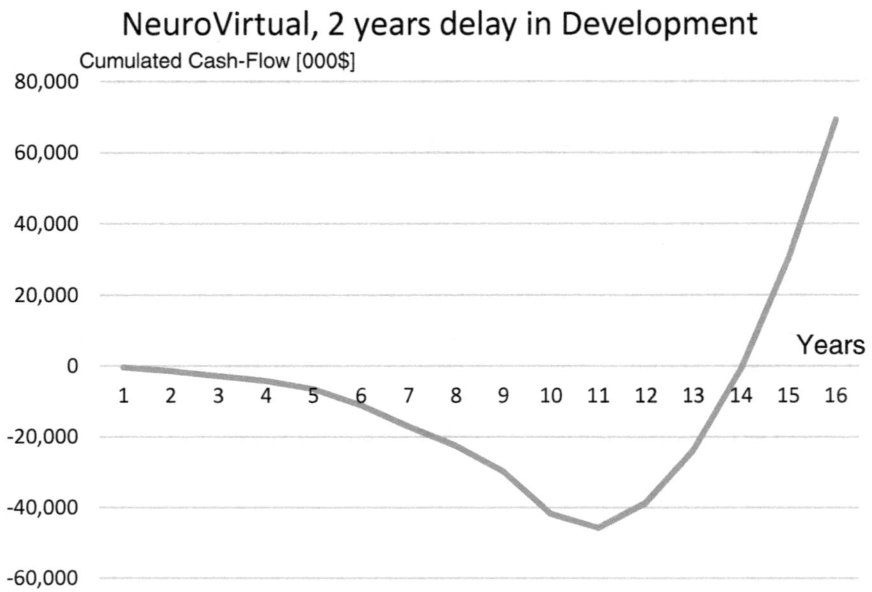

Fig. 10.2 Cumulated cash-flows with 2 years delay in the development phase

The FDA has taken the initiative and recently (February 2109) issued a first draft of a guidance "Implanted Brain-Computer Interface (BCI) Devices for Patients with Paralysis or Amputation – Non-clinical Testing and Clinical Considerations, Draft Guidance for the Industry and Food and Drug Administration Staff" [7]. As a draft, it is not yet intended for implementation and contains non-binding recommendations. When finalized, this guidance will represent the current thinking of the Food and Drug Administration (FDA) on the topic.

The FDA is conscious of the emergence and importance of BCI system. Extract of the Draft Guidance is as follows: "This draft guidance document provides draft recommendations for Q-Submissions and Investigational Device Exemptions (IDE) for implanted Brain-Computer Interface (BCI) devices for patients with paralysis or amputation. The field of implanted BCI is progressing rapidly from fundamental neuroscience discoveries to translational applications and market access. Implanted BCI devices have the potential to bring benefit to people with severe disabilities by increasing their ability to interact with their environment, and consequently, providing new independence in daily life. For the purposes of this draft guidance document, implanted BCI devices are neuro-prostheses that interface with the central or peripheral nervous system to restore lost motor and/or sensory capabilities in patients with paralysis or amputation. FDA's Center for Devices and Radiological Health (CDRH) believes it is important to help stakeholders (e.g. manufacturers, health-care professionals, patients, patient advocates, academia, and other government agencies) navigate the regulatory landscape for medical devices. Towards this goal, on November 21, 2014, CDRH held an open public workshop [8]

Table 10.2 Details with 2 years delay during the development phase

		2 years delay in development							
Yr	Q	Expenses	Revenues	Cash-flow	Cumul	Investm.	Capital	FTE	Phase
1	1	50		−50	−50	200	200	1	Seed funding
	2	70		−70	−120		200		
	3	100		−100	−220	300	500	2	2nd seed funding
	4	150		−150	−370		500		
2	1	200		−200	−570	1′000	1′500	3	Round A
	2	250		−250	−820		1′500		
	3	300		−300	−1′120		1′500	4	
	4	350		−350	−1′470		1′500		Start a redesign
3	1	350		−350	−1′820	5′000	6′500		Round B
	2	350		−350	−2′170		6′500		
	3	350		−350	−2′520		6′500		
	4	350		−350	−2′870		6′500		
4	1	350		−350	−3′220		6′500		
	2	350		−350	−3′570		6′500		
	3	350		−350	−3′920		6′500		
	4	350		−350	−4′270		6′500		
5	1	400		−400	−4′670		6′500	6	
	2	500		−500	−5′170	15′000	21′500		Round C
	3	700		−700	−5′870		21′500	8	Prototypes
	4	700		−700	−6′570		21′500		
6	1	1′000		−1′000	−7′570		21′500	10	Animals
	2	1′000		−1′000	−8′570		21′500		V&V
	3	1′000		−1′000	−9′570		21′500	12	
	4	1′500		−1′500	−11′070		21′500		Serie 0
7	1	1′000		−1′000	−12′070		21′500	14	FIM
	2	2′000		−2′000	−14′070		21′500		Serie 1
	3	1′500		−1′500	−15′570		21′500		Clinical
	4	1′500		−1′500	−17′070		21′500		
8	1	1′500		−1′500	−18′570		21′500		
	2	1′500		−1′500	−20′070		21′500		
	3	1′500		−1′500	−21′570		21′500		Application CE
	4	1′000		−1′000	−22′570	15′000	36′500		Round D
9	1	2′000		−2′000	−24′570		36′500	18	Prepare launch
	2	1′500		−1′500	−26′070		36′500		Approval
	3	2′000	50	−1′950	−28′020		36′500	20	*First sales*
	4	2′000	200	−1′800	−29′820		36′500		
10	1	3′000	400	−2′600	−32′420		36′500		Ramp-up
	2	3′500	600	−2′900	−35′320	10′000	46′500		Round E
	3	4′000	800	−3′200	−38′520		46′500		
	4	4′500	1′200	−3′300	−41′820		46′500		

(continued)

Table 10.2 (continued)

		2 years delay in development							
Yr	Q	Expenses	Revenues	Cash-flow	Cumul	Investm.	Capital	FTE	Phase
11	1	5'000	2'000	−3'000	−44'820		46'500		
	2	5'500	4'000	−1'500	−46'320		46'500		
	3	6'000	6'000	0	−46'320		46'500		End loosing
	4	6'500	7'000	500	−45'820		46'500		First profit
12	1	7'000	8'000	1'000	−44'820		46'500		
	2	7'500	9'000	1'500	−43'320		46'500		
	3	8'000	10'000	2'000	−41'320		46'500		
	4	8'500	11'000	2'500	−38'820		46'500		
13	1	9'000	12'000	3'000	−35'820		46'500		
	2	9'500	13'000	3'500	−32'320		46'500		
	3	10'000	14'000	4'000	−28'320		46'500		
	4	10'500	15'000	4'500	−23'820		46'500		
14	1	11'000	16'000	5'000	−18'820		46'500		
	2	11'500	17'000	5'500	−13'320		46'500		
	3	12'000	18'000	6'000	−7'320		46'500		
	4	12'500	19'000	6'500	−820		46'500		Payback
15	1	13'000	20'000	7'000	6'180		46'500		
	2	13'500	21'000	7'500	13'680		46'500		
	3	14'000	22'000	8'000	21'680		46'500		
	4	14'500	23'000	8'500	30'180		46'500		
16	1	15'000	24'000	9'000	39'180		46'500		
	2	15'500	25'000	9'500	48'680		46'500		
	3	16'000	26'000	10'000	58'680		46'500		
	4	16'500	27'000	10'500	69'180		46'500		

on its White Oak, MD campus with the aim of fostering an open discussion on the scientific and clinical considerations associated with the development of BCI devices. FDA considered the input provided during this workshop to develop the recommendations provided in this draft guidance document for implanted BCI devices. This draft guidance is issued for comment purposes only."

The deadline for comments was set to April 26, 2019, and is visible online [9]. The Wyss Center for Bio and Neuroengineering contributed remarks [10] which can be summarized as:

- The implanted BCI draft guidance should consider other technologies than radio frequency (RF) that can be used to transmit data from the human body to external equipment.
- Throughout the document, BCI devices are only described as recording electrical signals and interfacing through electrical stimulation.
- The guidance should anticipate rapidly developing technologies like light emission or ultrasounds.

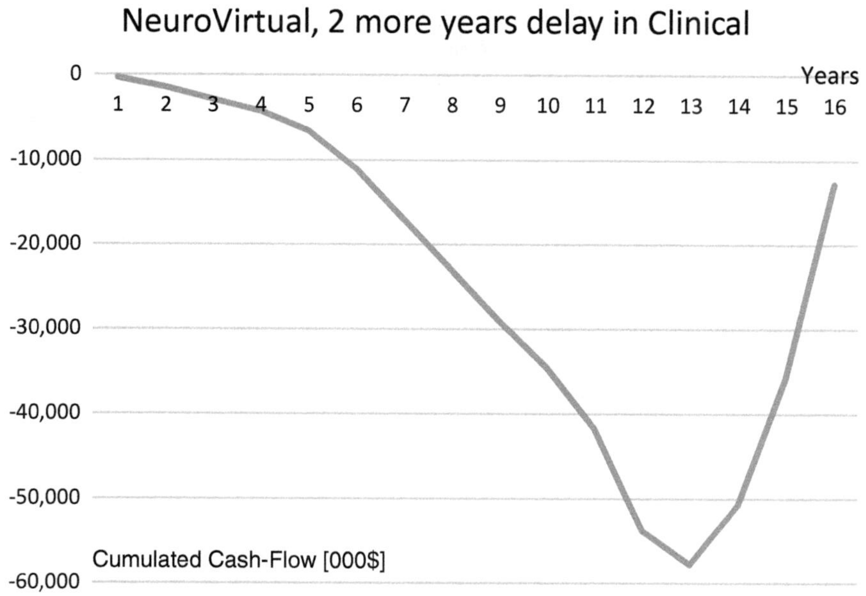

Fig. 10.3 Cumulated cash-flows with additional 2 years of delay in the clinical phase

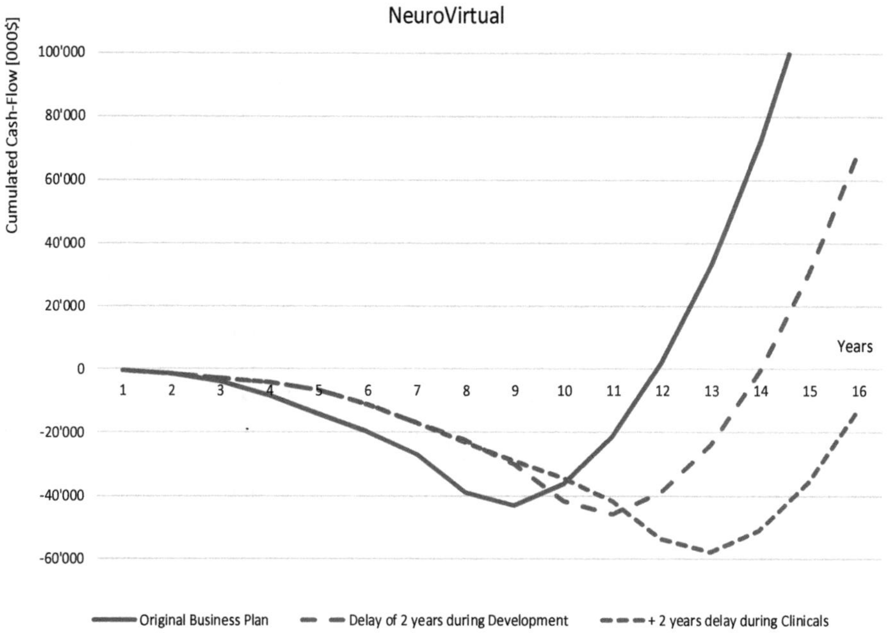

Fig. 10.4 Evolution of the cumulated cash-flows due to delays

Table 10.3 Details with additional 2 years delay during the clinical phase

		2 more years delay in clinical/approval							
Yr	Q	Expenses	Revenues	Cash-flow	Cumul	Investm.	Capital	FTE	Phase
1	1	50		−50	−50	200	200	1	Seed funding
	2	70		−70	−120		200		
	3	100		−100	−220	300	500	2	2nd seed funding
	4	150		−150	−370		500		
2	1	200		−200	−570	1′000	1′500	3	Round A
	2	250		−250	−820		1′500		
	3	300		−300	−1′120		1′500	4	
	4	350		−350	−1′470		1′500		Start a redesign
3	1	350		−350	−1′820	5′000	6′500		Round B
	2	350		−350	−2′170		6′500		
	3	350		−350	−2′520		6′500		
	4	350		−350	−2′870		6′500		
4	1	350		−350	−3′220		6′500		
	2	350		−350	−3′570		6′500		
	3	350		−350	−3′920		6′500		
	4	350		−350	−4′270		6′500		
5	1	400		−400	−4′670		6′500	6	
	2	500		−500	−5′170	15′000	21′500		Round C
	3	700		−700	−5′870		21′500	8	Prototypes
	4	700		−700	−6′570		21′500		
6	1	1′000		−1′000	−7′570		21′500	10	Animals
	2	1′000		−1′000	−8′570		21′500		V&V
	3	1′000		−1′000	−9′570		21′500	12	
	4	1′500		−1′500	−11′070		21′500		Serie 0
7	1	1′000		−1′000	−12′070		21′500	14	FIM
	2	2′000		−2′000	−14′070		21′500		Serie 1
	3	1′500		−1′500	−15′570		21′500		Clinical
	4	1′500		−1′500	−17′070		21′500		
8	1	1′500		−1′500	−18′570		21′500		
	2	1′500		−1′500	−20′070	15′000	36′500		Round D
	3	1′500		−1′500	−21′570		36′500		
	4	1′500		−1′500	−23′070		36′500		
9	1	1′500		−1′500	−24′570		36′500		
	2	1′500		−1′500	−26′070		36′500		
	3	1′500		−1′500	−27′570		36′500		
	4	1′500		−1′500	−29′070		36′500		
10	1	1′500		−1′500	−30′570		36′500		
	2	1′500		−1′500	−32′070		36′500		
	3	1′500		−1′500	−33′570		36′500		Application CE
	4	1′000		−1′000	−34′570	25′000	61′500		Round E

(continued)

Table 10.3 (continued)

		2 more years delay in clinical/approval							
Yr	Q	Expenses	Revenues	Cash-flow	Cumul	Investm.	Capital	FTE	Phase
11	1	2′000		−2′000	−36′570		61′500	18	Prepare launch
	2	1′500		−1′500	−38′070		61′500		Approval
	3	2′000	50	−1′950	−40′020		61′500	20	First sales
	4	2′000	200	−1′800	−41′820		61′500		
12	1	3′000	400	−2′600	−44′420		61′500		Ramp-up
	2	3′500	600	−2′900	−47′320		61′500		Round E
	3	4′000	800	−3′200	−50′520		61′500		
	4	4′500	1′200	−3′300	−53′820		61′500		
13	1	5′000	2′000	−3′000	−56′820		61′500		
	2	5′500	4′000	−1′500	−58′320		61′500		
	3	6′000	6′000	0	−58′320		61′500		End loosing
	4	6′500	7′000	500	−57′820		61′500		First profit
14	1	7′000	8′000	1′000	−56′820		61′500		
	2	7′500	9′000	1′500	−55′320		61′500		
	3	8′000	10′000	2′000	−53′320		61′500		
	4	8′500	11′000	2′500	−50′820		61′500		
15	1	9′000	12′000	3′000	−47′820		61′500		
	2	9′500	13′000	3′500	−44′320		61′500		
	3	10′000	14′000	4′000	−40′320		61′500		
	4	10′500	15′000	4′500	−35′820		61′500		
16	1	11′000	16′000	5′000	−30′820		61′500		
	2	11′500	17′000	5′500	−25′320		61′500		
	3	12′000	18′000	6′000	−19′320		61′500		
	4	12′500	19′000	6′500	−12′820		61′500		Not yet payback

- Closed-loop systems should also be included as they have specific requirements in terms of safety and control.
- Considering testing, the guidance should recommend methods for accelerating lifetime testing, adding reactive oxygen species.
- Moisture contents inside the casing should also be verified and controlled.
- The exclusion criterion related to reliance on ventilation support excludes de facto the use of BCI systems for ALS patients.

As a general comment, we would have liked the draft guidance to be applicable to a broader field of applications. Paralysis and amputation are only a subset of the possible uses of implanted BCIs. However, applications not strictly covered by the future guidance could refer to it by analogy.

The Draft Guidance also states: "Non-clinical testing methods may not be available or may not sufficiently provide the information needed to advance to a final version of an implanted BCI device under development. Therefore, if your device is still under development, we recommend that you consider performing an early feasibility study (EFS) through an IDE to collect early clinical evaluation of your device to provide proof of the principle and initial safety data." The IDE procedure includes a pre-submission optional but highly recommended preliminary step. It allows project owners to get an early non-binding feedback from the FDA.

This Draft Guidance is a major contribution to the reflection process we should engage in the preparation of the BCIs of the future.

References

1. https://en.wikipedia.org/wiki/Risk_management
2. Hubbard D (2009) The failure of risk management: why it's broken and how to fix it. Wiley and Sons, Inc, Hoboken, New Jersey. pp 46
3. https://en.wikipedia.org/wiki/Risk
4. Cline PB (2015) The merging of risk analysis and adventure education. Wilderness Risk Manag 5:43–45, March 3rd, 2015
5. https://en.wikipedia.org/wiki/ISO_14971
6. https://www.iso.org/standard/38193.html
7. https://www.fda.gov/regulatory-information/search-fda-guidance-documents/implanted-brain-computer-interface-bci-devices-patients-paralysis-or-amputation-non-clinical-testing
8. http://wayback.archive-it.org/7993/20170112091055/http:/www.fda.gov/MedicalDevices/NewsEvents/WorkshopsConferences/ucm410261.htm
9. https://www.regulations.gov/docketBrowser?rpp=25&so=DESC&sb=commentDueDate&po=0&dct=PS&D=FDA-2014-N-1130
10. https://www.regulations.gov/document?D=FDA-2014-N-1130-0011

Index

C. Clément, *Brain-Computer Interface Technologies*,
https://doi.org/10.1007/978-3-030-27852-6

MIX
Papier aus verantwortungsvollen Quellen
Paper from responsible sources
FSC® C105338

If you have any concerns about our products, you can contact us on
ProductSafety@springernature.com

In case Publisher is established outside the EU, the EU authorized representative is:
Springer Nature Customer Service Center GmbH
Europaplatz 3, 69115 Heidelberg, Germany

Printed by Libri Plureos GmbH
in Hamburg, Germany